Current Topics in Neuroendocrinology Volume 11

Current Topics in Neuroendocrinology

Vol. 4: Neurobiology of Vasopressin
1985. 53 figures. V, 203 pages. ISBN 3-540-11351-7
Biosynthesis of Vasopressin · Electrophysiological Studies of the Magnocellular Neurons · Volume Regulation of Antidiuretic Hormone Secretion · Vasopressin, Cardiovascular Regulation and Hypertension · Neuroanatomical Pathways Related to Vasopressin

Vol. 5: Actions of Progesterone on the Brain
1985. 61 figures. V, 216 pages. ISBN 3-540-13433-6
Antiprogestins · Progesterone Receptors in Brain and Hypophysis · Effects of Estrogen and Progesterone as Revealed by Neurophysiological Methods · Progesterone Receptors in CNS Correlated with Reproductive Behavior · Estradiol-Progesterone Interactions in the Reproductive Behavior of Female Rats · Behavioral Effects on Humans of Progesterone-Related Compounds Durring Development and in the Adult

Vol. 6: Neurobiology of Oxytocin
1986. 38 figures. V, 175 pages. ISBN 3-540-15341-1
Biosynthesis of Oxytocin in the Brain and Peripheral Organs · Regulation of Oxytocin Release · Proteolytic Conversion of Oxytocin, Vasopressin, and Related Peptides in the Brain · Oxytocin and Behavior · Oxytocin as an Ovarian Hormone · Oxytocin and the Milk-Ejection Reflex

Vol. 7: Morphology of Hypothalamus and Its Connections
1986. 94 figures. VI, 314 pages. ISBN 3-540-16919-9
The Pituitary Portal System · Functional Ultrastructure of Gonadotropes: a Review · Ultrastructure of Anterior Pituitary Cells · Catecholamine-Peptide Interactions in the Hypothalamus · Neuroendocrine Projections to the Median Eminence · Afferents onto Neuroendocrine Cells · Interconnectedness of Steroid Hormone-Binding Neurons: Existence and Implications · Ultrastructure of Regulatory Neuroendocrine Neurons and Functionally Related Structures · Synaptogenesis and Neuronal Plasticity to Gonadal Steroids: Implications for the Development of Sexual Dimorphism in the Neuroendocrine Brain

Vol. 8: Neuroendocrinology of Mood
1988. 80 figures. VI, 335 pages. ISBN 3-540-17892-9
Principles for the Hormone Regulation of Wiring Transmission and Volume Transmission in the Central Nervous System · Clinical Studies with Corticotropin Releasing Hormone: Implications for Hypothalamic-Pituitary-Adrenal Dysfunction in Depression and Related Disorders · Biological Rhythms and Mood Disorders · Recurrent Affective Disorders: Lessons from Limbio Kindling · The Mechanisms of Action of Antipsychotics and Antidepressant Drugs · Catecholamines and Mood: Neuroendocrine Aspects · Serotonin and Mood: Neuroendocrine Aspects · Cholinergic Mechanisms in Mood: Neuroendocrine Aspects · The Psychobiology of Neurotensin · Cholecystokinin and Mood · Opioid Peptides and Moods: Neuroendocrine Aspects · The Neuroendocrinology of Anorexia Nervosa · Effects of Peripheral Thyroid Hormones on the Central Nervous System: Relevance to Disorders of Mood

Vol. 9: Stimulus-Secretion Coupling in Neuroendocrine Systems
1988. 69 figures. V, 256 pages. ISBN 3-540-19043-0
Cellular Reorganization in Neuroendocrine Secretion · Stimulus-Secretion Coupling in the Oxytocin System · Coupling of Electrical Activity and Hormone Release in Mammalian Neurosecretory Neurons · The Bag Cell Neuroendocrine System of *Aplysia* · Electrophysiological Characteristics of Peptidergic Nerve Terminals Correlated with Secretion · Changes in Information Content with Physiological History in Peptidergic Secretory Systems · Insect Neuropeptides · Stimulus-Secretion Coupling in the Pancreatic B Cell

Vol. 10: Behavioral Aspects of Neuroendocrinology
1990. 56 figures. VI, 295 pages. ISBN 3-540-52801-6
Behavioral Effects of Corticotropin-Releasing Factor · Corticotropin-Releasing Factor: Central Regulation of Autonomic Nervous and Visceral Function · Oxytocin as Part of Stress Responses · Behavioral Effects of Vasopressin · Neuropeptide Control of Parental and Reproductive Behavior · Estrogen Regulation of mRNAs in the Brain and Relationship to Lordosis Behavior · The Neuroendocrinology of Thirst: Afferent Signaling and Mechanisms of Central Integration · Neuropeptide Y: A Novel Peptidergic Signal for the Control of Feeding Behaviour · Neuropeptide Mechanisms Affecting Temperature Control

H. Imura (Ed.)

Recombinant DNA Technologies in Neuroendocrinology

Contributors

Y. Dong, A. Fukamizu, R.H. Goodman, J.-Å. Gustafsson, V. Höllt, M.G. Kaplitt, L.A. Kukstas, P.-M. Lledo, M.J. Low, K. Murakami, Y. Nakai, S. Okret, D.W. Pfaff, S.D. Rabkin, R. Rehfuss, D. Richter, T. Tsukada, J.-D. Vincent, K. Walton

With 42 Figures

Springer-Verlag Berlin Heidelberg New York
London Paris Tokyo Hong Kong Barcelona
Budapest

Editor

HIROO IMURA, M.D., Professor
Dept. of Internal Medicine
Kyoto University
Faculty of Medicine
54 Kawaharacho, Shogoin, Sakyo-ku
Kyoto 606
Japan

The picture on the cover has been taken from Nieuwenhuys R., Voogd J., van Huijzen Chr.: The Human Central Nervous System. 2nd Edition.
Springer-Verlag Berlin Heidelberg New York 1981

ISBN-13: 978-3-642-77505-5 e-ISBN-13: 978-3-642-77503-1
DOI: 10.1007/978-3-642-77503-1

Typesetting: FotoSatz Pfeifer GmbH, Gräfelfing/München
2121/3130-543210 – Printed on acid-free paper

Table of Contents

Axonal Messenger RNA Transport: Fact or Fiction?

D. RICHTER[1]

Contents

> **Dogma**: a: something held as an established opinion, esp: one or more definite and authoritative tenets. b: a code or systematized formulation of such tenets ... c: a point of view or alleged authoritative tenet put forth as dogma without adequate grounds ... *(Webster's Third New International Dictionary)*

A current dogma in neurobiology states that axons and nerve terminals – in contrast to dendrites – do not contain messenger RNA and hence lack the capacity to synthesize proteins (Lasek and Brady 1981). Yet, at least in invertebrates, local translation in the axoplasm has been reported (Guiditta et al. 1990, 1991; Rapallino et al. 1988). Factors required for axonal protein synthesis have been detected in squid giant axons (Guiditta et al. 1977); these include elongation factors, aminoacyl – tRNA synthetases, tRNAs, polyA+ RNA (Perrone Capano et al. 1987) and polysomes carrying nascent peptide chains (Guiditta et al. 1991). Furthermore, axons of *Lymnaea stagnalis* have been shown to contain mRNA encoding the egg-laying hormone (Dirks et al. 1989). Thus these data strongly support the notion of *de novo* protein synthesis in the axoplasm of invertebrates, contradicting the earlier glia-neuron protein transfer hypothesis (Gainer et al. 1977), which proposed the transfer of newly synthesized proteins from periaxonal glial cells to giant axons.

[1] Institut für Zellbiochemie und klinische Neurobiologie, UKE, Universität Hamburg, Martinistr. 52, 2000 Hamburg 20, FRG

Current Topics in Neuroendocrinology, Vol. 11

1 Vasopressin- and Oxytocin-Encoding mRNAs Are Present in the Posterior Pituitary

In vertebrates, the existence of axoplasmic synthesizing machinery has not been established simply because the morphology of the nervous system does not easily permit the resolution of axons and nerve terminals from the surrounding glial cells. In this respect, a report by Lehmann (1988) that mRNAs encoding the neuropeptide hormones vasopressin and oxytocin are present in rodent posterior pituitary raised considerable interest (Murphy et al. 1989; Lehmann et al. 1990; Mohr et al. 1990a,b; McCabe et al. 1990). These transcripts are present in the posterior, but are absent in the anterior lobe of the rodent pituitary gland. When cloned and sequenced, both vasopressin- and oxytocin-encoding cDNAs derived from rat posterior pituitary mRNA were identical in nucleotide sequence to their hypothalamic counterparts (Mohr et al. 1990a), eliminating possible alternatively spliced transcripts in the posterior pituitary.

Genes encoding the vasopressin and oxytocin precursors are normally expressed in the perikarya of the hypothalamic magnocellular neurons which project their axons down to the posterior pituitary (for review, see Richter 1988). Anatomically, the latter consist of axons extending into nerve terminals and astrocyte-like cells, the pituicytes. Thus the transcripts detected in the posterior pituitary could be located in nerve terminals (including axons) or pituicytes.

2 Expression in Pituicytes?

A number of arguments seemed to support that the hormone transcripts present in the posterior pituitary were derived from pituicytes expressing the respective genes. For instance, vasopressin- and oxytocin-encoding mRNAs isolated from rat posterior pituitary were shorter compared to the corresponding hypothalamic transcripts (Murphy et al. 1989; Mohr et al. 1990a). Digestion assays using RNase H and oligo d(T) indicated that the differences in mRNA size were due to the presence of truncated polyA tails, a post-transcriptional modification step quite often found to occur with vasopressin and oxytocin mRNAs (Richter et al. 1991). It was speculated that if the two transcripts originated exclusively in the hypothalamus, having been axonally transported to the posterior pituitary, a differential degradation or stabilization process for magnocellular "short" versus "long" vasopressin and oxytocin transcripts would be required. Since this appeared to be rather unlikely, derivation of the transcripts from pituicytes was considered plausible. On the other hand, truncated polyA tails might reflect the turnover of this mRNA during transport.

Furthermore, rats that were exposed to saline solutions for several days responded with a three to five-fold increase in the levels of hypothalamic vasopressin and oxytocin mRNAs compared to those of controls. In contrast vasopressin mRNA level was elevated about 17-fold in the posterior pituitary, whereas that of oxytocin-encoding transcripts increased only two to three-fold, as observed in the hypothalamus (Murphy et al. 1989; Mohr et al. 1991; Richter et al. 1991). A simple explanation for the difference in vasopressin mRNA levels found in the two tissues could be that the signal "osmotic shock" increased cell-specifically the transcription rate of the vasopressin gene, with this being relatively higher in pituicytes than in magnocellular neurons. Alternatively, the turnover rate of vasopressin mRNA in nerve terminals (and axons) might be reduced as a function of the osmotic state, leading to a relative accumalation of that mRNA in the posterior pituitary. Since the vasopressin mRNA levels in the neural stalk showed the same relative increase in response to osmotic shock as those in the hypothalamus (three to five-fold increase; Mohr, unpublished data), axonal mRNA transport is apparently not accelerated – an observation which is in line with the properties of axonal transport of neurosecretory vesicles (Castel et al. 1984).

The possibility that, due to the osmotic state, the presence of vasopressin and oxytocin transcripts in axons and nerve terminals was the result of an "overflow" of mRNA from the perikarya into the axoplasm appears unlikely since the levels of the two mRNAs in the posterior pituitary were different (see above).

3 Expression During Fetal Rat Development

To gain further insights into the origin of the peptide hormone mRNAs, transcription of the vasopressin and oxytocin genes was studied during fetal rat development. RNase protection assays revealed that vasopressin transcripts were present in the hypothalamus and in the pituitary gland anlage as early as day 15 of gestation; these levels appeared to be slightly higher in the pituitary gland than in the hypothalamus itself (Mohr et al. 1991). The latter observation may be explained by slight variations in the yield of polyA+ RNA from different preparations or by accelerated axonal transport rates during early embryonic development as compared with that at later stages. Due to inconsistent reports in the literature concerning the initial embryonic onset of axonal outgrowth connecting the hypothalamus with the pituitary gland anlage, correlation of the transcripts with either glial cells or with axons was not possible. According to Galabov and Schiebler (1978), neurosecretory axons and the pituitary gland are connected at embryonic day 15, whereas Dellmann et al. (1978) observed this event later, at embryonic days 16 to 17.

In 16- and 17-day-old embryos, increasing amounts of vasopressin mRNA were detected in both the hypothalamus and the pituitary. In contrast, oxytocin transcripts could not be identified in the hypothalamus and the posterior pituitary until embryonic day 17, which has been reported to mark the onset of oxytocin gene transcription in the hypothalamus during fetal rat development (Laurent et al. 1989).

4 Evidence for the Presence of mRNA in Axons and Nerve Terminals of the Hypothalamo-Neurohypophyseal Tract

The finding that the two mRNAs were no longer detectable in the posterior pituitary following either transection of the hypothalamo-hypophyseal tract (Mohr et al. 1990a) or colchicine treatment (Levy et al. 1990) seemed to favor the concept of axonal mRNA transport. Rats that had paired electrical lesions disconnecting the hypothalamus from the pituitary continued to transcribe the vasopressin and oxytocin genes in the hypothalamus. However, the two transcripts were no longer detectable in the posterior pituitary, even when the animals were subjected to chronic intermittent salt loading, implying an axonal mRNA transport mechanism from the hypothalamic perikarya to the nerve terminals of the posterior pituitary. Alternatively, expression of the two genes could have taken place in pituicytes, but this was dependent on contacts with axons of magnocellular neurons and/or on neural inputs or factors deriving from other parts of the brain, and these had been destroyed by the electrical lesion.

The concept of axonal mRNA transport was further supported when studying the morphological distribution of the vasopressin-encoding mRNA in osmotically stressed rats. *In situ* hybridization experiments using coronal brain sections in the region of the median eminence, the infundibular stem, and the neural lobe indicated that vasopressin – specific signals were present in areas where mainly axons are located, particularly in the median eminence over the *zona interna* (Mohr et al. 1991). In contrast, the *zona externa*, consisting of other nerve endings and glial cells, was essentially unlabeled. In the infundibular stem, most of the signal was present over the internal part of the structure in which the axons are located. Sections through the neural, intermediate, and anterior lobes revealed positive signals only in the neural lobe (Murphy et al. 1989; Mohr et al. 1990a,b). However, *in situ* hybridization signals were not obtained in the above-mentioned tissues when using an oxytocin-specific probe, presumably due to the relatively low levels of this mRNA species even in osmotically stressed rats (e.g., a three-fold increase in the oxytocin mRNA level compared to a 17-fold stimulation of vasopressin mRNA levels).

5 Pituicytes Lack Primary Vasopressin Gene Transcripts

The possibility that the vasopressin and oxytocin genes may be expressed in glial cells of both the posterior pituitary and the neural stalk has recently been ruled out by a number of experiments designed to detect primary gene transcripts in rat posterior pituitary RNA preparations. If these genes were expressed in pituicytes or tanycytes, astrocyte-like cells found in the neural stalk, intron-containing primary gene transcripts should be present in the nuclei of these cells. Analysis of transcripts was carried out using three different approaches, including nuclear run-on experiments, RNase protection, and amplification of intronic sequences by the polymerase chain reaction (PCR), a particularly sensitive method for detecting minute amounts of RNA (Mohr et al. 1991).

When nuclear run-on experiments were performed with nuclei isolated from posterior pituitary cells, the newly synthesized ^{32}P-labeled mRNAs did not hybridize to either vasopressin or oxytocin cDNA probes, although a strong hybridization signal with a chicken B-actin cDNA was evident (Mohr et al. 1991). Earlier experiments using nuclei isolated from the hypothalamus showed that newly synthesized vasopressin and oxytocin transcripts were present in hypothalamic magnocellular neurons (Murphy and Carter 1990).

RNase protection experiments using hypothalamic or posterior pituitary primary gene transcripts and ^{32}P-labeled *in vitro*-synthesized RNA corresponding to part of the intron I sequence of the vasopressin gene yielded a protected fragment of the expected size (269 nucleotides) with hypothalamic, but not with posterior pituitary RNA (Fig 1, left panels). In contrast, when using a ^{32}P-labeled antisense RNA corresponding to exon A of the vasopressin gene, both the posterior pituitary and the hypothalamic RNAs gave identical protected bands; these were absent when using anterior pituitary RNA (Mohr et al. 1991).

In another set of experiments (Fig. 1, right panels), hypothalamic and posterior pituitary primary gene transcripts were converted into single-stranded cDNAs, which were then amplified by the PCR. Successful amplification of intronic sequences would, of course, depend on the first-strand cDNA synthesis, which, in the case of posterior pituitary RNA, may cause some technical difficulty due to the relatively low levels there of vasopressin transcripts. McCabe et al. (1990) calculated that in normal rats (not exposed to saline solutions) the vasopressin mRNA level in the posterior pituitary corresponds to about 1/100 of that in the hypothalamus. Because of this, the following precautions were taken: (a) posterior pituitary RNA was prepared from osmotically shocked rats, which respond with a 17-fold increase in the level of vasopressin mRNA (Murphy et al. 1989; Mohr et al. 1991); (b) in contrast, hypothalamic RNA was obtained from rats that were not exposed to saline solution; and (c) first-strand cDNA synthesis was carried out using RNA preparations isolated from the posterior pituitaries of three to four rats, whereas the hypothalamic RNA was prepared from only one tenth of a rat hypothalamus. Thus the posterior pituitary RNA preparation used for first-strand cDNA synthesis contained at least six times more vasopressin transcripts than were present in the hypothalamic RNA preparation.

For PCR amplification, forward and reverse primers were designed using sequences of either intron I or II of the vasopressin gene (Richter 1988). Southern blot hybridization of the amplified products with probes specific for intron I or II revealed the expected bands of 346 bp and 223 bp in hypothalamic but not in posterior pituitary cDNA (Fig. 1, right panels; Mohr et al. 1991), indicating that the glial cells of the posterior pituitary do not express the vasopressin or oxytocin genes and that the hormonal mRNAs must be of axonal and nerve terminal origin. Consequently, these transcripts are probably synthesized in the perikarya of hypothalamic neurons and then axonally transported to the posterior pituitary.

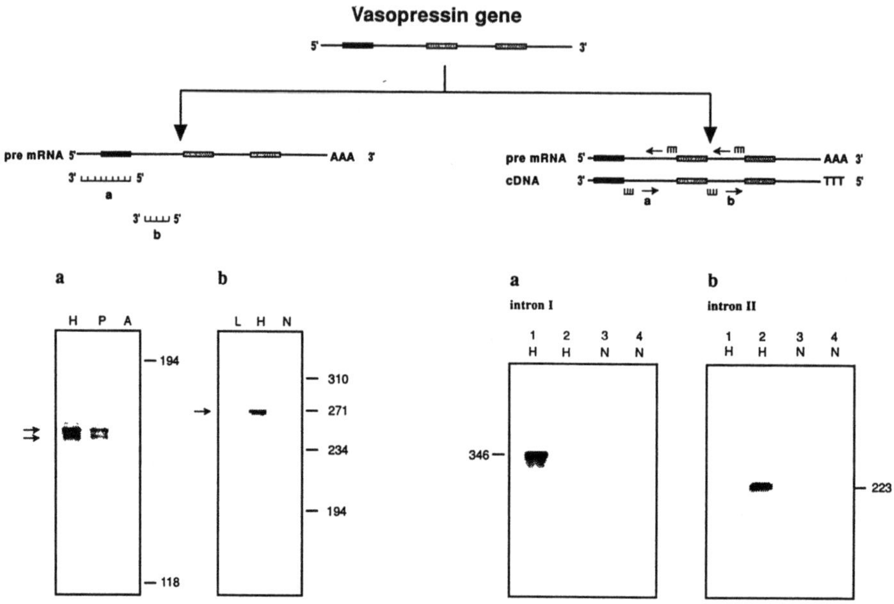

Fig. 1. RNase protection analysis (*left panels*) and polymerase chain reaction amplifications (*right panels*) of vasopressin-encoding primary gene transcripts.

Left panels: Total RNAs (for details, see Mohr et al. 1991) were analyzed by an RNase protection assay after hybridization to ^{32}P-labeled RNA corresponding to part of the 5' untranslated region and exon A (*a*) or to part of intron I (*b*) of the rat vasopressin gene. Liver and hypothalamic RNAs were derived from control animals whereas neural lobe RNA was isolated from osmotically stressed rats (2% saline for 7 days). In *a*, two protected fragments of 161 and 163 nucleotides in length, corresponding to the two mapped transcription initiation sites, are detected with hypothalamic and pituitary gland RNA. In *b*, the protected fragment of the expected length (269 nucleotides) can only be detected when the assay is performed with hypothalamic RNA, indicating the presence there of vasopressin primary gene transcripts. *H*, hypothalamus; *L*, liver; *N*, neural lobe; *A*, anterior lobe; *P*, pituitary gland.

Right panels: Southern blot analysis of DNA fragments generated by the polymerase chain reaction, using pairs of primers specific for either the vasopressin gene intron I (*a; lanes 1,3*) or intron II (*b; lanes 2,4*); single-stranded cDNAs were derived from hypothalamus (*H*) and neural lobe (*N*). For primer sequences, see Mohr et al. (1991). Southern blot hybridization was carried out using a ^{32}P-labeled 267 bp *Kpn*I/*Kpn*I DNA fragment specific for the rat vasopressin gene intron I (Mohr et al. 1991) (*a*). The same blot shown in *a* was stripped and hybridized with a ^{32}P-labeled 40-base oligonucleotide (Mohr et al. 1991) specific for intron II of the vasopressin gene (*b*).

6 Selectivity of Axonal mRNA Transport?

A number of questions remain to be answered. For instance, is the observed presence of mRNAs in the axons of hypothalamic neurons a general phenomenon, i.e., are mRNAs also found in axons projecting to other brain areas? What determines whether the mRNA species remain within the perikaryon or are transported to the neurohypophysis?

Preliminary data show that mRNAs encoding other hypothalamic neuropeptides and enzymes (dynorphin, proenkephalin, neuron-specific enolase) are present in the neural stalk as well as in the neurohypophysis (Mohr and Richter 1992). However, these tissues also contain the corresponding primary gene transcripts as indicated by PCR amplification using intron-specific primers, suggesting that glial cells express these genes. Hence, the detected mRNAs in the neural stalk and in the neurohypophysis are not necessarily of axonal or nerve terminal origin. Expression of proenkephalin- and dynorphin-encoding genes in the neurointermediate lobe is in line with other reports; cells of this lobe contain dynorphin-like immunoreactive substances (Watson et al. 1982) and proenkephalin encoding mRNA (Schäfer et al. 1990).

According to the literature, neuron-specific enolase has been considered to be a specific metabolic marker for neurons (Schmechel et al. 1978, 1980). However, when analyzing the expression of this gene by PCR amplification and Northern blotting, the presence of primary transcripts was revealed not only in neurons but also in non-neuronal cells and in organs such as liver (Mohr and Richter 1992). Again, the data do not permit the distinction between transcripts derived from axons or (and) glial cells.

A more suitable marker for studying the specificity of axonal mRNA transport appeared to be the transcript of the gene encoding neurofilament L, an intermediate filament protein present in neurons but not in glial cells (Lazarides 1980; Julien et al. 1985). Preliminary data show that the corresponding mRNA is detectable in the neural stalk as well as in the posterior pituitary (Fig. 2). Since the rat neurofilament L gene has not yet been sequenced, it has not been possible to use intron-specific sequences for the amplification of primary gene transcripts. Since neurofilament L is a neuronspecific protein, its mRNA represents another

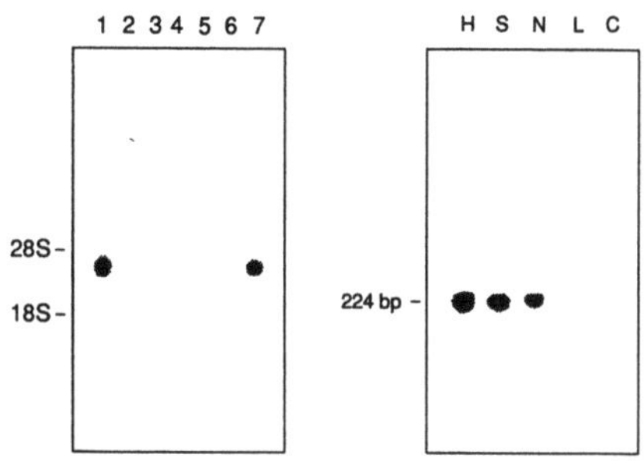

Fig. 2. Detection of rat neurofilament L (NF-L) encoding mRNA by Northern blot hybridization (*left panel*) and polymerase chain reaction (PCR) amplification (*right panel*). Northern blot hybridization (*left panel*): 25 μg of RNA extracted from different rat tissues or cell lines (*lane 1*, hypothalamus; *lane 2*, GH_3 cells; *lane 3*, anterior pituitary; *lane 4*, posterior pituitary; *lane 5*, liver; *lane 6*, rat fibroblast cells; *lane 7*, PC12 cells) was hybrizided to a ^{32}P-labeled NF-L specific 45-base antisense oligonucleotide (Julien et al. 1985). PCR amplification (*right panel*): RNAs isolated from rat hypothalamus (*H*), neural stalk (*S*), posterior pituitary (*N*), and liver (*L*) were converted into single-stranded cDNAs and subjected to PCR amplification using forward and reverse primers (20-mers) deduced from the rat NF-L mRNA sequence (Julien et al. 1985). Amplified products were analyzed by Southern blot hybridization with the same NF-L specific antisense oligonucleotide used for Northern blot analysis. *C*, control amplification without the addition of cDNA.

example of axonal transport from the perikarya of the hypothalamic neurons to the posterior pituitary.

Does the presence of mRNAs in axons and nerve terminals reflect the existence of a specific transport mechanism? The presence of oxytocin mRNA in neurosecretory vesicles from nerve endings of the rat posterior pituitary has recently been reported (Jirikowski et al. 1990). Although these data would be in line with the axonal transport hypothesis, mRNA transport within neurosecretory vesicles would raise the question of how this mRNA might have entered the vesicles in the perikaryon of a magnocellular neuron. At present, it appears more plausible that the presence of hypothalamic mRNAs in axons and nerve terminals is due to a transport mechanism similar to that postulated for dendrites (Garner et al. 1988).

7 Prospects

Major questions emerging from these studies will certainly concern the possible function(s) of axonal translation products. Local protein synthesis within nerve terminals and axons would require a complete protein synthesizing and secretory machinery which has not yet been observed in mammalian axons. Local translation, on the other hand, could be of advantage for the elongated cell, allowing it to respond immediately and more flexibly to external stimuli, as is the case for dendrites (Garner et al. 1988). If proteins are synthesized locally, then the functional significance of such synthesis remains to be established. It is rather unlikely that rough endoplasmic reticulum and Golgi structures have been overlooked in axons and nerve terminals, compartments which are known to be required for the packaging, processing, and secretion of peptide hormone precursors. Modulating activities by the peptides in response to external signals, as has been observed for neurons projecting to other brain areas (Swanson and Sawchenko 1983), or local modulation of their own release from nerve terminals (Mathison and Lederis 1978) may be possible functions. This would require precursor processing and secretion of the mature peptides by hitherto unknown mechanisms (Rubartelli et al. 1990). Certainly, sensitive methods will have to be applied in order to identify either of these processes.

Acknowledgments. Helpful discussions with Drs. E. Mohr, H. Schmale, and M. Darlison are gratefully acknowledged. This work was supported in part by grants from the Deutsche Forschungsgemeinschaft and the Nato collaborative research grant program.

References

Castel M, Gainer H, Dellmann HD (1984) Neuronal secretory systems. Int Rev Cytol 88:302–459

Dellmann HD, Castel M, Linner JG (1978) Ultrastructure of peptidergic neurosecretory axons in the developing neural lobe of the rat. Gen Comp Endocrinol 36:477–486

Dirks RW, Raap AK, van Minnen J, Vreughdenhil E, Smit AB, van der Ploeg M (1989) Detection of mRNA molecules coding for neuropeptide hormones of the pond snail *Lymnaea stagnalis* by radioactive and non-radioactive *in situ* hybridization: a model study for mRNA detection. J Histochem Cytochem 37:7–14

Gainer H, Tasaki I, Lasek RJ (1977) Evidence for the glia neuron transfer hypothesis from intracellular perfusion studies of squid giant axons. J Cell Biol 74:524–530
Galabov P, Schiebler TH (1978) The ultrastructure of the developing neural lobe. Cell Tissue Res 189:313–329
Garner CC, Tucker R, Matus A (1988) Selective localization of messenger RNA for cytoskeletal protein MAP2 in dendrites. Nature 336:674–677
Guiditta A, Metafora S, Felsani A, del Rio A (1977) Factors for protein synthesis in the axoplasm of squid giant axons. J. Neurochem 28:1393–1395
Guiditta A, Menichini E, Castigli E, Perrone Capano C (1990) Protein synthesis in the axonal territory. In: Stella AMG, de Vellis J, Perez-Polo JR (eds), Regulation of gene expression in the nervous system. Wiley-Liss, New York, pp 205–218 (Neurology and Neurobiology, vol 59)
Guiditta A, Menichini E, Perrone Capano C, Langella M, Martin R, Castigli E, Kaplan BB (1991) Active polysomes in the axoplasm of the squid giant axon. J Neurosci Res 28.18–28
Haimoto H, Takahashi Y, Koshikawa T, Nagura H, Kato K (1985) Immunohistochemical localization of γ-enolase in normal human tissues other than nervous neuroendocrine tissues. Lab Invest 52:257–263
Jirikowski GF, Sunna PP, Bloom FE (1990) mRNA coding for oxytocin is present in axons of the hypothalamo-neurohypophysial tract. Proc Natl Acad Sci USA 87:7400–7404
Julien JP, Ramachandran K, Grosveld F (1985) Cloning of a cDNA encoding the smallest neurofilament protein from the rat. Biochem Biophys Acta 825:398–404
Lasek RJ, Brady ST (1981) The axon: a prototype for studying expressional cytoplasm. Cold Spring Harbor Symp Quant Biol 46:113–124
Laurent FM, Hindelang C, Klein MJ, Stoeckel ME, Felix JM (1989) Expression of the oxytocin and vasopressin genes in the rat hypothalamus during development: an in situ hybridization study. Dev Brain Res 46:145–154
Lazarides E (1980) Intermediate filaments as mechanical integrators of cellular space. Nature 283:249–256
Lehmann E (1988) Untersuchungen zur Expression des Vasopressingens in der Ratte. Doctoral thesis, University of Heidelberg
Lehmann E, Hänze J, Pauschinger M, Ganten D, Lang RE (1990) Vasopressin mRNA in the neurolobe of the rat pituitary. Neurosci Lett 111:170–175
Levy A, Lightman SL, Carter DA, Murphy D (1990) The origin and regulation of posterior pituitary vasopressin ribonucleic acid in osmotically stimulated rats. J Endocrinol 2:329–334
Mathison R, Lederis K (1978) Modulation of vasopressin release from the neurohypophysis by the intact neuron. Proc. West Pharmacol Soc 21:257–260
McCabe JT, Lehmann E, Chastrette N, Hänze J, Lang RE, Ganten D, Pfaff D (1990) Detection of vasopressin mRNA in the neurointermediate lobe of the rat pituitary. Mol Brain Res 8:325–329
Mohr E, Zhou A, Thorn NA, Richter D (1990a) Rats with physically disconnected hypothalamo-pituitary tracts no longer contain vasopressin-oxytocin gene transcripts in the posterior pituitary lobe. FEBS Lett 263:332–336
Mohr E, Morley SD, Richter D (1990b) Evolution, expression and regulation of the vasopressin-oxytocin gene family. In: Schwartz TW, Hilsted LM, Rehfeld JF (eds) Neuropeptides and their receptors. Munksgaard, Copenhagen, pp 74–94 (Alfred Benzon symposium 29)
Mohr E, Fehr S, Richter D (1991) Axonal transport of neuropeptide encoding mRNAs within the hypothalamo-hypophyseal tract of rats. EMBO J 10:2419–2424
Mohr E, Richter D (1992) Diversity of mRNAs in the axonal compartment of peptidergic neurons in rat. The Europ. J. of Neuroscience, in press
Murphy D, Carter D (1990) Vasopressin gene expression in the rodent hypothalamus: transcriptional and posttranscriptional responses to physiological stimulation. Mol Endocrinol 4:1051–1059
Murphy D, Levy A, Lightman S, Carter D (1989) Vasopressin RNA in the neural lobe of the pituitary: dramatic accumulation in response to salt loading. Proc Natl Acad Sci USA 86:9002–9005
Perrone Capano C, Guiditta A, Castigli E, Kaplan BB (1987) Occurence and sequence complexity of polyadenylated RNA in squid axoplasm. J. Neurochem 49:698–704

Rapallino MV, Cupello A, Giuditta A (1988) Axoplasmic RNA species synthesized in the isolated giant axon. Neurochem Res 13:625–631

Richter D (1988) Molecular events in expression of vasopressin and oxytocin and their cognate receptors. Am J Physiol 255:F207–F219

Richter D, Mohr E, Schmale H (1991) Molecular aspects of the vasopressin gene family: evolution, expression and regulation. In: Jard S, Jamison R (eds) Vasopressin Libbey Eurotext, London, pp 3–11 (Colloque INSERM, Vol 208)

Rubaretelli A, Cozzolino F, Talio M, Sitia R (1990) A novel secretory pathway for interleukin-1β, a protein lacking a signal sequence. EMBO J 9:1503–1510

Schäfer MKH, Day R, Ortega MR, Akil H, Watson SJ (1990) Proenkephalin messenger RNA is expressed both in the rat anterior and posterior pituitary. Neuroendocrinology 51:444–448

Schmechel DE, Marangos PJ, Brightman MW, Goodwin FK (1978) Brain enolases as specific markers of neuronal glial cells. Science 199:313–315

Schmechel DE, Brightman MW, Marangos PJ (1980) Neurons switch from non-neuronal enolase to neuron-specific enolase during differentiation. Brain Res 190:195–214

Swanson LW, Sawchenko PE (1983) Hypothalamic integration: organization of the paraventricular and supraoptic nuclei. Annu Rev Neurosci 6:269–324

Watson SJ, Akil H, Fischli W, Goldstein A, Zimmermann E, Nilaver G, van Wimersma Greidanus TB (1982) Dynorphin and vasopressin: common localization in magnocellular neurons. Science 216:85–87

Molecular Mechanisms of Regulation of Gene Expression by Glucocorticoids*

Y. DONG, S. OKRET, and J.-Å. GUSTAFSSON[1]

Contents

---

* This work was supported by grants from the Swedish Medical Research Council (No. 13X-2819), the Swedish Cancer Society, the Magnus Bergvall Foundation and the Swedish Society of Medical Research.

[1] Department of Medical Nutrition, Karolinska Institute, Huddinge University Hospital F60, NOVUM, S-141 86 Huddinge, SWEDEN

Current Topics in Neuroendocrinology, Vol. 11

1 Introduction

1.1 Mechanism of Action of Glucocorticoid Hormones

The current model of glucocorticoid hormone action is summarized in Fig. 1. After synthesis, glucocorticoids are secreted into the blood stream and transported to target cells where they bind with high affinity ($K_d \sim 10^{-9} M$) and specificity to the intracellular glucocorticoid receptor (GR) protein. The subcellular localization of hormone-free GR is still a controversial issue. However, most data support the idea that unliganded GR is in the cytoplasmic compartment or loosely associated with the nucleus (Picard and Yamamoto 1987; Gustafsson et al. 1987 and references therein; LaFond et al. 1988; Gasc et al. 1989). Upon ligand binding, GR is activated into a form capable of interacting with DNA. The mechanism of GR activation probably involves a conformational change and dissociation from nonreceptor components, e.g., the 90-kDA heat shock protein (hsp90: Pratt et al. 1988; Bresnick et al. 1989; Denis and Gustafsson 1989). The subcellular location of activated GR has been firmly established to be inside the nucleus. In vivo, the hormone-receptor complex interacts with specific DNA

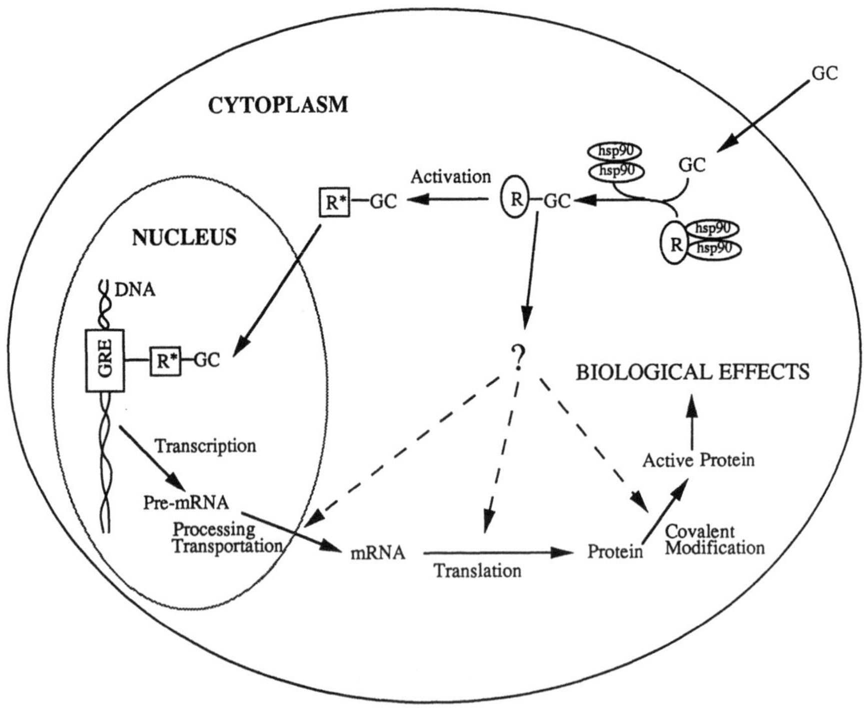

Fig. 1. Hypothetical mechanism of action of glucocorticoid hormones. *GC*, glucocorticoid hormone; *R*, glucocorticoid receptor; *R**, activated glucocorticoid receptor

sequences, termed glucocorticoid responsive element (GRE; see below and Becker et al. 1986), and elicits transcriptional activation or inhibition of target gene expression (reviewed in Yamamoto 1985) by a largely unknown mechanism. It is clear that GR is a central element in glucocorticoid hormone action and a defect at any step through the signal transduction pathway will cause glucocorticoid resistance.

Although transcriptional modulation of gene expression by glucocorticoids has been well documented, there is also an increasing body of evidence showing that glucocorticoids alter gene expression by posttranscriptional mechanisms, e.g., changes in mRNA stability (Paek and Axel 1987), mRNA splicing and processing (Vannice et al. 1984; Cote and Gagel 1986), and cytoplasmic transport of the mRNA (Fulton and Birnie 1985). Furthermore, translational and posttranslational regulatory levels have also been reported as possible mechanisms for glucocorticoid hormone action (McIntyre and Samuels 1985; Karlsen et al. 1986; Meyuhas et al. 1987). The mechanisms underlying posttranscriptional regulation by glucocorticoids are much less characterized compared to the transcriptional effects. Thus, in the following sections, the discussion concerning glucocorticoid hormone action will be mainly focused on modulation of transcriptional activities with special references to GR function and regulation.

1.2 Domain Structure of GR

The domain structure of GR was first defined by proteolytic cleavage pattern of the protein in vitro. Partial digestion of both crude and purified GR with proteases reveals two hypersensitive sites and separates the receptor into three functional domains: (1) an amino-terminal domain, (2) a central DNA-binding domain, and (3) a carboxy-terminal ligand-binding domain (reviewed in Gustafsson et al. 1987). Recent progress in GR cDNA cloning (Hollenberg et al. 1985; Miesfeld et al. 1986; Danielsen et al. 1986) has made it possible to map each domain by mutational analysis more precisely and define its function by transfection assays.

1.2.1 Hormone-Binding Domain

The hormone-binding domain lies in the carboxy-terminal part of GR. Contact points between glucocorticoids and GR have all been located within this domain (Simons et al. 1987; Carlstedt-Duke et al. 1988). In addition, the hormone-binding domain has been shown to repress transcriptional activation, since the removal of this region resulted in a receptor derivative which activates transcription constitutively in a hormone independent fashion (Danielsen et al. 1987; Hollenberg et al. 1987; Godowski et al. 1988; Picard et al. 1988). Furthermore, the ligand-binding region also has a function in mediating hormone-dependent nuclear localization of GR (Picard and Yamamoto 1987).

1.2.2 DNA-Binding Domain

The DNA-binding domain is located in approximately the central part of the receptor protein. The sequentially arranged cysteine residues and the predominating basic amino acids arginine and lysine are reminiscent of the zinc finger motif found in other DNA-binding proteins (reviewed in Berg 1986; Klug and Rhodes 1987). Based on the amino acid sequence of GR, two zinc fingers could theoretically be formed each of them with a zinc atom tetrahedrally coordinated to four cysteine residues (Giguere et al. 1986; Severne et al. 1988). Binding of zinc atoms by the DNA-binding domain of GR has been proven experimentally (Freedman et al. 1988). Deletions and insertions in regions encompassing the putative zinc fingers destroy the DNA-binding property of GR (Giguere et al. 1986; Rusconi and Yamamoto 1987; Danielsen et al. 1987; Hollenberg et al. 1987). Recently, the three-dimensional structure of the DNA-binding domain of GR has been determined using magnetic resonance spectroscopy and distance geometry (Härd et al. 1990). A model of the dimeric complex between the DNA-binding domain of GR and the GRE has been proposed (Härd et al. 1990).

1.2.3 *Trans*-Activation Domain

A *trans*-activation domain is defined as a protein region that does not bind DNA by itself, but in combination with the DNA-binding domain it increases transcriptional initiation by RNA polymerase II (for reviews see Sigler 1988; Guarente 1988). Two such regions have been identified in human GR; one lies in the amino-terminal part and the other lies in a region between the DNA- and hormone-binding domains (Hollenberg and Evens 1988). These regions are rich in acidic amino acids. The function of these highly negatively charged residues is probably to stimulate the formation or activity of a transcriptional preinitiation complex (Sigler 1988). Similar regions with comparable functions have also been mapped in the rat (Miesfeld et al. 1987) and mouse (Danielsen et al. 1987) GR.

1.3 Transcriptional Activation of Gene Expression by GR

The mechanism of GR mediated transcriptional enhancement of gene expression has been intensively studied during the past few years. It is generally accepted that GR, upon binding with hormone, interacts with enhancer-like DNA sequences of target genes, and thereby alters transcription of responsive genes (for review see Yamamoto 1985).

1.3.1 Consensus Sequence of GRE

Specific DNA recognition sites for GR have initially been identified and localized within the long terminal repeat of mouse mammary tumor virus (MMTVLTR) using in vitro DNA-binding techniques (Payvar et al. 1981; Scheidereit et al. 1983; Scheidereit and Beato 1984). Transfection experiments reveal that these viral

DNA fragments, when put in front of heterologous promoters in either orientation, confer glucocorticoid responsiveness to reporter genes (Lee et al. 1981; Chandler et al. 1983; Hynes et al. 1983; Majors and Varmus 1983; Ponta et al. 1985). Similar DNA sequences have been found at variable distances upstream from promoters or even within introns of several other glucocorticoid-responsive genes, e.g., human metallothionein IIA (Karin et al. 1984), rat tyrosine aminotransferase (TAT: Becker et al. 1986; Strähle et al. 1987), chicken lysozyme (Renkawitz et al. 1984; von der Ahe et al. 1985, 1986), rat tryptophan oxygenase (Danesch et al. 1987), phosphoenolpyruvate carboxykinase (Petersen et al. 1988), and growth hormone (Moore et al. 1985; Slater et al. 1985). Comparison of these GR binding sites revealed a 15-mer consensus GRE:

5'-GGTACAnnnTGTTCT-3'

which is an imperfect palindrome. Since these regulatory elements all share the characteristics of enhancers (Serfling et al. 1985; Johnson and Simon 1987), i.e., affect transcription independently of their orientation and position relative to homogolous or heterologous promoters, they are defined as hormone-dependent enhancer elements (Zaret and Yamamoto 1984; Ponta et al. 1985).

1.3.2. Interaction of GR with GRE

Transcriptional activation of genes occurs when *trans*-acting factors assemble on specific *cis*-acting elements, thereby altering initiation by RNA polymerase II (Ptashne 1986, 1988). According to the classical model of steroid hormone action, the hormone is required to change the structure of the receptor leading to acquisition of DNA binding activity. In vivo experiments showing that binding of GR to GRE requires binding of the receptor to the hormone are consistent with the classical model (Becker et al. 1986). Groyer et al. (1987) and Denis et al. (1988) have shown that association of GR with hsp90 stabilizes the receptor in its inactive form. Hormone is needed to release the receptor from hsp90 and triggers an activation process which results in the conversion of the receptor from a non-DNA-binding to a DNA-binding form.

The dyad symmetrical configuration of GRE suggests that the receptor binds to DNA as a dimer. In connection with this, Wrange et al. (1989) have shown that purified GR binds to DNA as a homodimer in vitro. Using a GR derivative that consists of the DNA-binding domain of GR only, Tsai et al. (1988) could show that the binding of one receptor to half of the 15-mer GRE facilitates the binding of another receptor to the other half of the GRE, implicating cooperativity of receptor dimerization.

1.3.3. Cooperativity in Function of GRE with Other *Cis*-Acting Elements

The upstream region of eukaryotic genes is generally composed of multiple regulatory elements. An important question concerns how the enhancer-like GRE functions together with other distinct transcriptionally regulatory elements. Induction of the rat tryptophan oxygenase gene by glucocorticoids requires both a GRE and a CACCC-box sequence implying cooperativity in function between GR and another transcription factor (Schüle et al. 1988). By in vivo exonuclease protection assay, Cordingley et al. (1987) have shown that activation at MMTV promoter by glucocorticoids is accompanied by the binding of nuclear factor-1 and probably other factors also, suggesting that the interaction of GR with GRE facilitates the interaction of a complex array of *trans*-acting factors at the promoter. Synergistic action between adjacent GREs or between GRE and other steroid responsive elements has also been reported in several models (Jantzen et al. 1987; Ankenbauer et al. 1988; Cato et al. 1988; Strähle et al. 1988; Tsai et al. 1989; Schmid et al. 1989).

1.4 Negative Control of Gene Expression by GR

In many physiological systems, negative feedback regulation is established to balance the stimulatory effects of hormones on gene transcription. Many genes are transcriptionally stimulated by glucocorticoids, but there are several genes that are transcriptionally repressed by glucocorticoids, e.g., proopiomelanocortin (POMC) (Charron and Drouin 1986), prolactin (Camper et al. 1985), chorionic gonadotropin alpha-subunit (Akerblom et al. 1988), and α-fetoprotein (Huang et al. 1985). Unlike transcriptional activation, very little is known about the mechanisms underlying GR-mediated *trans*-inhibition of gene expression. The interesting question is how it is possible for a single transcription factor to act in two opposite ways. Following the same procedure as for the investigation of gene activation, several groups have tried to study negative GREs (nGREs) and functional structural domains of GR with regard to gene suppression by glucocorticoids.

1.4.1 Negative GRE

Hypothetically, there are many possible mechanisms for hormonal repression of gene expression. For instance, nGRE, distinct in sequences from the well-known positive GRE, might alter the structure of receptor upon binding, thus, preventing its function as a transcriptional activator. Although there is still no solid evidence for a defined consensus sequence of nGRE, some studies point out clear differences in GR binding sites on negatively regulated genes. Results from the studies of the promoter region of the prolactin gene reveal multiple binding sites for GR between −51 and −562 bp which are apparently different from the consensus elements that confer positive regulation by glucocorticoids (Sakai et al. 1988). Since these sequences confer repression by glucocorticoids in vivo when linked to heterologous promoters, they are functionally opposing the positive GRE and thus, named nGRE (Sakai et al. 1988). By gene transfer and in vitro DNA binding studies, one GR binding site at position −63 in the promoter region of the POMC gene is found to be important for inhibition by glucocorticoids (Drouin et al. 1987). Even though the sequence of this binding site is similar to the GR binding sites of inducible genes, there are still remarkable differences in certain nucleotides between positive GRE and POMC GRE (Drouin et al. 1987). By comparing 11 binding sites for GR identified in negatively regulated genes, Beato et al. (1989) have recently proposed, by analogy to the 15-mer positive GRE, a 15-mer consensus for nGRE:

5'-ATYACnnTnTGATCn-3'

Whether this 15-mer sequence is common to all glucocorticoid-inhibited genes and whether the differences found in GR binding sequences are the real key for differentiating positive and negative effects remains to be determined.

1.4.2 Functional Domains of GR in Gene Inhibition

It is also possible that positive and negative regulation utilize functionally distinct domains of the receptor. Adler et al. (1988) reported that negative regulation of the prolactin gene by estrogens does not require the DNA-binding domain of estrogen receptor (ER). However, sequences within the "hinge" region (amino acids 251–314 of the human ER) are important suggesting that *trans*-repression and *trans*-activation are determined by different functional regions of the receptor. Since no DNA binding is required, a model of direct protein-protein interaction between a transcriptional activator and the ER is postulated for the ER-exerted inhibition of prolactin gene expression (Adler et al. 1988). Akerblom et al. (1988) showed that the human GR binds to the cAMP-inducible chorionic gonadotropin alpha-subunit promoter and possibly mediates negative regulation through interference with other enhancer binding factors. Subsequently, it has been shown that the inhibition of chorionic gonadotropin alpha-subunit promoter by human GR requires both the DNA- and ligand-binding domains of the receptor. However, the ligand-binding domain can be substituted for by a heterologous peptide and functions as a constitutive repressor suggesting that the human GR negatively regulates transcription via a steric hindrance mechanism (Oro et al. 1988). This type of interference would displace or mask the binding sites for enhancer/promoter binding factors.

1.5 Posttranscriptional Regulation: mRNA Stability

The steady-state level of cytoplasmic mRNA is determined by a balance between synthesis, processing, transport, and degradation. The rate of degradation of individual mRNA species, like the rate of synthesis, differs widely. The half-lives of mRNAs range from a few minutes (e.g., certain oncogenes) to hours or even days (e.g., β-globin). The stability of some mRNA species can be changed dramatically in response to hormonal induction, differentiation, and viral infection (summarized by Shapiro et al. 1987). Not until the early 1980s was the regulation of mRNA stability by cellular processes and hormonal stimuli recognized as important for gene expression. Unfortunately, still very little is known about the mechanisms underlying the regulation of mRNA turnover. Conceivably, there might be at least three determinants for this response: the intracellular signal, the

RNA degradation system, and structural characteristics on mRNA for degradative attack. Destruction of mRNA takes place in the cytoplasm by ribonucleases (RNases). Our knowledge concerning these degrading enzymes is very limited. To date, no site-specific RNases have been identified in eukaryotic cells. However, recent progress on the decay of mRNAs has provided evidence that the rate at which RNases degrade mRNAs is partly determined by the structural features of the mRNA molecule.

1.5.1 Sequence Elements that May Control mRNA Stability

A highly conserved AU-rich region within the 3' untranslated region is found in some polyadenylated mRNA species. When such an AT-rich sequence of 50–60 nucleotides from the 3' untranslated region of the human lymphokine gene was fused to 3' of the rabbit β-globin gene, the otherwise very stable β-globin mRNA became highly unstable (Shaw and Kamen 1986). Recently, Wilson and Treisman (1988) showed that the AU-rich sequence on the c-fos mRNA facilitates poly(A) shortening and consequently increased the degradation of c-fos mRNA. Conceivably, the AU-rich sequence functions as a signal element for nuclease attack on the poly(A) sequence. It has been shown that the first step in mRNA decay is the poly(A) shortening (Brewer and Ross 1988). A relation between the length of poly(A) and the rate of mRNA degradation has been observed in some mRNAs (Muschel et al. 1986; Paek and Axel 1987).

β-tubulin mRNA stability is autoregulated by the concentration of free tubulin subunits. The regulatory sequences responsible for this autoregulation reside within the first 13 nucleotides of β-tubulin mRNA. However, tubulin subunits destabilize its mRNA through interaction with newly synthesized amino-terminal β-tubulin tetrapeptides (encoded ny the first 13 nucleotides), rather than recognition of specific RNA sequences (Yen et al. 1988). It is not known how this cotranslational protein-protein interaction triggers β-tubulin mRNA degradation.

Stem-loop structures in the 3' untranslated region seem to determine the stability of iron-regulated human transferrin receptor mRNA (Müllner and Kühn 1988). These RNA stem-loops are termed iron-responsive elements (IREs; Casey et al. 1988). An IRE binding protein (IREB) has recently been purified from human liver (Rouault et al. 1989). The binding activity of IREB to IRE is decreased by iron and increased by iron chelator in parallel with the transferrin mRNA stability. Thus, the function of the IREB is to protect the transferrin receptor mRNA from degradation (Müllner et al. 1989).

Taken together, certain structural features within mRNA molecules are the determinants of mRNA turnover. It is obvious that mRNA turnover is a rather complex event and many cellular processes are involved in the determination of mRNA decay rates, e.g., polysome association (Pachter et al. 1987), translation (Graves et al. 1987), and *trans*-acting factors (e.g., IREB).

1.5.2 Hormonal Regulation of mRNA Stability

In many cases, cells control mRNA levels through selective changes of mRNA stability in response to hormonal stimuli. In *Xenopus* liver cells, the massive accumulation of vitellogenin mRNA in the presence of estrogens is achieved mainly by increasing the half-life of the mRNA (Brock and Shapiro 1983). The sequence element which is capable of conferring the estrogen-induced stabilization of vitellogenin mRNA lies within the 3' end of the transcript (Shapiro et al. 1987). In contrast to the vitellogenin mRNA, estrogen destabilizes albumin mRNA in xenopus liver cells (Kazmaier et al. 1985; Wolffe et al. 1985). Glucocorticoids enhance stability of human growth hormone mRNA (Diamond and Goodman 1985; Paek and Axel 1987). It has been observed that the poly(A) tract of human growth hormone mRNA increases in length with time after addition of glucocorticoids and decreases in length after hormone withdrawal, suggesting that poly(A) tailing protects the RNA from nuclease attack (Paek and Axel 1987). Although many more examples exist, the molecular mechanisms of hormonally regulated mRNA turnover are not yet understood.

1.6 Regulation of GR Expression

1.6.1 Regulation by Glucocorticoids

Based on ligand binding assays, there are numerous reports available documenting the effects of glucocorticoids on their own receptor. Svec and Rudis (1981) first reported on GR autoregulation in AtT-20 mouse pituitary tumor cells. Incubation of the cells with dexamethasone for 48–96 h results in a 75% diminution of total cellular GR content. Later, similar findings were reported in a number of different cell types including human HeLa S_3 cells (Cidlowski and Cidlowski 1981), normal human lymphocytes (Schlechte et al. 1982), murine GH_1 pituitary cells (Raaka and Samuels 1983), mouse thymoma-derived cell line W7 (Danielsen and Stallcup 1984), and human skin fibroblasts (Oikarinen et al. 1987). Also in intact animals, administration of glucocorticoids leads to a decreased number of hormone-binding sites (Ichii 1981; de Nicola et al. 1982; Sapolsky et al. 1984). In contrast, adrenalectomy elevates the level of GR in most, if not all, glucocorticoid target tissues, indicating an inhibitory effect of glucocorticoids on their own receptor expression (Gregory et al. 1976; Smith and Shuster 1984; Turner 1986). Thus, downregulation of GR by its cognate ligand is a common feature in a variety of glucocorticoid target cells. Furthermore, continuous stress reduces glucocorticoid-binding sites in certain brain areas, suggesting that downregulation of GR may constitute a physiological phenomenon (Sapolsky et al. 1984).

1.6.2 Regulation by Cyclic Adenosine Monophosphate

Modulation of GR activity by cyclic adenosine 3', 5'-monophosphate (cAMP) was first studied in cultured human skin fibroblast cells by Oikarinen et al. (1984). They showed that addition of cAMP to the culture medium induced the number of glucocorticoid-binding sites. Interestingly, the fluctuation of cellular cAMP concentration during cell growth paralleled the changes in GR concentration and glucocorticoid response. More direct evidence showing the involvement of cAMP-dependent protein kinase in regulating GR function came from studies of cAMP-resistant murine lymphoma W7 cell variants (Gruol et al. 1986a). Firstly, Gruol et al. (1986b) showed that cAMP or cAMP-inducing agents increased the number of glucocorticoid-binding sites per cell in wild-type but not in cAMP-resistant W7 cells that carried defects in the kinase activity. Secondly, the frequency of acquisition of glucocorticoid resistance was much higher in cAMP-resistant phenotype than in wild-type cells indicating a connection between the kinase activity and GR function (Gruol et al. 1986b).

The increase in cellular glucocorticoid-binding sites by cAMP could have two possible explanations: (1) conversion of the receptor from a non-hormone-binding to a hormone-binding form, and (2) de novo synthesis of receptor protein. In early experiments, Munck and Brink-Johnson (1968) found that GR binding capacity disappeared in ATP-depleted cells and rapidly reappeared when ATP levels were restored. This process was protein synthesis-independent (Bell and Munck 1973), leading to the proposal that GR exists in two functionally distinct forms, the non-hormone-binding and the hormone-binding form. The existence of non-hormone-binding receptor in the nuclei of ATP-depleted cells was confirmed by Mendel et al. (1986). Furthermore, it was shown that GR is a phosphoprotein (Housley and Pratt 1983) and undergoes ligand-dependent phosphorylation in the cytoplasm and dephosphorylation in the nucleus (Orti et al. 1989). It is conceivable that cAMP-dependent protein kinase could play a role in the catalytic conversion of non-hormone-binding to hormone-binding receptor via phosphorylation. In line with this, Singh and Moudgil (1985) demonstrated phosphorylation of purified GR by cAMP-dependent protein kinase in vitro. Whether this is also true in vivo remains to be determined. At this stage, it is difficult to distinguish between the two possible mechanisms mentioned above by which cAMP modulates GR function. Nevertheless, it is clear that cAMP is an important effector in the determination of cellular GR levels and, ultimately, glucocorticoid responsiveness.

1.7 Glucocorticoid Resistance

1.7.1 Resistance Due to Defects in GR

Glucocorticoid resistant cell lines are generally derived from cell culture systems. Most of the resistant clones isolated so far carry either defective receptor or low receptor activity (for reviews see Yamamoto et al. 1976; Stevens et al. 1983; Bourgeois and Gasson 1985; Gehring 1987; Thompson et al. 1988). The receptor deficient (r^-) phenotype has low or undetectable hormone binding activity (Sibley and Tomkins 1974). The nuclear transfer deficient (nt^-) cells contain GR which binds the hormone but has reduced affinity to both nonspecific (Gehring and Tomkins 1974) and specific (Pfahl 1982) DNA. Recently, a mutant receptor protein which does not bind the hormone has been detected in the nt^- S49 variant cells by a monoclonal antibody against GR (Northrop et al. 1985). This mutant receptor is actually encoded by a cDNA carrying a single amino acid substitution (Glu-546 to Gly) in the hormone-binding domain (Danielsen et al. 1986). Another cDNA isolated from the same cell encodes the classical nt^- type receptor and the lesion has been mapped to a single amino acid substitution (Arg-484 to His) located in the DNA-binding domain (Danielsen et al. 1986). The increased nuclear transfer (nt^i) variant has a receptor with increased affinity for nonspecific DNA after binding the hormone (Yamamoto et al. 1976). This mutant receptor has a molecular weight of 40kDa (Dellweg et al. 1982; Gehring and Hotz 1983) and is translated from a 5' truncated receptor mRNA (Miesfeld et al. 1984), probably due to a splicing defect (Miesfeld 1989).

1.7.2 Resistant Cells Contain Functional GR

Defective receptor and insufficient amounts of receptor are certainly important, but not the only determinants of hormone resistance in target cells. The lysis-resistant human leukemic clonal cell line CEM-C1 has a normal amount of functional GR as assessed by hormone binding and induction of glutamine synthetase (Zawydiwski et al. 1983). Somatic hybrids of CEM-C1 and a receptor-deficient variant clone gained sensitivity to glucocorticoid-induced cell lysis, indicating that a nonreceptor defect prevents the lymphocytolytic process in CEM-C1 (Yuh and Thompson 1987). The mouse lymphoma SAK-8 cell line has characteristics similar to CEM-C1, i.e., sensitive to hormonal stimulation of responsive genes, but resistant to lysis (Gasson and Bourgeois 1983). It has been suggested that the lysis resistance in SAK-8 cells might be caused by DNA methylation of lethality gene(s) since 5-azacytidine (an inhibitor of DNA methylation) can alter lysis resistance in the cell (Bourgeois and Gasson 1985).

Even when cells retain very low levels of GR, they are not necessarily resistant to all glucocorticoid inducible cellular responses. An MMTV-infected W7 cell variant has only 25% of the GR level found in wild-type cells and is resistant to the cytolytic effects of glucocorticoids, but expresses increased levels of MMTV in response to the hormone (Rabindran et al. 1987). This may indicate that different responses to glucocorticoids have different threshold level requirements regarding the cellular content of GR. However, the magnitude of the response might be correlated to the level of functional GR. A resistant rat hepatoma cell line 6.10.2 following transfection with a GR expression vector, is sensitive to transcriptional regulation of gene expression by glucocorticoids (Miesfeld et al. 1986). Moreover, the magnitude of response is proportional to the level of GR expressed (Vanderbilt et al. 1987). Thus, the presence of functional receptors in the cell is a prerequisite for glucocorticoid responsiveness, but it is not sufficient to guarantee glucocorticoid sensitivity as other factors might be involved in the determination of certain responses. In cases where GR is the limiting factor, glucocorticoid sensitivity seems to correlate with cellular GR levels.

2 Results and Discussion

2.1 Mechanisms of GR Autoregulation

Administration of glucocorticoids either in vitro or in vivo led to a decrease in cellular glucocorticoid binding sites (see Sect. 1). We have used a cDNA clone for the rat GR (Miesfeld et al. 1986) and monoclonal anti-GR antibodies (Okret et al. 1984) to study the mechanism of regulation of GR mRNA and protein levels. We could show that both in rat hepatoma tumor culture (HTC) cells and liver, GR mRNA levels were decreased by 60%–80% in response to treatment with dexamethasone. In parallel, a similar decrease in immunoreactive GR was also observed. The downregulation of GR by dexamethasone was dose-dependent and a maximal response was achieved at 10–20 n*M* close to the concentration for receptor saturation (see Okret et al. 1986; Dong et al. 1988a).

Glucocorticoid-induced cellular responses are mediated by several distinct pathways including both transcriptional and posttranscriptional mechanisms (for a review see Ringold 1985). We carried out experiments to elucidate possible mechanisms of action of glucocorticoids in GR downregulation. The stability of GR mRNA in the presence or absence of dexamethasone was studied by inhibition of transcription with actinomycin D. The half-life of GR mRNA was found to be approximately 4–5h in both control and hormone-treated HTC cells indicating that dexamethasone has no effect on the turnover rate of GR mRNA (Fig. 2). In line with this, a half-life of 4 h of GR mRNA has been determined in both hormone-treated and nontreated IM-9 cells (Rosewicz et al. 1988).

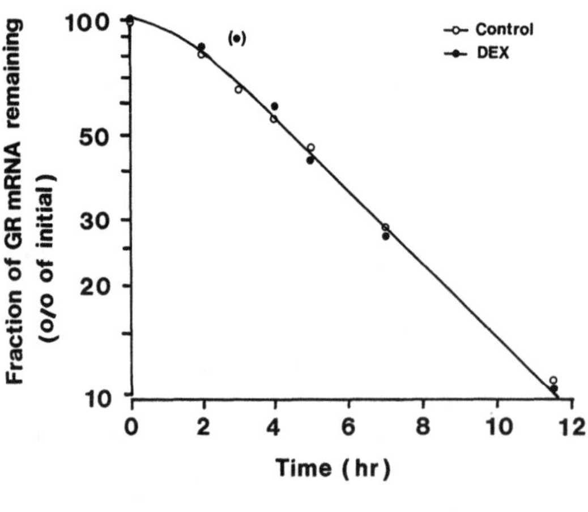

Fig. 2. Decay or GR mRNA in the presence or absence of glucocorticoids. HTC cells were grown in the presence or absence of 0.5-μ*M* dexamethasone (*DEX*) for 24 h. Thereafter, 5 μg/ml actinomycin D were added to the cells for indicated times. The rate of GR mRNA decay was determined by measuring the GR mRNA levels by RNA blot-hybridization (for details see Dong et al. 1988)

We also determined whether dexamethasone affected the turnover rate of GR protein in HTC cells after inhibition of translation with cycloheximide. A half-life of about 25 h for GR protein was estimated in the absence of hormone (Fig. 3). In the presence of hormone, the degradation of GR protein occurred with a biphasic decay pattern; an initial rapid phase was followed by a slower decay rate later. The fast decay of GR protein showed a half-life of about 11 h, which is 2.5-fold faster than the corresponding rate in nontreated cells (Fig. 3). Using a dense amino acid labeling method, McIntyre and Samuels (1985) have shown half-lives of GR to be 19 h in the absence and 9.5 h in the presence of triamcinolone acetonide in GH_1 cells. Their data agree well with our immunochemically determined half-lives of GR protein following inhibition of translation. Furthermore, Hoeck et al. (1989) have recently shown that the turnover rate of GR is increased 2.7-fold by dexamethasone in NIH3T3 cells. The faster decay rate of GR in the presence of glucocorticoids is agonist-dependent since no change in receptor half-life was observed following treatment of IM-9 cells with the antiglucocorticoid RU486 (Rajpert et al. 1987). The molecular mechanism for the ligand-induced increase in the turnover of GR protein is not known. It has been shown that ligand-dependent receptor phosphorylation is not involved in the downregulation of GR (Hoeck et al. 1989), suggesting that phosphorylation-induced covalent modification of GR is not responsible for the increased turnover of GR in the presence of glucocorticoids. It is possible that activated receptor may have a structure which is more susceptible to proteolytic attack, e.g., due to dissociation from hsp90.

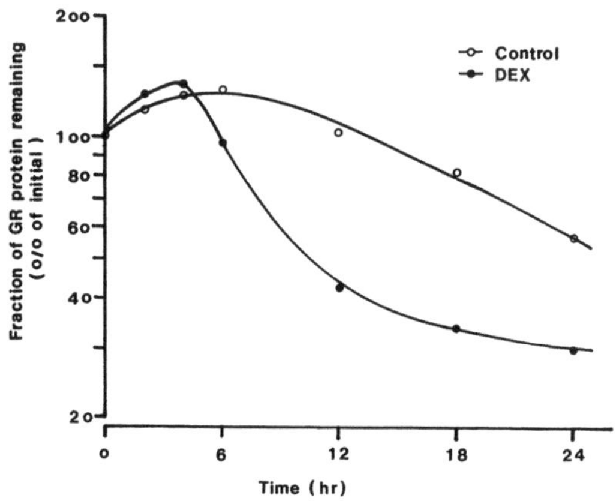

Fig. 3. Decay of GR protein in the presence or absence of glucocorticoids. HTC cells were grown in the presence or absence of 0.5-μ*M* dexamethasone (*DEX*) for 48 h. Thereafter, cells were exposed to 1,5 μg/ml cycloheximide. The rate of GR protein decay was determined by measuring total cellular GR protein levels by western immunoblotting (for details see Dong et al. 1988)

Using a nuclear run-on assay, we found that downregulation of GR mRNA levels in rat liver was caused by a decrease in the transcription rate of the GR gene (Fig. 4). Consistent with our observation, the decrease in GR mRNA levels by glucocorticoids in IM-9 cells has been reported to be due to a reduced GR gene transcription (Rosewicz et al. 1988). Also, in rat colon and intestine, a similar observation has been made by Walsh et al. (1987). Several lines of evidence have accumulated showing that transcriptional modulation of gene expression by glucocorticoids is mediated by the receptor protein itself (Yamamoto 1985). In the mutant nonresponsive rat hepatoma 6.10.2 cells, which contain very low levels of GR protein, glucocorticoid-induced downregulation of GR mRNA could not be detected (see below and Dong et al. 1990). This result might indicate that the decrease in GR mRNA levels by glucocorticoids seen in wild-type cells is a GR-mediated response. Furthermore, dexamethasone-dependent downregulation of GR mRNA occurs even in the presence of cycloheximide (see below), suggesting the involvement of preexisting regulatory factor (s), e.g., GR.

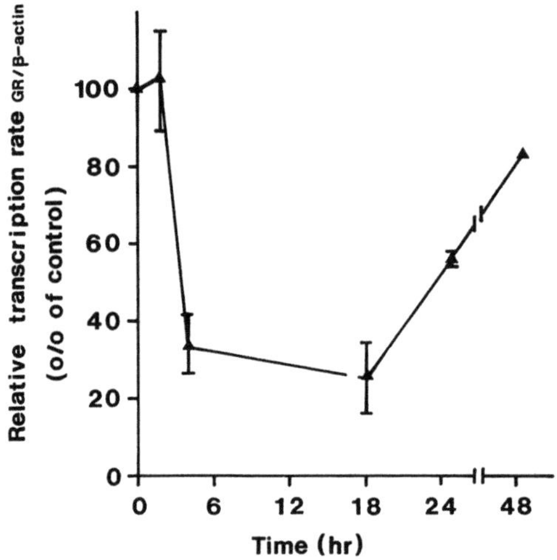

Fig. 4. Changes in transcription of GR gene in rat liver by glucocorticoids. Rats were adrenalectomized 9 days before treatment with dexamethasone-1-phosphate (4mg/kg). Cell nuclei were isolated from the livers and nuclear run-on transcription experiments were performed (for details see Dong et al. 1988)

2.2 Mechanism of cAMP-Induced GR Expression

In contrast to glucocorticoids, cAMP has been reported to increase cellular glucocorticoid-binding capacity (Oikarinen et al. 1984; Gruol et al. 1986). To further understand regulation of cellular GR concentrations by cAMP, we performed experiments elucidating the relationship between hormone-binding activity, GR protein concentrations, and GR mRNA levels. Our results indicate that treatment of HTC cells with forskolin (a cAMP-inducing agent) raised the number of glucocorticoid-binding sites by a factor of 2 as determined by a whole cell binding assay. Immunochemical determination of GR concentrations in forskolin-treated cells also revealed a parallel increase. Moreover, the steady-state level of GR mRNA was about two-fold higher in cells treated with forskolin than in control cells (Dong et al. 1989). Taken together, an increased receptor biosynthesis seemingly accounts for the increased hormone-binding activity. A cAMP analogue, 8-bromo cAMP was also used in order to show a direct involvement of cAMP in the induction of GR expression. Our data also show that the increased steady-state level of GR mRNA by cAMP was accompanied by decreased degradation of the receptor messenger providing evidence for posttranscriptional regulation of gene expression by cAMP (Fig. 5).

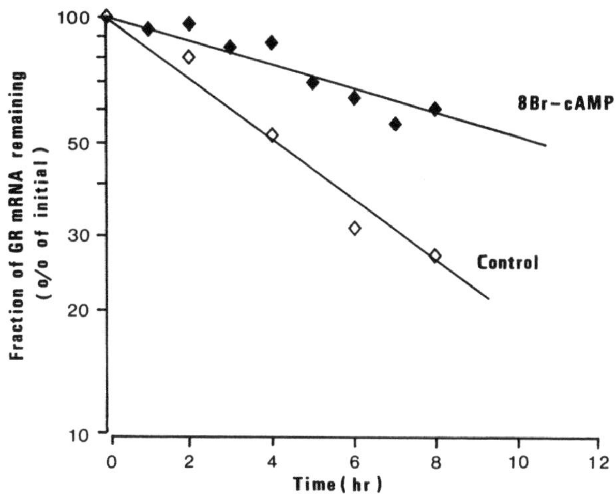

Fig. 5. Effect of cAMP on GR mRNA stability. HTC cells were grown with or without 50-μ*M* 8-bromo-cAMP (*8Br-cAMP*) for 16 h; then 5 μg/ml actinomycin D was added to the cell culture. The GR mRNA levels were determined by RNA blot-hybridization (for details see Dong et al. 1989)

2.3 Correlation Between Glucocorticoid Receptor Concentration and Responsiveness

The biological responses to steroid hormones are influenced by multiple factors. Changes in cellular receptor concentrations represent one potential mechanism which could eventually influence cellular hormone responsiveness. A variety of experimental approaches have been used to study the relationship between GR level and hormonal inducibility of target genes. In many systems, the degree of response has been directly related to the concentration of receptor. Prolonged exposure of cells to glucocorticoids often leads to a downregulation of GR followed by reduced hormone sensitivity (Shirwany et al. 1986; Yang et al. 1989). We show here that dexamethasone-induced depletion of GR was prevented by a simultaneous treatment with forskolin which concomitantly amplified the hormone-induced expression of both TAT and MMTV (Fig. 6). This result supports the notion that the cellular GR concentration is a limiting factor in glucocorticoid sensitivity of responsive genes.

2.4 Glucocorticoid Resistance

Steroid hormone resistance is often related to defective receptor or low receptor quantity. However, this is not the case in all glucocorticoid-resistant cell variants. It has been reported that a hormone-unresponsive cell line possesses not only functional but also sufficient amounts of receptors for both androgens and glucocorticoids (Darbre and King 1987). This implies that progression to steroid insensitivity can occur irrespectively of the presence of functional steroid receptors. We have studied the mechanisms of glucocorticoid resistance in the rat hepatoma cell variant 6.10.2 cells. The glucocorticoid resistance of 6.10.2 cells was defined by its inability to induce transcriptional activation of MMTV and TAT as well as transcriptional suppression of GR mRNA. Characterization of GR in 6.10.2 cells showed that the cells contain GR with normal physiochemical properties as assessed by both hormone- and DNA-binding assays, although the total number of receptor molecules is only 20% of that in wild-type rat hepatoma 762 cells. The mRNA level of GR is also only 20% of that in 762 cells, indicating that the lesion influenced expression of the GR gene and/or stability of the GR mRNA (Dong et al. 1990).

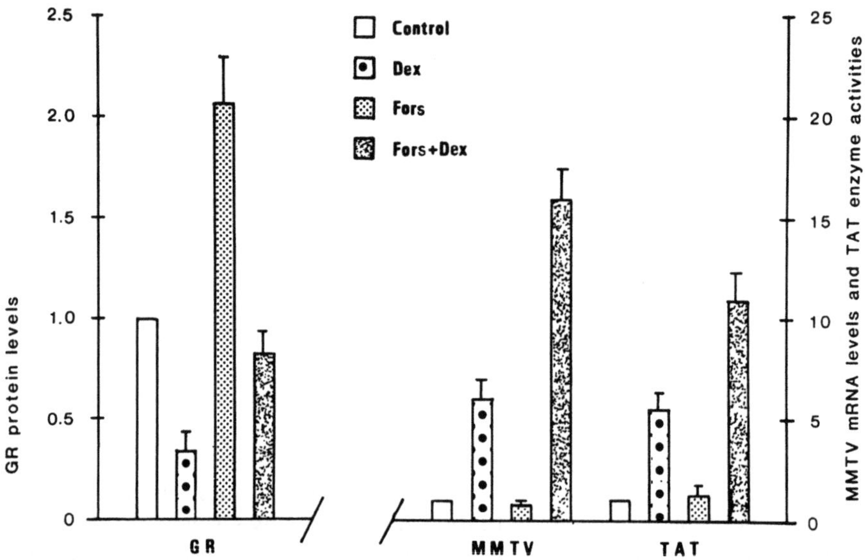

Fig. 6. Correlation between cellular GR protein level and the inducibility of MMTV and TAT by glucocorticoids. Rat hepatoma 762 cells were exposed to 0.5-μM dexamethasone (*DEX*) in the presence or absence of 25-μM forskolin (*Fors*) for 18 h. The level of GR protein relative to control was measured by western immunoblotting. The mRNA amount of MMTV was determined by RNA blot-hybridization. The enzyme activity of TAT relative to control was studied in isolated cell extract (Dong et al. 1989)

To finally investigate the basis for glucocorticoid resistance in 6.10.2 cells, it is important to analyze the biological activity of the residual number of GR present in the cells. To achieve this, one has to increase the endogenous receptor number since the quantity of receptor present under normal culture conditions was insufficient to elicit hormone-dependent responses. Therefore, we utilized the effect of cAMP on GR expression and increased the GR level about two-fold in 6.10.2 cells (Fig. 7). This slight increase in receptor levels altered the cellular glucocorticoid sensitivity as indicated by inductions of both MMTV and TAT, and downregulation of GR mRNA (Fig. 7). These findings strongly suggest the exist-

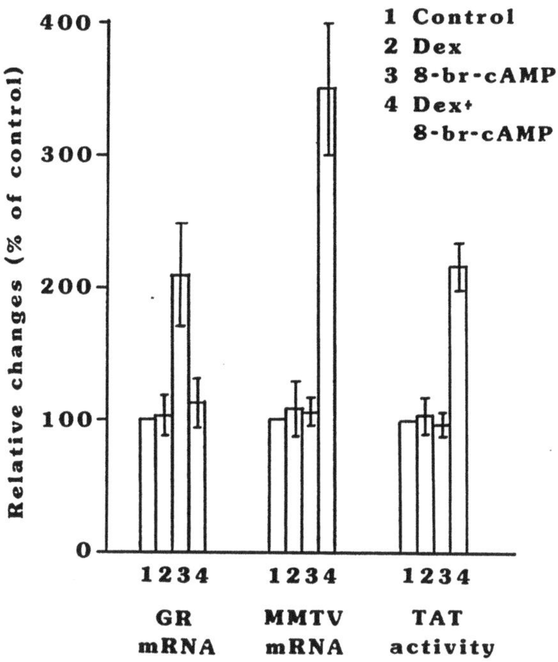

Fig. 7. Effect of cAMP on glucocorticoid-mediated cellular responses. Rat hepatoma 6.10.2 cells were pretreated with 50-μM 8-bromo-cAMP (*8-br-cAMP*) for 4 h and then exposed to 0.5-μM dexamethasone (*Dex*) for 18 h. Total cellular RNA was isolated and analyzed for GR mRNA and MMTV mRNA. Parallel dishes of cells with the same treatment were used for the measurement of TAT activity (Dong et al. 1990)

ence of biologically functional GR protein in 6.10.2 cells. We propose that the basis for glucocorticoid resistance in 6.10.2 cells may be a GR level below the threshold for hormone-dependent responses to occur.

2.5 Effects of Cycloheximide

The effects of cycloheximide on transcriptional modification of certain genes have in several cases led to the hypothesis that labile regulatory proteins may exist in mammalian cells (Elder et al. 1984; Israel et al. 1985; Dinter and Hauser 1987; Klein et al. 1988). In HTC cells, treatment with cycloheximide increased GR

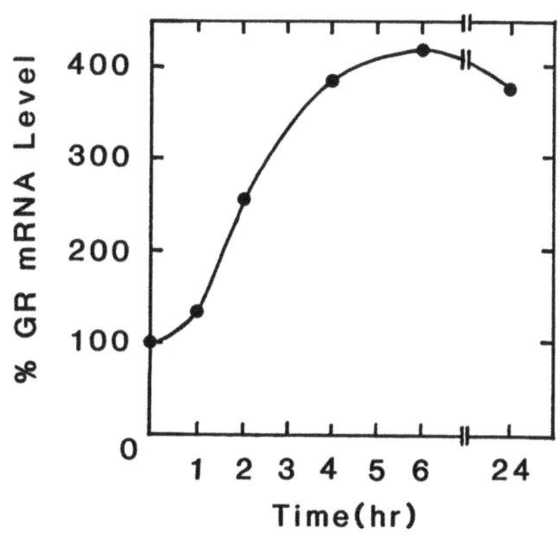

Fig. 8. Effect of cycloheximide on GR mRNA levels. HTC cells were incubated with 1.5 μg/ml cycloheximide. After various times/as indicated in the Fig.), total cellular RNA was isolated and the level of GR mRNA was analyzed for each sample

mRNA level about four-fold. This effect rapidly reached a maximal response after 4 h and was maintained for at least 24 h (Fig. 8). In glucocorticoid-resistant 6.10.2 cells, cycloheximide also increased GR mRNA about four-fold (Fig. 9). It is tempting to speculate that a short-lived repressor-like protein is controlling the basal level of expression of the GR gene in rat hepatoma cells. Removal of this protein by cycloheximide increases the transcriptional activity of the GR gene. Alternatively, cycloheximide may increase GR mRNA levels by decreasing its turnover. Further experiments are required to establish the mechanism underlying cycloheximide-induced GR mRNA levels.

In analogy to the effects on GR mRNA levels, treatment of 6.10.2 cells with cycloheximide increased the mRNA of MMTV by a factor of 2 (Fig. 9). Thus, it seems likely that the basal level of expression of MMTV is also under negative control. In line with this, it has been shown that a 96-bp DNA segment of the MMTV-LTR 5' to the GRE represses the basal activity of the proviral promoter in the absence of glucocorticoids (Morley et al. 1987). Moreover, Langer and Ostrowski (1988) have reported results showing that a *cis*-acting element distinct from the GR binding sites of MMTV represses transcription from its homologous promoter in vitro and heterologous promoter in vivo.

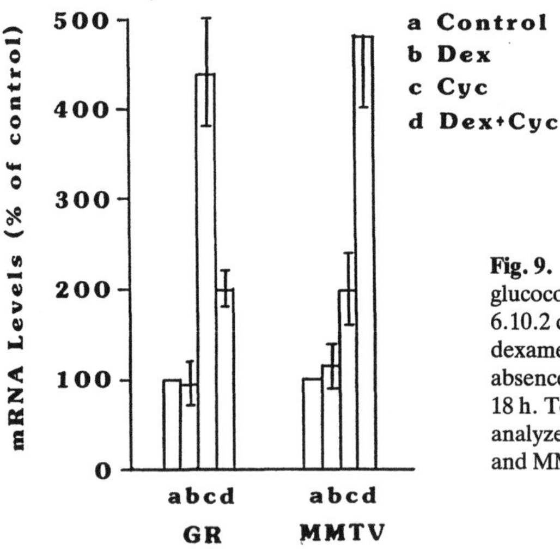

Fig. 9. Effect of cycloheximide on glucocorticoid-mediated cellular responses. 6.10.2 cells were treated with 0.5-μM dexamethasone (*Dex*) in the presence or absence of 1.5 μg/ml cycloheximide (*Cyc*) for 18 h. Total cellular RNA was extracted and analyzed for the levels of mRNA of both GR and MMTV (Dong et al. 1990)

We have shown that it is possible to restore glucocorticoid sensitivity in 6.10.2 cells by an increase in GR levels. In addition, in the presence of cycloheximide, it is also possible to regain glucocorticoid-dependent responses in 6.10.2 cells as assessed by transcriptional repression of the GR gene as well as the transcriptional induction of the MMTV gene (Fig. 9). Thus, either removal of a short-lived protein(s) or an increase in the GR concentration may be equally efficient in the restoration of glucocorticoid sensitivity in 6.10.2 cells. It is possible that the threshold level of GR required for certain glucocorticoid responses is eliminated by cycloheximide causing glucocorticoid sensitivity also in the absence of increased GR level.

2.6 Regulation of Alcohol Dehydrogenase Gene Expression by Glucocorticoids

Alcohol dehydrogenase (ADH) catalyzes the rate-limiting reaction in the metabolism of a wide variety of alcohols. We have used a cDNA probe encoding the human class I ADH to study the expression of rat ADH in HTC cells. The level of ADH mRNA was increased two-to four fold in the presence of dexamethasone (Dong et al. 1988b). Similar results have been reported by Hittle and Crabb (1988) using the same cell line. They showed that both ADH activity and mRNA levels were induced by dexamethasone. The turnover rate of ADH mRNA was measured after inhibition of transcription by actinomycin D. The data showed that the half-life of ADH mRNA was about 5–6 h in nontreated cells (Fig. 10). In cells treated with dexamethasone, the half-life of ADH mRNA was increased several-fold (Fig. 10) which might, at least partly, account for the increase in the steady-state levels of ADH mRNA.

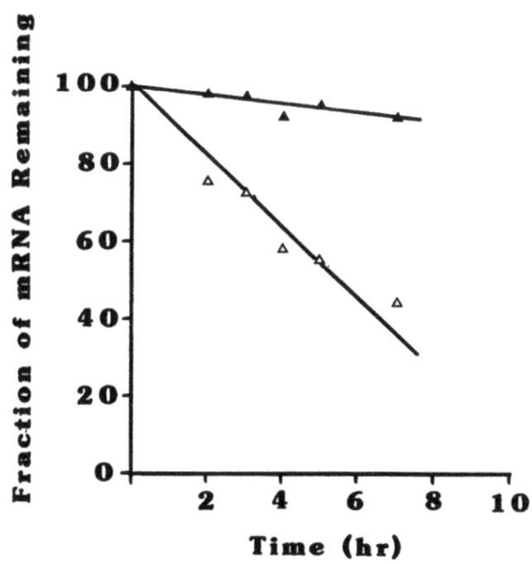

Fig. 10. Effect of glucocorticoid on class I ADH mRNA stability. HTC cells were grown in the absence (–Δ–) or presence (–▲–) of 0.5-μ*M* dexamethasone for 24 h. Thereafter, the cells were treated with actinomycin D (5 μg/ml) for the indicated times (Dong et al. 1988)

Both human and mouse ADH genes have been reported to contain sequences with a high degree of homology to the consensus GRE in the 5' flanking regions (Duester et al. 1986; Ceci et al. 1987). Using in vitro DNase I footprinting assays, several GR binding sites were detected within the 5' flanking region of the human β-ADH gene (one of the class I ADH genes). When DNA fragments containing these GR binding sites were linked to either homologous or heterologous promoters, they conferred glucocorticoid-induced expression of reporter genes as assessed by transfection assays (Winter et al. 1990). These results indicate that the human β-ADH gene contains *cis*-acting elements functionally identical to the GRE.

3 Summary and Conclusions

Research on hormonal control of gene expression has mainly been focused on promoter activation or transcriptional initiation. Our knowledge concerning the molecular mechanisms of gene activation by steroid hormones as well as cAMP-dependent hormones has increased considerably during the past few years. The present study has clearly shown that even posttranscriptional control levels (e.g., mRNA stability) of gene expression are important for the determination of cellular responses to hormonal stimuli. In many cases, more than one control mechanism is involved in the regulation of expression of a single target gene. The biological significance of such a complex network of gene regulation is not understood. However, such a control system results in an extremely fine adjustment of cellular responsiveness to endocrine stimuli.

The data from the present study show that the magnitude of glucocorticoid responses closely correlates with the cellular GR levels. Thus, the GR appears to be a limiting factor in the determination of glucocorticoid responsiveness. However, other factors may play a role, as hormone resistance can occur irrespectively of the presence of functional receptors (Darbre and King 1987; Wu and Pfahl 1988; John et al. 1989). In line with this contention, it has been clearly demonstrated that GR functions cooperatively with other transcription factors (Cordingley et al. 1987; Schüle et al. 1988). Negative *cis*-acting elements have been found in several steroid hormone inducible genes (Morley et al. 1987; Gaub et al. 1987; Langer and Ostrowski 1988). These negative elements control basal level of gene expression and inhibit enhancer activity of positive elements in the absence of an inducer. It indicates that *trans*-acting negative factors may be important in determining the hormonal sensitivity. The overall glucocorticoid responsiveness may therefore be a combined effect of promoter modulation by GR in the presence of hormone and promoter repression by negative factors in the absence of hormone. Glucocorticoid sensitivity could then depend on both the intracellular levels of GR and negative factors.

Acknowledgements. Some of the experiments presented in this review were performed in collaboration with Dr. Lorenz Poellinger, to whom we are most grateful.

References

Adler S Waterman ML, He X, Rosenfeld MG (1988) Steroid receptor-mediated inhibition of rat prolactin gene expression does not require the receptor DNA-binding domain. Cell 52:685–695

Akerblom IE, Slater EP, Beato M, Baxter JD, Mellon PL (1988) Negative regulation by glucocorticoids through interference with a cAMP responsive enhancer. Science 241:350–353

Ankenbauer W, Strähle U, Schütz G (1988) Synergistic action of glucocorticoid and estradiol responsive elements. Proc Natl Acad Sci USA 85:7526–7530

Beato M, Chalepakis G, Schauer M, Slater EP (1989) DNA regulatory elements for steroid hormones. J Steroid Biochem 32:737–748

Becker PB, Gloss B, Schmid W, Strähle U, Schütz G (1986) In vivo protein-DNA interactions in a glucocorticoid response element require the presence of the hormone. Nature 324:686–688

Bell A, Munck A (1973) Steroid-binding properties and stabilization of cytoplasmic glucocorticoid receptors from rat thymus cells. Biochem J 136:97–107
Berg JM (1986) Potential metal-binding domains in nucleic acid binding proteins. Science 232:485–487
Bourgeois S, Gasson JC (1985) Genetic and Epigenetic bases of glucocorticoid resistance in lymphoid cell lines. In: Littwack G Biochemical actions of hormones, vol. 12. (ed) Academic, New York pp 311–350
Bresnick EH, Dalman FC, Sanchez ER, Pratt WB (1989) Evidence that the 90-kDa heat shock protein is necessary for the steroid binding conformation of the L cell glucocorticoid receptor. J Biol Chem 264:4992–4997
Brewer G, Ross J (1988) Poly (A) shortening and degradation of the 3' A+U-rich sequences of human c-myc mRNA in a cell-free system. Mol Cell Biol 8:1697–1708
Brock ML, Shapiro DJ (1983) Estrogen stabilizes vitellogenin mRNA against cytoplasmic degradation. Cell 34:207–214
Casey JL, Hentze MW, Koeller DM, Caughman SW, Rouault TA, Klausner RD, Harford JB (1988) Iron-responsive elements:regulatory RNA sequences that control mRNA levels and translation. Science 240:924–928
Camper SA, Yao YAS, Rottman FM (1985) Hormonal regulation of the bovine prolactin promoter in rat pituitary tumor cells. J Biol Chem 260:12246–12251
Carlstedt-Duke J, Strömstedt PE, Persson B, Cederlund E, Gustafsson J-Å, Jörnvall H (1988) Identification of hormone-interacting amino acid residues within the steroid-binding domain of the glucocorticoid receptor in relation to other steroid hormone receptors. J Biol Chem 263:6842–6846
Cato ACB, Heitlinger E, Ponta H, Klein-Hitpass L, Ryffel GU, Bailly A, Rauch C, Milgrom E (1988) Estrogen and progesterone receptor-binding sites on the chicken vitellogenin II gene: synergism of steroid hormone action. Mol Cell Biol 8:5323–5330
Ceci JD, Zheng Y-W, Felder MR (1987) Molecular analysis of mouse alcohol dehydrogenase: nucleotide sequence of the Adh-1 gene and genetic mapping of a related nucleotide sequence to chromosome 3. Gene 59:171–182
Chandler VL, Maler BA, Yamamoto KR (1983) DNA sequences bound specifically by glucocorticoid receptor *in vitro* render a heterologous promoter hormone responsive *in vitro*. Cell 33:489–499
Charron J, Drouin J (1986) Glucocorticoid inhibition of transcription from episomal proopiomelanocortin gene promoter. Proc. Natl Acad Sci USA 83:8903–8907
Cidlowski JA, Cidlowski NB (1981) Regulation of glucocorticoid receptors by glucocorticoids in cultured Hela S3 cells. Endocrinology 109:1975–1982
Cordingley MG, Riegel AT, Hager GL (1987) Steroid-dependent interaction of transcription factors with the inducible promoter of mouse mammary tumor virus *in vivo*. Cell 48:261–270
Cote GJ, Gagel RF (1986) Dexamethasone differentially affects the levels of calcitonin and calcitonin gene-related peptide mRNAs expressed in a human medullary thyroid carcinoma cell line. J Biol Chem 261:15524–15528
Danesch U, Gloss B, Schmid W, Schütz G, Schüle R, Renkawitz R (1987) Glucocorticoid induction of the rat tryptophan oxygenase gene is mediated by two widely separated glucocorticoid-responsive elements. EMBO J 6:625–630
Danielsen M, Stallcup MR (1984) Down-regulation of glucocorticoid receptors in mouse lymphoma cell variants. Mol Cell Biol 4:449–453
Danielsen M, Northrop JP, Ringold GM (1986) The mouse glucocorticoid receptor: mapping of functional domains by cloning, sequencing and expression of wild-type and mutant receptor proteins. EMBO J 5:2513–2522
Danielsen M, Northrop JP, Jonklaass J, Ringold GM (1987) Domains of the glucocorticoid receptor involved in specific and nonspecific deoxyribonucleic acid binding, hormone activation, and transcriptional enhancement. Mol Endocrinol 1:816–822.
Darbre PD, King RJB (1987) Progression to steroid insensitivity can occur irrespective of the presence of functional steroid receptors. Cell 51:521–528
De Nicola AF, Tornello S, Coirini H, Heller C, Orti E, White A, Marusic ET (1982) Regulation of receptors for gluco and mineralocorticoids. In: De Nicola AF, Blaquier JA (eds) Physiopathology of hypophysial disturbances and diseases of reproduction. Liss, New York, pp 61–85

Dellweg H-G, Hotz A, Mugele K, Gehring U (1982) Active domains in wild-type and mutant glucocorticoid receptors. EMBO J 3:285–289
Denis M, Gustafsson J-Å (1989) The Mr~90,000 heat shock protein: important modulator of ligand and DNA-binding properties of the glucocorticoid receptor. Cancer Res 49:2245–2281
Denis M, Poellinger L, Wikström A-C, Gustafsson J-Å (1988) Requirement of hormone for thermal activation of the glucocorticoid receptor to a DNA binding state. Nature 333:686–688
Diamond DJ, Goodman HW (1985) Regulation of growth hormone messenger RNA synthesis by dexamethasone and triiodothyronine; transcriptional rate and mRNA stability changes in pituitary tumor cells. J Mol Biol 181:41–62
Dinter H, Hauser H (1987) Superinduction of the human interferon-β promoter. EMBO J 6:599–604
Dong Y, Poellinger L, Gustafsson J-Å, Okret S (1988a) Regulation of glucocorticoid receptor expression:evidence for transcriptional and posttranscriptional mechanisms. Mol Endocrinol 2: 1256–1264
Dong Y, Poellinger, Okret S, Höög J-O, von Bahr-Lindström H, Jörnvall H, Gustafsson J-Å (1988b) Regulation of gene expression of class I alcohol dehydrogenase by glucocorticoids. Proc Natl Acad Sci USA 85:767–771
Dong Y, Aronsson M, Gustafsson J-Å, Okret S (1989) The mechanism of cAMP-induced glucocorticoid receptor expression: correlation to cellular glucocorticoid response. J Biol Chem 264: 13679–13683
Dong Y, Cairns W, Okret S, Gustafsson J-Å (1990) A glucocorticoid-resistant rat hepatoma cell variant contains functional glucocorticoid receptor. J Biol Chem 265:7526–7531
Drouin J, Charron J, Gagner J-P Jeanotte L, Nemer M, Plante RK, Wrange Ö (1987) Pro-opiomelanocortin gene: a model for negative regulation of transcription by glucocorticoids. J Cell Biochem 35:293–304
Duester G, Smith M, Bilanchone V, Hatfield GW (1986) Molecular analysis of the human class I alcohol dehydrogenase gene family and nucleotide sequence of the gene encoding the β-subunit. J Biol Chem 261:2027–2033
Elder PK, Schmidt LJ, Ono T, Getz MJ (1984) Specific stimulation of actin gene transcription by epidermal growth factor and cycloheximide. Proc Natl Acad Sci USA 81:7476–7480
Freedman LP, Luisi BF, Korszun ZR, Basavappa R, Sigler PB, Yamamoto KR (1988) The function and structure of the metal coordination sites within the glucocorticoid receptor DNA binding domain. Nature 334:543–546
Fulton R, Birnie G (1985) Post-transcriptional regulation of rat liver gene expression by glucocorticoids. Nucleic Acids Res 13:6467–6482
Gasc J-M, Delahaye F, Baulieu E-E (1989) Compared intracellular localization of the glucocorticosteroid and progesterone receptors: an immunocytochemical study. Exp Cell Res 181: 492–504
Gasson JC, Bourgeois S (1983) A new determinant of glucocorticoid sensitivity in lymphoid cell lines. J Cell Biol 96:409–415
Gaub M-P, Dierich A, Astinotti D, Touitou I, Chambon P (1987) The chicken ovalbumin promoter is under negative control which is relieved by steroid hormones. EMBO J 6:2313–2320
Gehring U (1987) Wild-type and mutant glucocorticoid receptors of mouse lymphoma cells. In: Moudgil VK (ed) Recent advances in steroid hormone action. De Gruyter, Berlin, pp 427–442
Gehring U, Hotz A (1983) Photoaffinity labeling and partial proteolysis of wild-type and variant glucocorticoid receptors. 22:4013–4018
Gehring U, Tomkins GM (1974) A new mechanism for steroid unresponsiveness: loss of nuclear binding activity of steroid hormone receptor. Cell 3:301–306
Giguere V, Hollenberg SM, Rosenfeld MG, Evans RM (1986) Functional domains of the human glucocorticoid receptor. Cell 46:645–652
Godowski PJ, Picard D, Yamamoto KR (1988) Signal transduction and transcriptional regulation by glucocorticoid receptor-LexA fusion proteins. Science 241:812–816
Graves RA, Pandey NB, Chodchoy N, Marzluff WF (1987) Translation is required for regulation of histone mRNA degradation. Cell 48.615–626
Gregory MC, Duval D, Meyer P (1976) Changes in cardiac and hepatic glucocorticoid receptors after adrenalectomy. Clin Sci Mol Med 51:487–493

Groyer A, Schweizer-Groyer G, Cadepond F, Mariller M, Baulieu E-E (1987) Antiglucocorticosteroid effects suggest why steroid hormones are required for receptors to bind DNA in vivo but not in vitro. Nature 328:624–626
Gruol DJ, Ashby MN, Faith Campbell N, Bourgeois S (1986a) Isolation of new types of dexamethasone-resistant variants from a cAMP-resistant lymphoma. J Steroid Biochem 24:255–258
Gruol DJ, Faith Campbell N, Bourgeois S (1986b) Cyclic AMP-dependent protein kinase promotes glucocorticoid receptor function. J Biol Chem 261:4909–4914
Guarente L (1988) UASs and enhancers: common mechanism of transcriptional activation in yeast and mammals. Cell 52:303–305
Gustafsson J-Å, Carlstedt-Duke J, Poellinger L, Okret S, Wikström A-C, Brönnegård M, Gillner M, Dong Y, Fuxe K, Cintra A, Härfstrand A, Agnati L (1987) Biochemistry, molecular biology, and physiology of the glucocorticoid receptor. Endocr Rev 8:185–234
Härd T, Kellenbach E, Boelens R, Maler BA, Dahlman K, Freedman LP, Carlstedt-Duke J, Yamamoto KR, Gustafsson J-Å, Kaptein R (1990) Solution structure of the glucocorticoid receptor DNA-binding domain. Science 249:157–160
Hittle JB, Crabb DW (1988) The molecular biology of alcohol dehydrogenase: implications for the control of alcohol metabolism. J Lab Med 112:7–15
Hoeck W, Rusconi S, Groner B (1989) Down-regulation and phosphorylation of glucocorticoid receptors in cultured cells. J Biol Chem 264:14396–14402
Hollenberg SM, Evans RM (1988) Multiple and cooperative trans-activation domains of the human glucocorticoid receptor. Cell 55:899–906
Hollenberg SM, Weinberger C, Ong ES, Cerelli G, Oro A, Lebo R, Thompson EB, Rosenfeld MG, Evans RM (1985) Primary structure and expression of a functional human glucocorticoid receptor cDNA. Nature 318:635–641
Hollenberg SM, Giguere V, Segul P, Evans RM (1987) Colocalization of DNA-binding and transcriptional activation functions in the human glucocorticoid receptor. Cell 49:39–46
Housely P, Pratt WB (1983) Direct demonstration of glucocorticoid phosphorylation by intact L-cells. J Biol Chem 258:4630–4635
Huang DP, Cote GJ, Massari RJ, Chiu JF (1985) Dexamethasone inhibits alpha-fetoprotein gene transcription in neonatal rat liver and isolated nuclei. Nucleic Acids Res 13:3873–3890
Hynes N, van Ooyen AJJ, Kennedy N, Herrlich P, Ponta H, Groner B, (1983) Subfragments of the larger terminal repeat cause glucocorticoid-responsive expression of mouse mammary tumor virus and of an adjacent gene. Proc Natl Acad Sci USA 80:3637–3641
Ichii S (1981) Depletion and replenishment of glucocorticoid receptor in cytosols of rat tissues after administration of various glucocorticoids. Endocrinol Japan 28(3):293–304
Israel DI, Estolano MG, Galeazzi DR, Whitlock JP Jr (1985) Superinduction of cytochrome P1–450 gene transcription by inhibition of protein synthesis in wild type and variant mouse hepatoma cells. J Biol Chem 260:5648–5653
Jantzen H-M, Strähle U, Gloss B, Stewart F, Schmid W, Boshart M, Miksicek R, Schütz G (1987) Cooperativity of glucocorticoid response elements located far upstream of the tyrosine aminotransferase gene. Cell 49:29–38
John NJ, Bravo DA, Firestone GL (1989) Glucocorticoid responsiveness of mouse mammary tumor virus (MMTV) promoters in a down-transcription hepatoma tissue culture (HTC) variant. Mol Cell Endocrinol 61:57–68
Johnson RC, Simon MI (1987) Enhancer of site-specific recombination in bacteria. Trends Genet 3:262–267
Karin M, Haslinger A, Holtgreve H, Richards RI, Krauter P, Westphal HM, Beato M (1984) Characterization of DNA sequences through which cadmium and glucocorticoid hormones induce human metallothionein-IIA gene. Nature 308:513–519
Karlsen K, Vallerga AK, Hone J, Firestone GL (1986) A distinct glucocorticoid hormone response regulates phosphoprotein maturation in rat hepatoma cells. Mol Cell Biol 6:574–585
Kazmaier M, Brüning E, Ryffel GU (1985) Post-transcriptional regulation of albumin gene expression in Xenopus liver. EMBO J 4:1261–1266
Klein ES, DiLorenzo D, Posseckert G, Beato M, Ringold GM (1988) Sequences downstream of the glucocorticoid regulatory element mediate cycloheximide inhibition of steroid induced

expression from the rat alpha-acid glycoprotein promoter: evidence for a labile transcription factor. Mol Endocrinol 2:1343–1351

Klug A, Rhodes D (1987) Zinc fingers': a novel protein motif for nucleic acid recognition. Trends Biochem Sci 12:464–469

LaFond RE, Kennedy SW, Harrison RW, Villee CA (1988) Immunocytochemical localization of glucocorticoid receptors in cells, cytoplasts, and nucleoplasts. Exp Cell Res 175:52–62

Langer SJ, Ostrowski MC (1988) Negative regulation of transcription in vitro by a glucocorticoid response element is mediated by transacting factor. Mol Cell Biol 8:3872–3881

Lee F, Mulligan R, Berg P, Ringold G (1981) Glucocorticoids regulate expression of dihydrofolate reductase cDNA in mouse mammary tumor virus chimaeric plasmids. Nature 294:228–232

Majors J, Varmus HE (1983) A small region of the mouse mammary tumor virus long terminal repeat confers glucocorticoid hormone regulation on a linked heterologous gene. Proc Natl Acad Sci USA 80:5866–5870

McIntyre WR, Samuels HH (1985) Triamcinolone acetonide regulates glucocorticoid-receptor levels by decreasing the half-life of the activated nuclear-receptor form. J Biol Chem 260:418–427

Mendel DB, Bodwell JE, Munck A (1986) Glucocorticoid receptors lacking hormone-binding activity are bound in nuclei of ATP-depleted cells. Nature 324.478–480

Meyuhas O, Thompson EA, Jr, Perry RP (1987) Glucocorticoids selectively inhibit translation of ribosomal protein mRNAs in P1798 lymphosarcoma cells. Mol Cell Biol 7:2691–2699

Miesfeld R (1989) The structure and function of steroid receptor proteins. Crit Rev Biochem Mol Biol 24(2):101–117

Miesfeld R, Okret S, Wikström A-C, Wrange Ö, Gustafsson J-Å, Yamamoto KR (1984) Characterization of a steroid hormone receptor gene and mRNA in wild-type and mutant cells. Nature 312:779–781

Miesfeld R, Rusconi S, Godowski PJ, Maler BA, Okret S, Wikström A-C, Gustafsson J-Å, Yamamoto KR (1986) Genetic complementation of a glucocorticoid receptor deficency by expression of cloned receptor cDNA. Cell 46:389–399

Miesfeld R, Godowski PJ, Maler BA, Yamamoto KR (1987) Glucocorticoid receptor mutants that define a small region sufficient for enhancer activation. Science 236:423–427

Moore DD, Marks AR, Buckley DI, Kapler G, Payvar F, Goodman HM (1985) The first intron of the human growth hormone gene contains a binding site for glucocorticoid receptor. Proc Natl Acad Sci USA 82:699–702

Morley KL, Toohey MG, Peterson DO (1987) Transcriptional repression of a hormone-responsive promoter. Nucleic Acids Res 15:6973–6989

Müllner EW, Kühn LC (1988) A stem-loop in the 3' untranslated region mediates iron-dependent regulation of transferrin receptor mRNA stability in the cytoplasm. Cell 53:815–825

Müllner EW, Neupert B, Kühn LC (1989) A specific mRNA binding factor regulates the iron-dependent stability of cytoplasmic transferrin receptor mRNA. Cell 58:373–382

Munck A, Brinck-Johnson T (1968) Specific and nonspecific physicochemical interactions of glucocorticoids and related steroids with rat thymus cells. J Biol Chem 243:5556–5565

Muschel R, Khoury G, Reid LM (1986) Regulation of insulin mRNA abundance and adenylation: dependence on hormones and matrix substrata. Mol Cell Biol 6:337–341

Northrop JP, Gametchu B, Harrison RW, Ringold GM (1985) Characterization of wild type and mutant glucocorticoid receptors from rat hepatoma and mouse lymphoma cells. J Biol Chem 260: 6398–6403

Oikarinen J, Hämäläinen L, Oikarinen A (1984) Modulation of glucocorticoid receptor activity by cyclic nucleotides and its implications on the regulation of human skin fibroblasts growth and protein synthesis. Biochim Biophys Acta 779:158–165

Oikarinen A, Oikarinen H, Meeker CA, Tan EML, Uitto J (1987) Glucocorticoid receptors in cultured human skin fibroblasts: evidence for down-regulation of receptor by glucocorticoid hormone. Acta Derm Venereol (Stockh) 67:461–468

Okret S, Wikström A-C, Wrange Ö, Andersson B, Gustafsson J-Å (1984) Monoclonal antibodies against the rat liver glucocorticoid receptor. Proc Natl Acad Sci USA 81:1609–1613

Okret S, Poellinger L, Dong Y, Gustafsson J-Å (1986) Down-regulation of glucocorticoid receptor mRNA by glucocorticoid hormones and recognition by the receptor of a specific binding sequence within a receptor cDNA clone. Proc Natl Acad Sci USA 83:5899–5903

Oro AE, Hollenberg SM, Evans RM (1988) Transcriptional inhibition by a glucocorticoid receptor-β-galactosidase fusion protein. Cell 55:1109–1114
Orti E, Mendel DB, Smith LI, Munck A (1989) Agonist-dependent phosphorylation and nuclear dephosphorylation of glucocorticoid receptors in intact cells. J Biol Chem 264:9728–9731
Pachter JS, Yen TJ, Cleveland DW (1987) Autoregulation of tubulin expression is achieved through specific degradation of polysomal tubulin mRNAs. Cell 51:283–292
Paek I, Axel R (1987) Glucocorticoids enhance stability of human growth hormone mRNA. Mol cell Biol 7:1496–1507
Payvar F, Wrange Ö, Carlstedt-Duke J, Okret S, Gustafsson J-Å, Yamamoto KR (1981) Purified glucocorticoid receptors bind selectively *in vitro* to a cloned DNA fragment whose transcription is regulated by glucocorticoids *in vivo*. Proc Natl Acad Sci USA 78:6628–6632
Petersen DD, Magnuson MA, Granner DK (1988) Location and characterization of two widely separated glucocorticoid response elements in the phosphoenolpyruvate carboxykinase gene. Mol Cell Biol 8:96–104
Pfahl M (1982) Specific binding of the mouse glucocorticoid-receptor complex to mouse mammary tumor proviral promoter region. Cell 31:475–482
Picard D, Salser SJ, Yamamoto KR (1988) A movable and regulable inactivation function within the steroid binding domain of the glucocorticoid receptor. Cell 54:1073–1080
Picard D, Yamamoto KR (1987) Two signals mediate hormone-dependent nuclear localization of the glucocorticoid hormone receptor. EMBO J 6:3333–3340
Ponta H, Kennedy N, Skroch P, Hynes NE, Groner B (1985) Hormonal response region in the mouse mammary tumor virus long terminal repeat can be dissociated from the proviral promoter and has enhancer properties. Proc Natl Acad Sci USA 82:1020–1024
Pratt WB, Jolly DJ, Pratt DV, Hollenberg SM, Giguere V, Cadepond FM, Schweizer-Groyer G, Catelli M-G, Evans RM, Baulieu E-E (1988) A region in the steroid binding domain determines formation of the non-DNA-binding, 9S glucocorticoid receptor complex. J Biol Chem 263:267–273
Ptashne M (1986) Gene regulation by protein acting nearby and at a distance. Nature 322:697–700
Ptashne M (1988) How eukaryotic transcriptional activators work. Nature 335:683–689
Raaka BM, Samuels HH (1983) The glucocorticoid receptor in GH_1 cells. J Biol Chem 258:417–425
Rabindran SK, Danielsen M, Stullcup (1987) Glucocorticoid-resistant lymphoma cell variants that contain functional glucocorticoid receptor. Mol Cell Biol 7:4211–4217
Rajpert EJ, Lemaigre FP, Eliard PH, Place M, Lafontaine DA, Economidis IV, Belayew A, Martial JA, Rousseau GG (1987) Glucocorticoid receptors bound to the antagonist RU486 are not downregulated despite their capacity to interact in vitro with defined gene regions. J Steroid Biochem 26:513–520
Renkawitz R, Schütz G, von der Ahe D, Beato M (1984) Sequences in the promoter region of the chicken lysozyme gene required for steroid regulation and receptor binding. Cell 37:503–510
Ringold GM (1985) Steroid hormone regulation of gene expression. Annu Rev Pharmacol Toxicol 25:529–566
Rosewicz S, McDonald AR, Maddux BA, Goldfine ID, Miesfeld RL, Logsdon CD (1988) Mechanism of glucocorticoid receptor down-regulation by glucocorticoids. J Biol Chem 263:2581–2584
Rouault TA, Hentze MW, Haile DJ, Harford JB, Klausner RD (1989) The iron-responsive element binding protein: a method for the affinity purification of a regulatory RNA-binding protein. Proc Natl Acad Sci USA 86:5768–5772
Rusconi S, Yamamoto KR (1987) Functional dissection of the hormone and DNA binding activities of the glucocorticoid receptor. EMBO J 6.1309–1315
Sakai DD, Helms S, Carlstedt-Duke J, Gustafsson J-Å, Rottman FM, Yamamoto KR (1988) Hormone-mediated repression: a negative glucocorticoid response element from the bovine prolactin gene. Genes Dev 2:1144–1154
Sapolsky RM, Krey LC, McEwen BS (1984) Stress down-regulates corticosterone receptors in a site-specific manner in the brain. Endocrinology 114:287–292
Scheidereit C, Beato M (1984) Contacts between hormone receptor and DNA double helix within a glucocorticoid regulatory element of mouse mammary tumor virus. Proc Natl Acad Sci USA 81:3029–3033

Scheidereit C, Geisse S, Westphal HM, Beato M (1983) The glucocorticoid receptor binds to defined nucleotide sequences near the promoter of mouse mammary tumor virus. Nature 304:749–752
Schlechte JA, Ginsberg BH, Sherman BM (1982) Regulation of the glucocorticoid receptor in human lymphocytes. J Steroid Biochem 16:69–74
Schmid W, Strähle U, Schütz G, Schmitt J, Stunnenberg H (1989) Glucocorticoid receptor binds cooperatively to adjacent recognition sites. EMBO J 8:2257–2263
Schüle R, Müller M, Otsuka-Murakami H, Renkawitz R (1988) Cooperativity of the glucocorticoid receptor and the CACCC-box binding factor. Nature 332:87–90
Serfling E, Jasin M, Schaffner W (1985) Enhancers and eukaryotic gene transcription. Trends Genet 224–230
Severne V, Wieland S, Schaffner W, Rusconi S (1988) Metal binding 'finger' structures in the glucocorticoid receptor defined by site-directed mutagenesis. EMBO J 7:2503–2508
Shapiro DJ, Blume JE, Nielsen DA (1987) Regulation of messenger RNA stability in eukaryotic cells. BioEssays 6(5):221–226
Shaw G, Kamen R (1986) A conserved AU sequence from the 3' untranslated region of GM-CSF mRNA mediates selective mRNA degradation. Cell 46:659–667
Shirwany TA, Hubbard JR, Kalimi M (1986) Glucocorticoid regulation of hepatic cytosolic glucocorticoid receptors in vivo and its relationship to induction of tyrosine aminotransferase. Biochim Biophysica Acta 886:162–168
Sibley CH, Tomkins GM (1974) Mechanisms of steroid resistance. Cell 2:221–227
Sigler PB (1988) Acid blobs and negative noodles. Nature 333:210–212
Simons SS Jr, Pumphrey JG, Rudikoff S, Eisen HJ (1987) Identification of cysteine 656 as the amino acid of hepatoma tissue culture cell glucocorticoid receptors that is covalently labeled by dexamethasone 21-mesylate. J Biol Chem 262:9676–9680
Singh VB, Moudgil VK (1985) Phosphorylation of rat liver glucocorticoid receptor. J Biol Chem 260:3684–3690
Slater EP, Rabenau O, Karin M, Baxter JD, Beato M (1985) Glucocorticoid receptor binding and activation of a heterologous promoter by dexamethasone by the first intron of the human growth hormone gene. Mol Cell Biol 5:2984–2992
Smith K, Shuster S (1984) Effect of adrenalectomy and steroid treatment on rat skin cytosol glucocorticoid receptor. J Endocrinol 102:161–165
Stevens J, Stevens Y-W, Haubenstock H (1983) Molecular basis of glucocorticoid resistance in experimental and human leukemia. In: Littwack G (ed) Biochemical actions of hormones, vol. 10. Academic, New York, pp 383–446
Strähle U, Klock G, Schütz G (1987) A DNA sequence of 15 base pairs is sufficient to mediate both glucocorticoid and progesterone induction of gene expression. Proc Natl Acad Sci USA 84:7871–7875
Strähle U, Schmid W, Schütz G (1988) Synergistic action of the glucocorticoid receptor with transcription factors. EMBO J 7:3389–3395
Svec F, Rudis M (1981) Glucocorticoids regulate the glucocorticoid receptor in the AtT-20 cell. J Biol Chem 256:5984–5987
Thompson EB, Yuh Y-S, Gametchu AB, Linder M, Harmon JM (1988) Molecular genetic analysis of glucocorticoid actions in human leukemic cells. In Lippman M E (ed) Growth regulation of cancer. Liss, New York pp 221–237
Tsai SY, Carlstedt-Duke J, Weigel NL, Dahlman K, Gustafsson J-Å, Tsai M-J, O'Malley BW (1988) Molecular interaction of steroid hormone receptor with its enhancer element: evidence for receptor dimer formation. Cell 55:361–369
Tsai SY, Tsai M-J, O'Malley BW (1989) Cooperative binding of steroid hormone receptors contributes to transcriptional synergism at target enhancer elements. Cell 57:443–448
Turner AB (1986) Tissue differences in the up-regulation of glucocorticoid-binding proteins in the rat. Endocrinology 118:1211–1216
Vanderbilt JN, Miesfeld R, Maler BA, Yamamoto KR (1987) Intracellular receptor concentration limits glucocorticoid-dependent enhancer activity. Mol Endocrinol 1:68–74
Vannice JL, Taylor JM, Ringold GM (1984) Glucocorticoid-mediated induction of alpha$_1$-acid glycoprotein: evidence for hormone-regulated RNA processing. Proc Natl Acad Sci USA 81: 4241–4245

Von der Ahe D, Janich S, Scheidereit C, Renkawitz R, Schütz G, Beato M (1985) Glucocorticoid and progesterone receptors bind to the same sites in two hormonally regulated promoters. Nature 313:706–709

Von der Ahe D, Renoir J-M, Buchou T, Bauleiu E-E, Beato M (1986) Receptors for glucocorticoid and progesterone recognize distinct features of a DNA regulatory element. Proc Natl Acad Sci USA 83:2817–2821

Walsh MJ, LeLeiko NS, Sterling KM Jr (1987) Regulation of types I, III, and IV procollagen mRNA synthesis in glucocorticoid-mediated intestinal development. J Biol Chem 262:10814–10818

Wilson T, Treisman R (1988) Removal of poly(A) and consequent degradation of c-fos mRNA facilitated by 3' AU-rich sequences. Nature 336:396–399

Winter LA, Stewart MJ, Lou Shean M, Dong Y, Poellinger L, Okret S, Gustafsson J-Å, Duester G (1990) A hormone response element upstream of the human alcohol dehydrogenase gene ADH2 consists of three tandem glucocorticoid receptor binding sites. Gene 91:233–240

Wolffe AP, Glover JF, Martin SC, Tenniswood MPR, Williams JL, Tata JR (1985) Deinduction of transcription of Xenopus 74-kDa albumin genes and destabilization of mRNA by estrogen in vivo and in hepatocyte cultures. Eur J Biochem 146:489–496

Wrange Ö, Eriksson P, Perlmann T (1989) The purified activated glucocorticoid receptor is a homodimer. J Biol Chem 264:5253–5259

Wu KC, Pfahl M (1988), Variable responsiveness of hormone inducible hybrid genes in different cell lines. Mol Endocrinol 2:1294–1301

Yamamoto KR (1985) Steroid receptor regulated transcription of specific genes and gene networks. Annu Rev Genet 19:209–252

Yamamoto KR, Gehring U. Stampfer MR, Sibley CH (1976) Genetic approaches to steroid hormone action. Recent Prog Horm Res 32:3–32

Yang Y-L, Tan J-X, Xu R-B (1989) Down-regulation of glucocorticoid receptor and its relationship to the induction of rat liver tyrosine aminotransferase. J Steroid Biochem 32:99–104

Yen TJ, Machlin PS, Cleveland DW (1988) Autoregulated instability of β-tubulin mRNAs by recognition of the nascent amino terminus of β-tubulin. Nature 334:580–585

Yuh Y-S, Thompson EB (1987) Complementation between glucocorticoid receptor and lymphocytolysis in somatic cell hybrids of two glucocorticoid-resistant human leukemic clonal cell lines. Somatic Cell Mol Gene 13(1):33–46

Zaret KS, Yamamoto KR (1984) Reversible and persistent changes in chromatin structure accompany activation of a glucocorticoid-dependent enhancer element. Cell 38:29–38

Zawydiwski R, Harmon JM, Thompson EM (1983) Glucocorticoid-resistant human acute lymphoblastic leukemic cell line with functional receptor. Cancer Res 43:3865–3873

Regulation of Neuropeptide Gene Expression

R. H. Goodman, R. Rehfuss, K. Walton, and M. J. Low[1]

Contents

Understanding of the mechanisms of neuropeptide gene regulation has increased dramatically in the past 5 years. In large part, this increased understanding has resulted from advances in the technology of introducing reporter genes into intact cells in culture. This chapter addresses two methods for gene transfer that have been particularly important for studies of neuropeptide gene regulation: DNA-mediated gene transfer into mammalian cell lines and microinjection of foreign genes into the germline of transgenic mice. Two specific areas of gene regulation will be considered – transcriptional control by cAMP and tissue-specific expression of neuropeptides in the brain and pituitary gland.

[1] Vollum Institute for Advanced Biomedical Research, Oregon Health Sciences University, 3181 SW Sam Jackson Park Road, Portland, OR 97201-3098

Current Topics in Neuroendocrinology, Vol. 11

1 Transcriptional Regulation by cAMP

Cyclic adenosine 3', 5' monophosphate (cAMP) is an important regulator of neuropeptide gene expression. The regulatory actions of cAMP on mammalian gene transcription occur through the activation of protein kinase A (PKA), a tetrameric kinase containing two regulatory and two catalytic subunits. The catalytic subunits of PKA dissociate from the inhibitory regulatory subunits upon treatment with cAMP, and thereby become activated. The catalytic subunit of PKA is proposed to stimulate gene expression by altering the ability of specific transcription factors to activate RNA polymerase II. Many of the details of how this transcriptional activation occurs are currently unknown, however. This section reviews some of the recent advances in the understanding of neuropeptide gene regulation through the cAMP second messenger pathway. The term "advances" is used loosely, because it appears that cAMP-induced gene transcription is considerably more complex than previously imagined.

cAMP-regulated transcriptional enhancers fall into three general categories: (a) sequences related to the "classical" cAMP-regulated enhancer (the cAMP-responsive element; CRE), containing one ore more copies of the consensus sequence 5'-CGTCA-3'; (b) sequences related to the AP-2 binding site, 5'-CCCCAGGC-3'; and (c) DNA elements which have no obvious sequence similarity to either of these sites. *Trans*-acting factors which act at the CRE and AP-2 sites have been identified; factors that act at "novel" cAMP-regulatory sites have yet to be characterized. The family of CRE-binding proteins appears to be growing rapidly, however, and it is likely that hundreds of factors may have the potential to mediate transcriptional regulation through the CRE site. Below, we describe the *cis*-acting sequences responsible for transcriptional regulation by cAMP and the *trans*-acting factors that bind to the CRE sequences.

1.1 The "Classical" CRE

The CRE was first identified in the genes for phosphoenolpyruvate carboxykinase (Wynshaw-Boris et al. 1984; Short et al. 1986) and the neuropeptides proenkephalin (Comb et al. 1986) and somatostatin (Montminy et al. 1986). Delineation of the critical sequences in this element was achieved independently by these investigators through functional studies of fusion genes introduced into heterologous mammalian cells in culture. Examination of the nucleotide sequences in these three genes indicated that the cAMP-responsive regions were highly related. A wide variety of other eukaryotic genes have since been shown to contain similar sequences which, in many cases, have been shown to confer cAMP-responsiveness (*Table 1*). Whether cAMP is the primary regulator of all CREs is unknown. The demonstration that a CRE-containing gene can be regulated by cAMP does not address whether the element serves this purpose in intact cells. Although CRE sequences generally are located within a few hundred base pairs (bp) of the TATA box, the element fulfills all of the requirements of a classical enhancer in that it is both orientation- and position-independent. Some genes contain functional CRE sequences that are located at a great distance from the promoter (Boshart et al. 1990).

Table 1: Nucleotide sequence of various cellular CREs

Gene	*Position*	*Sequence*	*Reference*
Proenkephalin, human	− 96 to − 82	GGCCTGCGTCAGCTG CCGGACGCAGTCGAC	Comb et al. (1986)
Somatostatin, rat	− 51 to − 37	TGGC**TGACGTCA**GAGA ACCG**ACTGCAGT**CTCT	Montminy et al. (1986)
Fibronectin, human rat	−176 to −161 −164 to 159	CCCG**TGACGTCA**CCCG GGGC**ACTGCAGT**GGGC	Dean et al. (1988)
Vasoactive intestinal peptide, human	− 90 to − 68	TGGCCGTCATACTGTGACGTCTT ACCGGCAGTATGACACTGCAGAA	Tsukada et al. (1987)
c-fos, mouse	− 58 to − 71	CCAGTGACGTAGG GGTCACTGCATCC	Gilman et al. (1986)
p-Enolpyruvate carboxykinase, rat	− 91 to − 79	CTTACGTCAGAGC GAATGCAGTCTCG	Short et al. (1986)
Parathyroid hormone, bovine	− 79 to − 64	GGAG**TGACGTCA**TCTG CCTC**ACTGCAGT**AGAC	Weaver et al. (1984)

Nucleotide structure of cAMP-responsive elements (CRES) from various cellular promoters. The position of the CRE relative to the transcriptional start site is indicated. The core CRE motif 5' CGTCA 3' is indicated by a bold arrow.

The structural requirements for the CRE function are only partially understood. One of the best characterized CREs, 5'-TGACGTCA-3', was found within the promoter of the somatostatin gene. This element was used to purify the original CRE-binding protein (CREB; Montminy and Bilezikjian 1987), and was used in a recognition-site cloning strategy to clone the CREB cDNA (Hoeffler et al. 1988). Fink et al. (1988) proposed that the basic unit of the CRE was the 5-bp motif, 5'-CGTCA-3'. The somatostatin CRE contains an inverted, overlapping repeat of this motif. Other CREs, such as the one in the *c-fos* promoter, contain a single copy of this sequence. Point mutations within the motif diminish or totally eliminate the ability of the element to respond to cAMP (Deutsch et al. 1988a). However, in elements containing multiple copies of the motif, mutations in each CGTCA sequence are not equivalent. For example, while the sequences 5'-AGACGTCA-3' and 5'-TGACGTCT-3' each contain only the copy of the 5-bp motif, the former partially responds to cAMP, while the latter is completely unresponsive. Presumably, the recognition sites provided by the two CGTCA sequences are viewed differently by the DNA-binding proteins involved in the cAMP response.

Unlike many other regulatory transcriptional elements, the CGTCA motifs in the CRE do not need to be palindromic to maintain their function. For example, the CRE in the vasoactive intestinal peptide (VIP) gene contains an inverted repeat of the CGTCA sequence separated by five nucleotides, but changing the orientation of one of the motifs of the VIP gene has little effect on the ability of the element to respond to cAMP (Tsukada et al. 1987). As in the somatostatin CRE, point mutations within the two CGTCA motifs have differential effects. Mutation of the downstream motif affects cAMP stimulation more than mutation of the upstream motif (Fink et al. 1988; Hyman et al. 1988). All of these analyses may be overly simplistic however, as the recognition sites for DNA-binding proteins probably are not represented very well by linear DNA sequences. For example, McMurray et al. (1991) have suggested that the CREs in the enkephalin and other neuropeptide gene promoters assume a cruciform structure in solution and that this structure differs markedly from that inferred from the linear representation of the DNA sequence. The association of chromatin components may alter the "appearance" of CREs in specific genes still further.

The nucleotides adjacent to the CREs in different genes are poorly conserved, but appear to influence transcriptional activity significantly (Deutsch et al. 1988b). For example, the glycoprotein α-subunit, somatostatin, and parathyroid hormone genes each contain the same 8-bp CRE consensus sequence, but, when the sequences containing these elements are placed upstream from an identical promoter, each DNA fragment differs markedly in its ability to confer cAMP-mediated transcriptional induction. This finding is consistent with the results of DNase I footprinting assays, which suggest that DNA-binding proteins interact with a region that extends beyond the minimum consensus sequence (Delegeane et al. 1987; Montminy et al. 1987). Whether the flanking sequences contribute to the specification of individual CRE-binding complexes remains to be determined.

1.2 The AP-2 Site: An Alternative Mode of cAMP Regulation

The AP-2 site, first identified in simian virus 40 and human metallothionein IIA promoters (Haslinger and Karin 1985; Karin et al. 1987; Mitchell et al. 1987), was initially believed to be a basal transcriptional element. Subsequent studies determined that this site, when reiterated within a gene construction containing the human β-globin promoter, was responsive to both phorbol esters and forskolin (Imagawa et al. 1987). The classical CRE also appears to respond to both of these pathways (Fink at al. 1991).

AP-2 sites have been found in the growth hormone, c-myc, H-2Kb, and pro-enkephalin genes, among others (Imagawa et al. 1987; Comb et al. 1986). Unlike proteins that bind to the CRE, which appear to be fairly ubiquitous, AP-2 expression may be relatively restricted. For example, AP-2 footprinting activity, *trans*-activation, and messenger (m)RNA are apparent in HeLa cells, but not in HepG2 cells (Williams et al. 1988; Rickles et al. 1989). The mechanism of AP-2 activation by cAMP is unknown.

1.3 Novel cAMP-Responsive Elements: CREs in Disguise?

Several cAMP-regulated genes, such as the human steroid 21-hydroxylase (P-450_{C21}; Kagawa and Waterman 1990) and the bovine steroid 17 α-hydroxylase genes (Lund et al. 1990), contain neither CRE nor AP-2 sequences. Furthermore, fusion genes containing promoter elements from each of these genes have been constructed that clearly recapitulate the cAMP responsiveness of the native genes, but do not contain recognizable CRE or AP-2 sequences. Although oligonucleotides representing the classical CRE compete for binding to these promoter elements in gel mobility shift assays, they do so at very low affinity, suggesting that the hydroxylase genes utilize a novel element for cAMP regulation.

Differences between regulation through classical and nonclassical CRE sites are demonstrated by the α- and β-chorionic gonadotropin subunit genes. The α-subunit CRE is of the classical type, and contains two tandem copies of the 8-bp consensus sequence (Jameson and Lindell 1988; Milstead et al. 1987). The β-subunit contains no recognizable CRE sequences, but is also stimulated by cAMP. cAMP-mediated induction of the β-subunit is delayed relative to the α-subunit gene and is sensitive to protein synthesis inhibitors such as cycloheximide, indicating that the activation of transcription is due to the synthesis of new factors. Classical CRE-binding factors are typically not blocked by cycloheximide, again suggesting that the factors that mediate the α CRE response are different from those that activate the β-subunit.

Non classical CREs may be involved in the cAMP-mediated expression of other neuroendocrine genes such as proopiomelanocortin. It has been proposed that some of these nonclassical CREs may function primarily as calcium-responsive elements (Lorange et al. 1992).

2 Is the CRE a Basal Transcriptional Element?

Several studies have suggested that the CRE functions as a basal or tissue-specific enhancer in selected genes. As in regulated CRE function, the environment of a CRE affects its ability to activate basal expression. For example, deleting the CRE from the tissue-type plasminogen activator or insulin I promoters moderately decreases basal activity (Medcalf et al. 1990; Philippe and Missotten 1990), while removal of the CRE from the somatostatin or glycoprotein α-subunit promoters blocks basal transcriptional activity almost completely (Andrisani et al. 1987; Jameson et al. 1989a,b).

CRE sequences from the glucagon, parathyroid hormone, glycoprotein α-subunit, and somatostatin genes have been examined for their ability to activate the glycoprotein α-subunit promoter (Deutsch et al. 1988a). Although each of these CREs contain the identical 8-bp classical consensus sequence, the basal activity conferred by each CRE differed markedly. The glucagon and parathyroid hormone CREs increased basal activity only 1,5-fold, while the α subunit and somatostatin CREs increased basal activity seven-and tenfold, respectively (Deutsch at al.

1988a). Inclusion of an additional 5-bp adjacent to the glucagon CRE completely eliminated its ability to confer basal and stimulated responses. Whether the same transcription factors mediate both basal and stimulated expression of CRE-containing genes has not been determined.

The basal activation of transcription provided by the various CRE sequences also depends in part on cellular factors. For example, the α-subunit CRE increases transcription in JEG-3 cells (which normally express the α-subunit) 2.5-fold when placed upstream from the somatostatin promoter, while it produces a 36-fold increase in transcription in INRI-G9 cells (a rat islet cell line which normally produces somatostatin but not the α-subunit; Jameson at al. 1989a). When placed upstream from its own promoter, the α-subunit CRE increases basal activitiy tenfold in JEG-3 cells, but has no effect in INRI-G9 cells. These findings suggest that a complex interaction among CRE sequences, promoter elements, and cellular factors each contribute to the ability of the CRE to confer basal transcriptional activity. Recent studies by Boshart et al. (1991) suggest that these basal effects may depend on low levels of activated PKA.

3 Interaction of the CRE with Other Transcriptional Enhancers

Another level of complexity is added by the interaction between CRE sequences and adjacent enhancer sequences. Several examples are described below.

The CRE in the fibronectin gene appears to function in a tissue-specific manner. While the promoter is inducible by cAMP in JEG-3 cells (Dean et al. 1988), it is inhibited by cAMP in granulosa cells (Dorrington and Skinner 1986; Bernath at al. 1990). This cell-specific inhibition may be due to *cis*-acting sequences found upstream (within 300 bp) of the CRE. The precise sequences involved in this inhibition have yet to be fully elucidated, however.

The two CRE-like sequences in the proenkephalin gene have been designated ENKCRE-1 and ENKCRE-2 (Hyman et al. 1988; Comb et al.1988). While it has been shown that a common factor binds ENKCRE-2 and the VIP CRE, a novel factor, ENKTF-1, binds to the distal ENKCRE-1 site. Deletion of the upstream ENKCRE-1 site diminishes the cAMP response by about ten fold. Deletions of the more proximal ENKCRE-2 site completely eliminates all basal as well as cAMP-stimulated activity. Consistent with the idea that activity of the enkephalin enhancer requires the interaction of two distinct transcription factors, the spacing between the ENKCRE-1 and ENKCRE-2 motifs has been shown to be critical. These elements are normally 5 bases apart, i.e., one half turn to the helix. Adding 5 additional bases between two elements, which increases this distance to a full turn, decreases the cAMP response to 33% of control. Adding 5 more bases, which returns the two sites to the same side of the DNA helix, increases the activity to 66% of control. The proteins which regulate the VIP CRE, on the other hand, may not need to interact. In this case, increasing the distance between the two CGTCA motifs had essentially no effect in the response to cAMP (Fink et al. 1988).

An AP-2 site proximal to ENKCRE-2 in the enkephalin gene may also contribute to cAMP responsiveness (Hyman et al. 1988; Comb et al. 1988). Mutation of this sequence inhibits basal activity by 50%, and decreases cAMP responsiveness fourfold (Hyman et al. 1989). The factors AP-1 and AP-4 can also bind to sequences in and around ENKCRE-2, perhaps contributing to the ability of this promoter to respond to both the PKA and the protein kinase C (PKC) pathways.

The CRE located in the glycoprotein α-subunit interacts with both upstream and downstream elements. The α-subunit CRE consists of two identical 18-bp repeats, each of which contain a classical CRE. While a single copy of the 18-bp element is sufficient for mediating the response to cAMP, basal activitiy of genes containing the single element is reduced by about fourfold (Deutsch et al. 1987). Immediately upstream from the 18-bp repeats is a tissue-specific enhancer designated the upstream regulatory element (URE), which increases basal expression about fivefold and is dependent on the CRE for activity (Jameson et al. 1989a,b; Delegeane et al. 1987). Factors that interact with the URE have not yet been identified.

Mutational analysis of the region immediately downstream from the α-subunit gene CRE identified another element, which, when mutated, decreases basal transcription tenfold (Kennedy et al. 1990). This element binds a unique factor which appears to be tissue specific (Anderson et al. 1990). The α-subunit promoter can be repressed by activated glucocorticoid receptor and it has been suggested that this effect involves the direct binding of the receptor to sequences which overlap the CRE, thus blocking activity of the CRE-binding proteins (Akerbloom et al. 1988). However, other investigators have suggested that the inhibition may involve the physical interaction of the glucocorticid receptor with the CRE-binding proteins rather than with the CRE sequence (Chatterjee et al. 1991).

4 The Plot Thickens: The Multiplicity of CRE-Binding Proteins

Considerable evidence indicates that the CRE has different functions in different genes, and that these functions may vary depending on cell-specific factors. It is possible that the concept of the CRE as a single genetic element is no longer useful, given that the regulation of gene expression through this element is so variable. One explanation for the complex pattern of gene regulation through the CRE is that a multiplicity of different CRE-binding proteins might interact to control expression of individual target genes. Indeed, the initial isolation of cDNAs encoding some of these factors has confirmed the existence of a complex superfamily of CRE-binding proteins (see Table 2). These factors appear to bind to DNA as dimers and share common structural elements, including a carboxy-terminal basic DNA-binding region and a leucine zipper motif (Landschultz et al. 1988). The primary sequence of amino acids in the basic region varies slightly among the different CRE-binding proteins, but greater than 50% of the residues in this domain are either lysine or arginine. This region has been proposed to form an α-helix which, together with a second α-helix from the other half of the dimer, recognizes the CRE in a "scissors-grip" configuration. Dimerization of the CRE-binding proteins is mediated by thc leucine zipper, which consists of four or five leucines separated from each other by seven amino acids. The zipper regions in the two components of a dimer appear to interact with each other by forming a coiled-coil structure (O'Shea et al. 1989). All of the CRE-binding proteins have amino-terminal domains which, while poorly conserved overall, frequently have consensus phosphorylation sites for a variety of protein kinases. The amino-terminal domains are hypothesized to be involved in transcriptional activation.

Table 2: CRE-binding proteins

Factor	Molecular Mass (kDa)	Source	Reference	Comments
ATF-α	52	HeLa	Gaire et al. (1990)	N and C terminus
ATF-αΔ	50	HeLa	Gaire et al. (1990)	similar to CRE-BP1 Splice variant of ATF-α
ATF 1	28.2	MG63	Hai et al. (1989)	70% similar to CREB
ATF 2	54,5	MG63	Hai et al. (1989)	Identical to CRE-BP1
ATF 3	?	HeLa	Hai et al. (1989)	Dimerizes with
ATF 4	?	MG63	Hai et al. (1989)	CRE-BP1,Jun
ATF 6	?	HeLa	Hai et al. (1989)	Dimerizes with
ATF 7,8	?	MG63	Hai et al. (1989)	Fra-1, Fos, Jun
CREB 341	37	Rat brain	Gonzalez et al. (1989)	Activated by kinase A
CREB 327	35	Human placenta	Hoeffler et al. (1988)	Splice variant of CREB 341
CRE-BP1	54.5	Human fetal brain	Maekawa et al. (1989)	Forms CRE binding heterodimer with c-*Jun*
HB 16	?	Human B-cell	Kara et al. (1990)	Truncated CRE-BP1
mXBP/ CRE-BP2	?	Mouse spleen	Ivashkiv et al. (1990)	Splice variant of CRE-BP1
TREB 5	28.7	HUT 102	Yoshimura et al. (1990)	Basic region homology only
CREM α,β,γ	26	Mouse pituitary	Foulkes et al. (1991)	Suppressors of CREB activity

Partial list of known CRE binding proteins. The indicated molecular weight of the protein is based on the predicted weight from cDNA cloning experiments.
?, a partial cDNA isolate of unknown molecular weight, HUT 102, a human T-cell lymphoma, MG 63, a human osteosarcoma, HeLa, a human cervical carcinoma line.

4.1 Diversity Through Dimerization

Signals included in the leucine zippers of the various CRE-binding proteins determine which factors may interact with each other. All of the CRE-binding proteins can form homodimers and several are capable of forming heterodimers. The ability of these factors to dimerize in specific combinations greatly increases the number of discrete forms of DNA-binding complexes. For example, CRE-BP1 (also called ATF-2) readily forms dimers with c-*jun* that are capable of interacting with CRE sequences (Macgregor et al. 1990). This heterodimer combination is of particular interest for two reasons: First, CRE-BP1 appears to be abundant in all tissues and can bind to many different CREs, including those from the long terminal repeat (LTR) of HTLV I, the MHC A α-gene, and the ATF site in the adenovirus early gene promoter (Hai et al. 1989; Kara et al. 1990; Yoshimura et al. 1990). Secondly, this complex may provide a mechanism for crosstalk between the PKC and cAMP signal transduction pathways. When present in the AP-1 complex with c-*fos*, c-*jun* mediates transcriptional signals generated by PKC through the phorbol ester response element (TRE). When dimerized with CRE-BP1, the recognition site of the complex changes so that it exerts its effects through the CRE sequence. In addition, CRE-BP1 can form dimers with ATF-3 (Hai et al. 1989) and possibly c-*fos*, although the physiological relevance of these heterodimer combinations are unknown. Recent studies have shown that CREB can form dimers with ATF-1 (Rehfuss et al. 1991, c-*jun* with ATF-3 und ATF-4, and c-*fos* with ATF-4 (Hai and Curran 1991).

4.2 Diversity Through Alternative Splicing

Many of the CRE-binding proteins appear to use alternative RNA splicing to generate a wide diversity of isoforms (Yamamoto et al. 1990; Ivashkiv et al. 1990; Gaire et al. 1990). With the exception of the originally characterized CREB, functional differences between the various isoforms has not been explored.

CREB occurs in two isoforms which differ by the presence or absence of an 11-amino-acid region in the activator domain termed the α-peptide. Both CREB isoforms are expressed in a wide variety of tissues, with the larger isoform (CREB 341) being several times more abundant than the smaller one (CREB 327). Yamamoto et al. (1990) have reported that the larger form was ten times more active than the smaller one in *trans*-activating CRE-containing genes in F9 terato-carcinoma cells. Berkowitz und Gilman (1990) were unable to detect any difference between the two isoforms in their ability to impart cAMP responsiveness to the c-*fos* promoter in BALB/c 3T3 fibroblasts. The reasons for these differences are unknown.

In at least one instance, alternative splicing has been utilized to place different DNA-binding domains onto the same activator region (Foulkes et al. 1991). The mouse CRE-binding protein CREM consists of three known isoforms designated α, β, and γ. The DNA-binding domain of the α isoform (designated binding domain I) differs from that in β und γ (designated binding domain II) by about

20% in nucleotide sequence. Binding domains I and II are 95% and 75% identical, respectively, to that in CREB and probably can dimerize with CREB as well as with each other. This ability to form CREB/CREM heterodimers has interesting implications because it has been suggested that CREM inhibits CREB action (Foulkes et al. 1991). All three isoforms of CREM appear to block CREB-induced transcription of a CRE-reporter gene. The amino terminus of CREM contains regions that are homologous to the PDE boxes found in CREB (see below). The role of these elements in CREM function is only conjectural at this point, but it is possible that the activity of CREM, like CREB is regulated by phosphorylation. Whether CREM suppresses CREB activity by forming heterodimers with CREB or by preventing CREB from occupying the CRE is unknown. All three isoforms of CREM are produced in multiple tissues, including heart, pituitary, kidney, and brain. Whether the subtle differences in the CREM- and CREB-binding domains direct the factors to different types of CREs has not been determined. Rehfuss et al. (unpublished) have determined that the CREM gene can also generate isoforms which contain different activator regions spliced onto binding domains I and II.

5 Transcriptional Activity

CREB is clearly the most well-studied CRE-binding protein and provides the basis for much of the understanding of CRE regulation. Phosphorylation of CREB by PKA is necessary for complete transcriptional activity. Substitution of the serine at position 133 with either alanine, glutamic acid, or aspartic acid severely impairs the ability of CREB to activate transcription (Gonzalez et al. 1989; Lee et al. 1990). Phosphorylation of CREB by PKA is not sufficient to fully activate the transcription factor, however. Lee et al. (1990) showed that additional serine-rich sequences near the PKA site were also necessary for activity. Phosphopeptide mapping of mutant CREB proteins expressed *in vivo* indicate that these "phosphorylation boxes" are modified subsequent to the primary phosphorylation at the PKA site and further, that two of these elements (PDE 1 and PDE 2) are absolutely required for activity. At present it is not understood how any of these phosphorylation events actually produces an active transcription factor. Negatively charged domains have been suggested to act as potent transcriptional activators (Ma and Ptashne 1987a,b), and it is possible that the multiple phosphorylations of the PDE elements may serve to increase the net negative charge of this region of CREB. Alternatively, phosphorylation may induce conformational changes that reveal previously cryptic activator sequences. In either case, the observation that CREB requires multiple phosphorylations to be fully active implies that several different kinases, possibly acting through multiple signaling pathways, may impinge on the activity of CREB. Recent studies have demonstrated that some other CRE-binding proteins, such as ATF-1, can also mediate transcriptional regulation through cAMP (Rehfuss et al. 1991). It appears that CREB and probably other proteins in this family are well positioned to integrate inputs from many types of environmental cues. The suggestion that calcium signaling pathways may lead to the phosphorylation of CREB at position 133 (Sheng et al. 1991), the same position that is phosphorylated by PKA, supports this hypothesis.

While most of the information about cAMP-mediated transcriptional regulation concerns proteins that bind to the CRE sequence, some progress has been made in understanding cAMP regulation through the transcription factor AP-2 (Williams et al. 1989). AP-2 is completely dissimilar from CREB: it does not contain a leucine zipper and binds DNA through a helix-loop-helix (HLH) DNA-binding motif. This motif is similar to that in other transcription factors including myc, daughterless and E12 (Williams and Tjian 1991). The only consensus phosphorylation site for PKA is located just amino-terminal to the DNA-binding region. A role for PKA in mediating the function of AP-2 has not been determined.

6 Tissue-Specific Gene Expression

An essential characteristic of gene expression in multicellular organisms is the tightly regulated spatial and temporal control of gene transcription. Ultimately, it is the regulation of cell-specific and developmentally appropiate expression of each gene encoding the multitude of receptors, channels, neuropeptides, enzymes, and components of signal transduction pathways that determines the phenotype of individual neurons and the structure of the nervous system. Although many aspects of hormonally mediated regulation of neuropeptide gene expression, such as the cAMP regulation discussed above, have been analyzed to advantage in cell lines, the very nature of the questions concerning cell-specific and developmental regulation demands the use of an experimental model that retains the characteristics under study. For example, reporter genes containing as little as 50 bp of the 5'-flanking region of the somatostatin gene including the CRE are highly functional in medullary thyroid carcinoma cell line that normally expresses somatostatin and less efficiently transcribed in some cell lines that do not produce somatostatin (Andrisani et al. 1987). In contrast, reporter genes containing up to 4 kb of somatostatin 5'-flanking sequences are expressed inconsistently or ectopically in many areas of the CNS of transgenic mice and are not expressed at all in periventricular neurons of the hypothalamus, a major location of endogenous somatostatin production (our unpublished observations). These results suggest that DNA-regulatory elements that are essential for the correct *in vivo* expression of the somatostatin gene are quite distinct from and must supplement the CRE element that confers a high degree of basal and cAMP-regulated gene expression in transfected cell lines. This section of the chapter will focus on the use of transgenic mice to study neuropeptide gene regulation.

The pituitary gland of transgenic mice has been a particularly useful experimental paradigm for the study of cell-specific gene expression. The pituitary and nervous system share several important characteristics, including a diverse population of cellular phenotypes, an extensive network of intercellular signaling pathways, and coexpression of many neuropeptides and plasma membrane proteins. The pituitary is a less structurally complicated and more readily accessible organ that nonetheless accurately reflects the complexities of regulated gene expression in the CNS. Furthermore, both the somatotrophs and corticotrophs of the anterior lobe of the pituitary are also represented by extensively characterized immortalized cell lines, GH4 or GC and AtT20, respectively, that have been invaluable in parallel with transgenic mice for the study of growth hormone (GH), prolactin, and proopiomelanocortin (POMC) gene expression.

6.1 Tissue-Specific Expression of GH and Prolactin

GH and prolactin belong to a gene family that also includes chorionic somatomammotropin. Each member of the family arose by gene duplication from a single ancestral gene and shares significant homology with the other members, yet each has evolved its own specific pattern of regulation. The first step in defining the molecular basis of somatotroph-specific expression of the GH gene was the identification of the minimal nucleotide sequences that confer strict pituitary-specific transcription. Three laboratories independently delineated two binding sites upstream of the GH promoter that are essential for tissue-specific expression by means of deletion analyses with reporter fusion genes in transfected pituitary-derived cell lines (Nelson et al. 1986; Lefevre et al. 1987; West et al. 1987). Contained within these sequences is a common motif 5' A(A/T)TTATNCAT 3' that is the consensus binding site for a nuclear protein. Further evidence implicating the binding site as a transcriptional enhancer for GH gene expression was supplied by functional assays in both cell-free *in vitro* transcription reactions (Bodner and Karin 1987) and by cotransfection of the cloned transcription factor with a reporter gene driven by the enhancer element *in vivo* (Mangalam et al. 1989). Although the prolactin gene is coexpressed with GH in a population of bihormonal mammosomatotroph cells in the pituitary and some cell lines, there is a distinct pool of pituitary lactotrophs that do not express GH. It was of interest, therefore, to understand the functional importance of the repeats of the consensus site also present in the prolactin gene in terms of cell-specific regulation. Four of these clustered in a proximal domain and others are found in a distal domain (Nelson et al. 1986). *In vitro*, the two domains appear to be functionally equivalent in supporting transcription of reporter genes (Elsholtz et al. 1986; Nelson et al. 1986; Lufkin and Bancroft 1987). Studies in transgenic mice have helped to clarify the importance of the GH enhancer sequence in both genes but have also produced some surprising twists.

The most straightforward result was that reporter genes containing the 180 bp of 5'-flanking information from the rat GH gene expressed faithfully only in the pituitary gland and no other tissues (Lira et al. 1988). This 180-bp area of the rat and human GH genes contains the *cis*-active sequences that are essential for *in vitro* expression. Expression within the transgenic pituitaries was found not only in the expected GH cells, but also in a substantial proportion of thyroid stimulating hormone (TSH)-producing thyrotrophs and occasionally in lactotrophs, however. Similar studies have also been performed with 5'-flanking sequences from the rat prolactin gene (Crenshaw et al. 1989). 3 kb of 5' information ligated to reporter genes resulted in lactotroph-specific pituitary expression with a high penetrance of expression indicating that the prolactin sequences were entirely sufficient for expression of the prolactin gene. Either the distal or proximal enhancer domains alone also resulted in pituitary expression, but at significantly lower levels. Moreover, the distal element alone, or its positioning immediately adjacent to the proximal element with intervening sequences deleted, resulted in thyrotroph as well as lactotroph expression. None of the prolactin reporter genes ever expressed in somatotrophs. Like the CRE in the somatostatin gene, the common binding site present in both the GH and prolactin genes is only partially responsible for the exact cell-specific regulation of transcription. It has been postulated that a combination of specific activator and repressor sequences are necessary to achieve correct *in vivo* expression of the GH and prolactin genes (Crenshaw et al. 1989; Larsen et al. 1986; Tripputi et al. 1988; Castrillo et al. 1989).

6.2 Characterization of Pit-1 or GHF-1

The identification of the pituitary-specific enhancer sequence in the GH and prolactin genes rapidly led to strategies for the purification of the transcriptional factor termed alternatively Pit-1 (Ingraham et al. 1988) or GHF-1 (Bodner et al. 1988). Both groups identified a protein with a molecular mass of 33 kDa and identical cDNA sequence based on the different cloning strategies of either site-specific screening of a bacteria phage expression library or hybridization screening with a synthetic oligonucleotide sequence deduced from partial peptide sequence. Pit-1 was found to be expressed specifically in the adult pituitary in somatotrophs, lactotrophs, and thyrothrophs. Delevelopmental expression was found at very early stages in the neural tube, but was then confined to the pituitary starting at day 15 of embryonic life, which is earlier than expression of the GH or prolactin genes, but later than that of the TSH-β gene (He et al. 1989; Dollé et al. 1990). Both the Snell and Jackson dwarf mice have mutations in the Pit-1 gene that are responsible for their phenotype, which includes hypoplasia of somatotrophs, lactotrophs, and thyrotrophs and the absence of the respective hormones (Li et al. 1990). Thus Pit-1 appears to be both a transcription factor for specific pituitary hormone genes and an important regulator of mammalian pituitary development.

Pit-1 shares significant homology with a 60-amino-acid region known as the homeodomain that was originally described in developmentally important *Drosophila* genes. A second region of amino acid homology is known as the POU domain (Herr et al. 1988). One of the most important results of experiments originally focused on elucidating the transcriptional mechanism of GH regulation was the identification of a large gene family of POU-domain genes (Rosenfeld 1991). Many of these POU-domain proteins are expressed in early embryogenesis and during forebrain development and are likely to play important roles in the transcriptional regulation of CNS genes.

Recent studies have focused on the transcriptional regulation of the gene encoding Pit-1. The 5'-flanking region contains two Pit-1 binding sites and can be positively autoregulated (McCormick et al. 1990; Chen et al. 1990). The Pit-1 gene also contains two consensus CREs that are positively activated by CREB through the action of growth hormone releasing factor binding to its specific receptor and the subsequent elevation of cAMP. One possibility, therefore, is that developmental activation of Pit-1, which eventually leads to a cascade of developmental effects including the cell-specific activation of GH, is mediated in part by regulation through the CREB family of transcription factors. In addition, McCormick et al. (1991) have shown that the extinction of GH expression in somatic cell hybrids is correlated with the disappearance of GHF-1 protein and mRNA at the level of GHF-1 transcription. Cell-specific expression of GHF-1 is mediated by a 15-bp region that contains the TATA box and binds a candidate protein named PTF that is responsible for the differential activation of GHF-1 in pituitary cells.

6.3 Tissue-Specific Expression of POMC

The POMC gene is transcriptionally highly active in pituitary corticotrophs and melanotrophs and in a subpopulation of neurons in the arcuate nucleus of the hypothalamus (Civelli et al. 1982). Remarkable features of its regulated expression include stimulation by CRH and vasopressin and inhibition by glucocorticoids and dopamine, the latter factor interacting with D_2 receptors (Lundblad and Roberts 1988). The signaling pathways that mediate these responses are the subject of some debate. Although binding of CRH to its specific receptor is known to activate PKA, there is no consensus CRE in the POMC promoter and the limits of the CRH-response element are not known with precision (Roberts et al. 1987). Similarly, although several glucocorticoid binding sites have been identified within the rat POMC sequences between nucleotides −706 to +64 by gel retardation and DNase I protection assays, there remains controversy as to the location and mechanism of action of the functionally important regulatory sites (Drouin et al. 1989a). Drouin has identified a negative glucocorticoid receptor element (nGRE) between positions −50 to −68 and has suggested that activated glucocorticoid receptors binding to this site may interfere with transcription by competing for binding with transcriptional activators at closely adjacent sites (Drouin et al. 1989a, b). Replacement mutations within this area do not interfere with basal transcription of reporter genes in transfected cell lines, however, (Therrien and Drouin 1991). Recently Riegel et al. (1991) reported that another area of the POMC 5'-flanking sequence between −480 and −360 was required in addition to the nGRE for functional repression by glucocorticoids, based on experiments with transiently transfected AtT20 cells and *in vitro* transcription reactions. Furthermore, a binding site was identified in the 5'-flanking region at positions −15 and −5, between the TATA box and the transcriptional start site, the mutation of which leads to a significant loss of transcriptional activity of the POMC promoter in a number of expression systems (Riegel et al. 1990). A protein (or proteins) of an apparent molecular mass of 54–56 kDa, designated PO-B, has been partially purified that specifically binds to this site and is found in a variety of cell lines besides AtT20 (Wellstein et al. 1991). Although the exact identity of this protein is not yet known, it appears safe to conclude that it is not the factor that determines the cell-specific regulation of the POMC-gene, nor is it known what role PO-B plays in the regulated expression of POMC. Interestingly, phosphorylation of PO-B has been reported to greatly decrease the binding affinity of the protein to its cognate site in the POMC promoter (Wellstein et al. 1991).

Drouin's laboratory was the first to demonstrate accurate expression of the POMC gene in AtT20 cells (Jeanotte et al. 1987) and in transgenic mice (Tremblay et al. 1988). A reporter gene containing rat POMC sequences between −706 and +64 was ligated to coding sequences from the *neo* gene and expressed specifically in both the anterior and intermediate lobes of the pituitary gland and in the testis from the normally used start site at position +1. We subsequently reported that the identical regulatory sequences efficiently expressed β-Galactosidase from the *Escherichia coli lacZ* gene in transgenic mouse pituitaries (Hammer et al. 1990). β-Galactosidase expression was advantageous because it could be detected both enzymatically and immunochemically in single cells, and its activity could be quantified in individual pituitaries as an indirect measure of transcriptional activity of the promoter in response to *in vivo* hormonal manipulations.

β-Galactosidase colocalized with ACTH immunoreactivity only in pituitary corticotrophs and melanotrophs and none of the other hormone cell types. Thus, unlike the GH and prolactin promoter regions that produced expression in an inappropriate cell type, the POMC sequences apear to be absolutely cell-type specific. Developmental studies showed that the reporter enzyme was appropriately expressed in the mouse intermediate lobe as early as day 14.5 of embryonic life, corresponding to the normal transcriptional activation of the endogenous POMC gene (unpublished observations). Adrenalectomy caused a significant increase in anterior lobe β-Galactosidase expression that was completely suppressed by dexamethasone. Haloperidol, a D_2-receptor antagonist, caused a threefold increase in enzymatic activity specifically in the intermediate lobe, again paralleling the response of the endogenous gene to the same treatment. In summary, the 770 nucleotides between −706 and +64 are sufficient in transgenic mice to duplicate every aspect of pituitary expression of the POMC gene that has been tested.

Deletional and mutational analyses of the POMC promoter using transfected AtT20 cells have been performed to characterize the minimal DNA elements important for POMC expression. Although the results have not been entirely consistent among laboratories, in part due to the resistance of the cell line to transfection, a pattern has emerged that suggests that multiple elements are necessary in combination to explain all aspects of regulated POMC expression. Our own studies in transgenic mice have identified the sequences between −323 and −34 as essential and sufficient for all aspects of the spatial, temporal, and hormonally regulated expression of POMC in the pituitary (unpublished results). Individual, but not pairs of nuclear protein binding sites within this region can be deleted and pituitary expression in transgenic mice still be maintained. Furthermore, no tested deletions have dissociated cell-specific expression from in vivo hormonal regulation of the reporter transgenes.

Neither the POMC-*lacZ* nor the POMC-*neo* fusion genes were expressed in the hypothalamus of transgenic mice, sugesting that additional nucleotide signals are required for neuronal expression of the POMC gene. We have tested 5'-flanking sequences out to −4000, and although 50% of the transgenic lines produced have accurate pituitary expression, none have had arcuate nucleus expression (unpublished results). Almost all of the lines with additional 5'-flanking sequences have exhibited widely varying patterns of ectopic neuronal expression. however. Apparently, chromosomal integration positions that are favorable for the faithful recognition of POMC promoter elements by pituitary-specific transcription factors also permit access of more promiscuous neural factors to normally ignored elements upstream of position −706.

The demonstration of ectopic neural and/or pituitary expression of transgenes has been a relatively common occurrence and was first well documented in pedigrees of mice expressing a fusion gene consisting of the mouse metallothionein I promoter and human GH gene reporter sequences (Swanson et al. 1985). Unexpected expression of GH was found in many groups of neurons despite the fact that neither the metallothionein or GH genes are expressed in the ectopic sites. A more systematic study found similar levels of ectopic neural expression from a wide variety of hybrid fusion genes that juxtaposed neuropeptide promoter elements with reporter molecules (Russo et al. 1988). We reported that fusion genes containing the common element of 3'-flanking sequences from the hGH gene were selectively expressed in pituitary gonadotrophs of transgenic mice (Low et al. 1989). Together with many other examples of ectopic expression, these data suggest that cell-specific gene expression involves the combinatorial effect of multiple transcriptional factors that share similar or identical DNA-binding sites present in many unrelated genes. Combining the elements from different genes leads to novel expression patterns based on the new combination of transcriptional regulators interacting with the hybrid gene.

6.4 Neural-Specific Expression of Transgenes

Numerous examples of appropriate expression of neural-specific genes in transgenic mice now exist. One of the most dramatic was rescue of the hypogonadal phenotype by introduction of the rat gonadotropin-releasing hormone (GnRH) gene into *hpg* mice that contain a large deletion in their endogenous GnRH gene (Mason et al. 1986). Another neuropeptide, bovine vasopressin, has also been expressed correctly in the paraventricular, supraoptic, and accessory magnocellular nuclei of transgenic mice (Ang et al. 1991). Other notable successes include the human nerve growth factor (NGF) receptor (Patil et al. 1990), the human amyloid precursor protein (APP) gene (Wirak et al. 1991), mylein basic protein (Readhead et al. 1987), the Purkinje cell-specific L7 protein (Oberdick et al. 1990), and human neurofilament (Julien et al. 1987). A common thread tying together many of these transgenic studies has been the introduction of large pieces of genomic DNA. Wether the fragments contain "locus control sequences," as has been described for the globin genes, or simply contain multiple *cis* sequences spread over many kilobases of sequence is unknown. It is abundantly clear, however, that important regulatory sequences can exist in virtually any location relative to the transcriptional unit, including introns and 3'-flanking sequences. Conversely, the simple inclusion of large tracts of genomic DNA does not guarantee appropriate expression in transgenic mice. For example, a 15-kb mouse genomic somatostatin clone, including 10 kb and 3 kb of 5'- and 3'-flanking sequences, respectively, in addition to the entire transcribed sequences, was highly expressed in the hypothalamus of transgenic mice. The tissue specificity at the gross level was misleading because microscopic analysis revealed that all the transgene expression was ectopic and confined to magnocellular vasopressinergic and not periventricular parvicellular somatostatinergic neurons (Rubinstein et al. 1992). Unlike the situations discussed earlier, this form of ectopic expression is unique in that it occurred from an intact gene marked only with a 30-bp oligonucleotide sequence and not from the combination of disparate genetic elements.

7 Future Directions

While great strides have been made in the past few years in understanding the general mechanisms involved in second messenger signal transduction along the cAMP pathway, much still needs to be done. The nature of CRE-binding proteins indicates that cells must exert complex and dynamic control over signaling along this pathway. Given the central nature of cAMP to the control of metabolism and cell growth, this is perhaps not so surprising. Future work will focus on gaining a better understanding of exactly how many different binding proteins are involved in CRE regulation, how these proteins are controlled by specific kinases, and what role heterodimerization plays in modulating activity.

What have we learned from the neural expression of genes in transgenic mice? First, the fidelity of transgene expression in the CNS, as well as other somatic tissues, has varied markedly depending on the gene construction. The process of designing a transgene remains largely empirical and does not necessarily correlate well with transfection studies in cell lines. Large genomic fragments, including introns, and the use of more than one reporter gene have been more likely to express appropriately. With these caveats it is likely that transgenic mice will be increasingly useful for the precise identification of regulatory sequences in neural genes. A new phase of targeted delivery of specific protein products to selected neurons has directly resulted from the earlier descriptive studies and will contribute to an understanding of neurophysiology and pathology. Finally, the technique of homologous recombination in embryonic stem cells and production of mouse germline in transformants will be used to rigorously test the specific role of DNA-binding proteins, including members of the CREB or POU domain family, in transcriptional regulation and development of the CNS.

References

Akerbloom IE, Slater EP, Beato M, Baxter JD, Mellon PL (1988) Negative regulation by glucocorticoids through interference with a cAMP-responsive enhancer. Science 241:350–353

Anderson B, Kennedy GC Nilson JH (1990) A *cis*-acting element located between the cAMP-response elements and CCAAT box augments cell-specific expression of the glycoprotein hormone α-subunit gene. J Biol Chem 265:21874–21880

Andrisani OM, Hayes TE, Roos B, Dixon JE (1987) Identification of the promoter sequences involved in the cell-specific expression of the rat somatostatin gene. Nucleic Acids Res 15:5715–5728

Ang HL, Funkhower J, Carter DA, Ho MY, Murphy D (1991) Expression of bovine vasopressin in the hypothalamus of the transgenic mouse and its regulation during osmotic challenge. Soc Neurosci Abstr 17:1287

Benbrook DM, Jones NC (1990) Heterodimer formation between CREB and JUN proteins. Oncogene 5:295–302

Berkowitz LA, Gilman MZ (1990) Two distinct forms of active transcription factor CREB. Proc Natl Acad Sci USA 87:5258–5262

Bernath VA, Muro AF, Vitullo AD, Bley MA, Baranao JL, Kornblihtt AR (1990) Cyclic AMP inhibits fibronectin gene expression in a newly developed granulosa cell line by a mechanism that suppresses cAMP-responsive element-dependent transcriptional regulation. J Biol Chem 265: 18219–18226

Bodner M, Karin M (1987) A pituitary-specific *trans*-acting factor can stimulate transcription from the growth hormone promoter in extracts of nonexpressing cells. Cell 50:267–275

Bodner M, Castrillo J-L, Theill LE, Deerinck T, Ellisman M, Karin M (1988) The pituitary–specific transcription factor GHF-1 is a homeobox-containing protein. Cell 55:505–518

Boshart M, Weih F, Schmidt A, Fournier REK, Schutz G (1990) A cyclic AMP-responsive element mediates repression of tyrosine aminotransferase gene transcription by the tissue-specific extinguisher locus Tse-1. Cell 61:905–916

Boshart M, Weih F, Nichols M, Schutz G (1991) The tissue-specific extinguisher locus TSE1 encodes a regulatory subunit of cAMP-dependent protein kinase. Cell 66:849–859

Castrillo J-L, Bodner M, Karin M (1989) Purification of growth hormone-specific transcription factor GHF-1 containing homeobox. Science 243:814–817

Chatterjee VKK, Madison LD, Mayo S, Jameson JL (1991) Repression of the human glycoprotein hormone α-subunit gene by glycocorticoids: evidence for receptor interactions with limiting transcriptional activators. Mol Endocrinol 5:100–110

Chen R, Ingraham HA, Treacy MN, Albert VR, Wilson L, Rosenfeld MG (1990) Autoregulation of Pit-1 gene expression mediated by two *cis*-active promoter elements. Nature 346:583–586

Civelli O, Birnberg N, Herbert E (1982) Detection and quantitation of proopiomelanocortin mRNA in pituitary and brain tissues from different species. J Biol Chem 257:6783–6787

Comb M, Birnberg NC, Seasholtz A, Herbert E, Goodman HM (1986) A cyclic AMP- and phorbol ester-inducible DNA element. Nature 323:353–356

Comb M, Mermod N, Hyman SE, Pearlberg J, Ross ME, Goodman HM (1988) Proteins bound at adjacent DNA elements act synergistically to regulate human proenkephalin cAMP inducible transcription. EMBO J 73793–3805

Crenshaw EB, Kalla K, Simmons DM, Swanson LW, Rosenfeld MG (1989) Cell-specific expression of the prolactin gene in transgenic mice is controlled by synergistic interactions between promoter and enhancer elements. Genes Dev 3:959–972

Dean DC, Newby RF, Bourgeois S (1988) Regulation of fibronectin biosysnthesis by dexamethasone, transforming growth factor β, and cAMP in human cell lines. J Cell Biol 106:2159–2170

Delegeane AM, Ferland LH, Mellon PL (1987) Tissue-specific enhancer of the human glycoprotein hormone α-subunit gene: dependence on cyclic AMP-inducible elements. Mol Cell Biol 7:3994–4002

Deutsch PJ, Jameson JL, Habener JF (1987) Cyclic AMP-responsiveness of human gonadotropin-α gene transcription is directed by a repeated 18-base pair enhancer. J Biol Chem 262:12169–12174

Deutsch PJ, Hoeffler JP, Jameson JL, Lin JC, Habener JF (1988a) Structural determinants for transcriptional activation by cAMP-responsive DNA elements. J Biol Chem 263:18466–18472

Deutsch PJ, Hoeffler JP, Jameson JL, Habener JF (1988b) Cyclic AMP- and phorbol ester-stimulated transcription mediated by similar DNA elements that bind distinct proteins. Proc Natl Acad Sci USA 85:7922–7926

Dollé P, Castrillo J-L, Theill LE, Deerinck J, Ellisman M, Karin M (1990) Expression of GHF-1 protein in developing mouse pituitaries correlates both temporally and spatially with the onset of growth hormone gene activity. Cell 60:809–820

Dorrington JH, Skinner MK (1986) Cytodifferentiation of granulosa cells induced by gonadotropin-releasing hormone promoters fibronectin secretion. Endocrinology 118:2065–2071

Drouin J, Trifiro MA, Plante RK, Nemer M, Eriksson P, Wrange O (1989a) Glucocorticoid receptor binding to a specific DNA sequence is required for hormone-dependent repression of pro-opiomelanocortin gene transcription. Mol Cell Biol 9:5305–5314

Drouin J, Nemer M, Charron J, Gagner JP, Jeanotte L, Sun YL, Therrien M, Tremblay Y (1989b) Tissue-specific activity of the pro-opiomelanocortin (POMC) gene and repression by glycocorticoids. Genome 31:510–519

Elsholtz HP, Mangalam HS, Potten E, Albert VR, Supowit S, Evans RM, Rosenfeld MG (1986) Two different *cis*-active elements transfer the transcriptional effects of both EGF and phorbol esters. Science 234:1552–1557

Fink JS, Verhave M, Kasper S, Tsukada T, Mandel G, Goodman RH (1988) The CGTCA sequence motif is essential for biological activity of the vasoactive intestinal peptide gene cAMP-regulated enhancer. Proc Natl Acad Sci USA 85:6662–6666

Fink JS, Verhave M, Walton K, Mandel G Goodman RH (1991) Cyclic AMP- and phorbol ester-induced transcriptional activation are mediated by the same enhancer element in the human vasoactive intestinal peptide gene. J Biol Chem 266:3882–3887

Foulkes NS, Borelli E, Sassone-Corsi P (1991) CREM gene: use of alternative DNA-binding domains generates multiple antagonists of cAMP-induced transcription. Cell 64:739–749

Gaire M, Chatton B, Kedinger C (1990) Isolation of two novel, closely related ATF cDNA clones from HeLa cells. Nucleic Acids Res 18:3467–3473

Gilman MZ, Wilson RN, Weinberg RA (1986) Multiple protein-binding sites in the 5'-flanking region regulate c-*fos* expression. Mol Cell Biol 6:4305–4306

Gonzalez GA, Montminy MR (1989) Cyclic AMP stimulates somatostatin gene transcription by phosphorylation of CREB at serine 133. Cell 59:675–680

Gonzalez GA, Yamamoto KK, Fischer WH, Karr D, Menzel P, Biggs W, Vale WW, Montminy MR (1989) A cluster of phosphorylation sites on the cyclic AMP-regulated nuclear factor CREB predicted by its sequence. Nature 337:749–752

Hai T, Curran T (1991) Cross-family dimerization of transcription factors Fos/Jun and ATF/CREB alters DNA-binding specificity. Proc Natl Acad Sci USA 88:3720–3724

Hai T, Liu F, Coukos WJ Green MR (1989) Transcription factor ATF cDNA clones: and extensive family of leucine zipper proteins able to selectively form DNA-binding heterodimers. Genes Dev 3:2083–2090

Hammer GD, Fairchild-Huntress V, Low MJ (1990) Pituitary-specific and hormonally regulated gene expression directed by the rat proopiomelanocortin promoter in transgenic mice. Mol Endocrinol 4:1689–1697

Haslinger A, Karin M (1985) Upstream promoter element of the human metallothionein-IIA gene can act like an enhancer element. Proc Natl Acad Sci USA 82:8572–8576

He X, Treacy MN, Simmons DM, Ingraham HA, Swanson LW, Rosenfeld MG (1989) Expression of a large family of POU-domain regulatory genes in mammalian brain development. Nature 340:35–42

Herr W, Sturm RA, Clerc RG, Corcosan LM, Baltimore D, Sharp PA, Ingraham HA, Rosenfeld MG, Finney M, Ruvkun G, Horvitz R (1988) The POU domain: a large conserved region in the mammalian Pit-1, Oct-1, Oct-2, and Caenorhabditis elegans unc-86 gene products. Genes Dev 2: 1513–1516

Hoeffler JP, Meyer TE, Yun Y, Jameson JL, Habener JF (1988) Cyclic AMP-responsive DNA-binding structure based on a cloned placental DNA. Science 242:1430–1433

Hyman SE, Comb M, Lin YS, Pearlberg J, Green MR, Goodman HM (1988) A common *trans*-acting factor is involved in transcriptional regulation of neurotransmitter genes by cyclic AMP. Mol Cell Biol 8:4225–4233

Hyman SE, Comb M, Pearlberg J, Goodman HM (1989) An AP-2 element acts synergistically with the cyclic AMP- and phorbol ester-inducible enhancer of the human proenkephalin gene. Mol Cell Biol 9:321–324

Imagawa M, Chiu R, Karin M, (1987) Transcription factor AP-2 mediates induction by two different signal transduction pathways: protein kinase C and cAMP. Cell 51:251–260

Ingraham HA, Chen R, Mangalam HJ, Elsholtz HP, Flynn SE Lin CR, Simmons DM, Swanson L, Rosenfeld MG (1988) A tissue-specific transcription factor containing a homeodomain specifies a pituitary phenotype. Cell 55:519–529

Ivashkiv LB, Liou HC, Kara CJ, Lamph WM, Verma IM, Glimcher LH (1990) MXBP/CRE-BP-2 and c-*jun* form a complex which binds to the cyclic AMP, but not the 12-O-tetradecanoylphorbol-13-acetate, response element. Mol Cell Biol 10:1609–1621

Jameson JL, Lindell CM (1988) Isolation and characterization of the human chorionic gonadotropin β-subunit (CGβ) gene cluster: regulation of a transcriptionally active CGβ gene by cyclic AMP. Mol Cell Biol 8:5100–5107

Jameson JL, Powers AC, Gallagher GD, Habener JF (1989a) Enhancer and promoter element interactions dictate cyclic adenosine monophosphate mediated and cell-specific expression of the glycoprotein hormone α-gene. Mol Endocrinol 3:763–772

Jameson JL, Albanese C, Habener JF (1989b) Distinct adjacent protein-binding domains in the glycoprotein hormone α gene interact with a cAMP-responsive enhancer. J Biol Chem 264: 16190–16196

Jeannotte L, Trifiro MA, Plante RK, Chamberland M, Drouin J (1987) Tissue-specific activity of the pro-opiomelanocortin gene promoter. Mol Cell Biol 7:4058–4064

Julien JP, Tretjakoff I, Beaudet L, Peterson A (1987) Expression and assembly of a human neurofilament protein in transgenic mice provide a novel neuronal marking system. Genes Dev 1:1085–1086

Kagawa N, Waterman MR (1990) cAMP-dependent transcription of the human CYP21B ($P\text{-}450_{c21}$) gene requires a *cis*-regulatory element distinct from the consensus cAMP-regulatory element. J Biol Chem 265:11299–11305

Kara CJ, Liou HC, Ivashkiv LB, Glimcher LH (1990) a cDNA for human cyclic AMP-response element-binding protein which is distinct from CREB and expressed preferentially in brain. Mol Cell Biol 10:1347–1357

Karin M, Haslinger A, Heguy A, Dietlin T, Cooke T (1987) Metal-responsive elements act as positive modulators of human metallothionein-IIA enhancer activity. Mol Cell Biol 7:606–613

Kennedy GC, Anderson B, Nilson JH (1990) The human α-subunit glycoprotein hormone gene utilizes a unique CCAAT binding factor. J Biol Chem 265:6279–6285

Landschultz WH, Johnson PF, McKnight SL (1988) The leucine zipper: a hypothetical structure common to a new class of DNA-binding proteins. Science 240:1759–1764

Larsen PR, Harney JW, Moore DD (1986) Repression mediates cell-type specific expression of the rat growth hormone gene. Proc Natl Acad Sci USA 83:8283–8287

Lee CQ, Yun Y, Hoeffler JP, Habner JP (1990) Cyclic AMP-responsive transcriptional activation of CREB-327 involves interdependent phosphorylated subdomains. EMBO J 9:4455–4465

Levevre C, Imagawa M, Dana S, Grindlay J, Bodner M, Karin M (1987) Tissue-specific expression of the humen growth hormone gene is conferred in part by binding of a specific *trans*-acting factor. EMBO J:971–981

Li S, Crenshaw EB, Rawson EJ, Simmons DM, Swanson LW, Rosenfeld MG (1990) Dwarf locus mutants lacking three pituitary cell types result from mutations in the POU-domain gene Pit-1. Nature 347:528–534

Lira SA, Crenshaw EB, Glass CK, Swanson LW, Rosenfeld MG (1988) Identification of rat growth hormone genomic sequences targeting pituitary expression in transgenic mice. Proc Natl Acad Sci USA 85:4755–4759

Lorange D, Lundblad JR, Blum M, Roberts JL (1992) Calcium plays a major role in mediating CRH-regulation of proopiomelanocortin gene transcriptional regulation in AtT20 cells. Mol Cell Biol (in press)

Low MJ, Goodman RH, Ebert KM (1989) Cryptic human growth hormone gene sequences direct gonadotroph-specific expression in transgenic mice. Mol Endocrinol 3:2028–2033
Lufkin T, Bancroft C (1987) Identification by cell fusion of gene sequences that interact with positive *trans*-acting factors. Science 237:283–286
Lund J, Ahlgren R, Wu D, Kagimoto M, Simpson ER, Waterman MR (1990) Transcriptional regulation of the bovine CYP17 (P-45017a) gene. J Biol Chem 265:3304–3312
Lundblad JR, Roberts JL, (1988) Regulation of proopiomelanocortin gene expression in pituitary. Endocr Rev 9:135–158
Ma J, Ptashne M (1987a) A new class of yeast transcriptional activators. Cell 51:113–114
Ma J, Ptashne M (1987b) Deletion analysis of GAL 4 defines two transcriptional activator sequences. Cell 48:847–853
Macgregor PF, Abate C, Curran T (1990) Direct cloning of leucine zipper proteins: Jun binds cooperatively to the CRE with CRE-BP1. Oncogene 5:451–458
Maekawa T, Sakura H, Kanei-Ishii C, Sudo T, Yoshimura T, Fujisawa J, Mitsuaki Y, Ishii S (1989) Leucine zipper structure of the protein CRE-BP1 binding to the cyclic AMP-response element in brain. EMBO J 8:2023–2028
Mangalam HJ, Albert VR, Ingraham HA, Kapiloff M, Wilson L, Nelson C, Elsholtz H, Rosenfeld MG (1989) A pituitary POU domain protein, Pit-1, activates both growth hormone and prolactin promoters transcriptionally. Genes Dev 3:946–958
Mason AJ, Pihs SL, Nikoliu K, Szonyi W, Wilcox S, Seeburg PH, Stewart TA (1986) The hypogonadal mouse: reproductive functions restored by gene therapy. Science 234:1372–1378
McCormick A, Brady H, Theill LE, Karin M (1990) Regulation of the pituitary-specific homeobox gene GHF1 by cell-autonomous and environmental cues. Nature 345:829–832
McCormick A, Brady H, Fukushima J, Karin M (1991) The pituitary-specific regulatory gene GHF1 contains a minimal cell-type specific promoter centered around its TATA box. Genes Dev 5:1490–1503
McMurray CT, Wilson WD, Douglass JO (1991) Hairpin formation within the enhancer region of the human enkephalin gene Proc Natl Acad Sci USA 88:666–670
Medcalf RL, Ruegg M, Schleuning WD (1990) A DNA motif related to the cAMP-responsive element and an exon-located activator protein-2 binding site in the human tissue-type plasminogen activator gene promoter cooperate in basal expression and convey activation by phorbol ester and cAMP. J Biol Chem 265:14618–14626
Milstead A, Cox RP, Nilson JH (1987) Cyclic AMP regulates transcription of the genes encoding human chorionic gonadotropin with different kinetics. DNA 6:213–219
Mitchell PJ, Wang C, Tjian R (1987) Positive and negative regulation of transcription *in vitro*: enhancer-binding protein AP-2 inhibited by SV40 T antigen. Cell 50:847–861
Montminy MR, Bilezikjian LM (1987) Binding of a nuclear protein to the cyclic AMP-response element of the somatostatin gene. Nature 328:175–178
Montminy MR, Sevarino KA, Wagner JA, Mandel G, Goodman RH (1986) Identification of a cyclic AMP-responsive element within the rat somatostatin gene. Proc Natl Acad Sci USA 83: 6682–6686
Nelson C, Crenshaw EB, Franco R, Lira SA, Albert VR, Evans RM, Rosenfeld MG (1986) Discrete *cis*-active genomic sequences dictate the pituitary cell-type specific expression of rat prolactin and growth hormone genes. Nature 322:557–562
Nelson C, Albert VR, Elsholtz HP, Lu LIW, Rosenfeld MG (1988) Activation of cell-specific expression of rat growth hormone and prolactin genes by a common transcription factor. Science 239:1400–1405
Oberdick J, Smeyne RJ, Mann JR, Zackson S, Morgan JI (1990) A promoter that drives transgene expression in cerebellar purkinje and retinal bipolar neurons. Science 248:223–226
O'Shea EK, Rutkowski R, Kim PS (1989) Evidence that the leucine zipper is a coiled-coil. Science 243:538–542
Patil N, Lacy E, Chao MV (1990) Specific neuronal expression of human NGF receptors in the basal forebrain and cerebellum of transgenic mice. Neuron 4:437–447
Philippe J, Missotten M (1990) Functional characterization of a cAMP-responsive element of the rat insulin I gene. J Biol Chem 265:1465–1469
Philippe J, Drucker DJ, Knepel W, Jepeal L, Misulovin Z, Habener JF (1988) α cell-specific

expression of the glucagon gene is conferred to the glucagon promoter element by the interactions of DNA-binding proteins. Mol Cell Biol 8:4877–4888
Readhead C, Popko B, Takahashi N, Shine HD, Saavedra RA, Sidman RL, Hood L (1987) Expression of a myelin basic protein gene in transgenic shiverer mice: correction of the dysmyelinating phenotype. Cell 48:703–712
Rehfuss RP, Walton KM, Loriaux MM, Goodman RH (1991) The CRE-binding protein ATF-1 activates transcription in response to cAMP-dependent protein kinase A. J Biol Chem 266: 18431–18434
Rickles RJ, Darrow AL, Strickland S (1989) Differentiation-responsive elements in the 5' region of the mouse tissue plasminogen activator gene confer two-stage regulation by retinoic acid and cyclic AMP in teratocarcinoma cells. Mol Cell Biol 9:1691–1704
Riegel AT, Remenick J, Wolford RG, Berard DS Hager GL (1990) A novel transcriptional activator (PO-B) binds between the TATA box and cap site of the pro-opiomelanocortin gene. Nucleic Acids Res 18:4513–4521
Riegel AT, Lu Y, Remenick J, Wolford RG, Berard DS, Hager GL (1991) Proopiomelanocortin gene promoter elements required for constitutive and glococorticoid-repressed transcription. Mol Endocrinol 5:1973–1982
Roberts JL, Lundblad JR, Eberwine JH, Fremeau RT, Salton SRJ, Blum M (1987) Hormonal regulation of POMC gene expression in pituitary. Ann NY Acad Sci 512:275–285
Rosenfeld MG (1991) POU-domain transcription factors: pou-er-ful developmental regulators. Genes Dev 5:897–907
Rubinstein M, Liu B, Goodman RH, Low MJ (1992) Targeted expression of somatostatin in vasopressinergic magnocellular hypothalamic neurons of transgenic mice. Mol Cell Neurosci (in press)
Russo AF, Crenshaw EB, Lira SA, Simmons DM, Swanson LW, Rosenfeld MG (1988) Neuronal expression of chimeric genes in transgenic mice. Neuron 1:311–320
Sheng M, Thompson MA, Greenberg ME (1991) CREB: A Ca^{+2}-regulated transcription factor phosphorylated by calmodulin-dependent kinases. Science 252:1427–1430
Short JM, Wynshaw-Boris A, Short HP, Hanson RW (1986) Characterization of the phosphoenolpyruvate carboxykinase (GTP) promoter-regulatory region. J Biol Chem 261:9721–9726
Swanson LW, Simmons DM, Arriza J, Hammer R, Brinster R, Rosenfeld MG, Evans RM (1985) Novel developmental specificity in the nervous system of transgenic animals expressing growth hormone fusion genes. Nature 317:363–366
Therrien M, Drouin J (1991) Pituitary pro-opiomelanocortin gene expression requires synergistic interactions of several regulatory elements. Mol Cell Biol 11:3492–3503
Tremblay Y, Tretjakoff I, Peterson A, Antakly T, Zhang CX, Drouin J (1988) Pituitary-specific expression and glucocorticoid regulation of a proopiomelanocortin fusion gene in transgenic mice. Proc Natl Acad Sci USA 85:8890–8894
Tripputi P, Guerin SL, Moore DD (1988) Two mechanisms for the extinction of gene expression in hybrid cells. Science 241:1205–1207
Tsukada T, Fink JS, Mandel G, Goodman RH (1987) Identification of a region in the human vasoactive intestinal peptide gene responsible for regulation by cyclic AMP. J Biol Chem 262: 8743–8747
Weaver CA, Gordon DF, Kissel MS, Mead DA, Kemper B (1984) Isolation and complete nucleotide-sequence of the gene for bovine parathyroid-hormone. Gene 28:319–329
Wellstein A, Dobrenski AF, Radonovich MN, Brady JF, Riegel AT (1991) Purification of PO-B, a protein that has increased affinity for the pro-opiomelanocortin gene promoter after dephosphorylation. J Biol Chem 266:12234–12241
West BL, Catanzaro DF, Mellon SH, Cattini PA, Baxter JD, Reudelhuber TL (1987) Interaction of a tissue-specific factor with an essential rat growth hormone gene promoter element. Mol Cell Biol 7:1193–1197
Williams T, Tjian R (1991) Characterization of a dimerization motif in AP-2 and its function in heterologous DNA-binding proteins. Science 251:1067–1071
Williams T, Adom A, Luescher B, Tjian R (1988) Cloning and expression of AP-2, a cell-type specific transcription factor that activates inducible enhancer elements. Genes Dev 2:1557–1569

Wirak DO, Bayney R, Kundel CA, Lee A, Scangos GA, Trapp BD, Unterbeck AJ (1991) Regulatory region of human amyloid precursor protein (APP) gene promotes neuron-specific gene expression in the CNS of transgenic mice. EMBO J 10:289–296
Wynshaw-Boris A, Lugo TG, Short JM, Fournier REK, Hanson RW(1984) Identification of a cAMP regulatory region in the gene for rat cytosolic phosphoenolpyruvate carboxykinase (GTP). J Biol Chem 259:12161–12169
Yamamoto KK, Gonzalez GA, Menzel P, Rivier J, Montminy MR (1990) Characterization of a bipartite activator domain in transcription factor CREB. Cell 60:611–617
Yoshimura T, Fujisawa J, Yoshida M (1990) Multiple cDNA clones encoding nuclear proteins that bind to the tax-dependent enhancer of HTLV-I: all contain a leucine zipper structure and basic amino acid domain. EMBO J 9:2537–2542

Opioid Peptide Genes: Structure and Regulation

V. HÖLLT[1]

Contents

[1] Department of Physiology, University of Munich, Pettenkoferstraße 12, D-8000 München 40, FRG

Current Topics in Neuroendocrinology, Vol. 11

1 Introduction

All mammalian opioid peptides belong to one of three peptide families, each deriving from a distinct precursor: proopiomelanocortin (POMC), proenkephalin (PENK), and prodynorphin (PDYN). These precursor molecules are translation products from separate genes (for a review see Höllt 1986).

2 Proopiomelanocortin

POMC is the precursor of several biologically active peptides, such as β-endorphin, adrenocorticotropic hormone (ACTH), and various peptides with melanocyte-stimulating activity (melanocyte-stimulating hormones: α-MSH, β-MSH, γ-MSH) (Fig. 1).

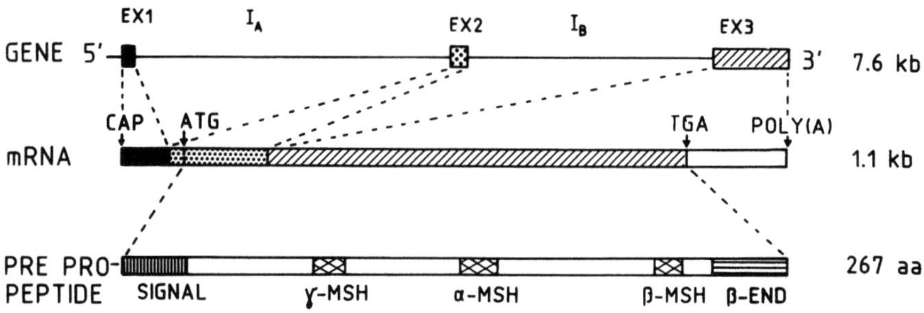

Fig. 1. Structural organization of the POMC gene, mRNA, and propeptide. *CAP*, transcription initiation site; *ATG*, start of translation; *TGA*, end of translation; *POLY(A)*, polyadenylation site. *EX*, exon; *I*, intron; β-END, β-endorphin

2.1 Structure

The complete sequence of the human gene (Takahashi et al. 1983) is 7665 bp long and is comprised as follows (Fig. 1): exon 1 (86bp), exon 2 (152 bp), exon 3 (833 bp); intron A (3708 bp); intron B (2886 bp). With the exception of two structurally different genes in *Xenopus laevis* (Martens 1986), there appears to exist only one functional gene in the other species investigated. The mouse has an additional pseudogene which does not produce a functional mRNA (Uhler et al. 1983; Notake et al. 1983). The human POMC gene is localized on chromosome 2 (Owerbach et al. 1981) and that of the mouse on chromosome 19 (Uhler et al 1983).

The size of POMC mRNA in the pituitary and the hypothalamus is 1100-1200 nucleotides. In some tissues (adrenal medulla, thyroid, etc.) shorter POMC transcripts have been found. These smaller mRNA species are derived from aberrant transcription initiations next to the 5'-end of exon 3 and, thus, do not contain any exon 1 or exon 2 sequences (Jingami et al. 1984; Jeanotte et al. 1987;

Kilpatrick et al. 1987). Diversity of POMC transcription also occurs by alternative *splicing*. Thus, in the intermediate pituitary of rats, a different splicing occurs between exon 1 and exon 2 giving rise to two POMC mRNAs which differ in 30 nucleotides within the 5'-untranslated portion of the POMC mRNA (Oates and Herbert 1984). Larger, 5'-extended POMC transcripts of about 1450 nucleotides have been observed in ectopic tumors. In the human POMC gene, these are derived from initiations at promoter sites located 100–350 nucleotides upstream (de Keyzer et al. 1989a).

2.2 Regulation

2.2.1 Adenohypophysis

In the anterior pituitary, POMC gene expression is under the negative feedback control of adrenal steroids. Adrenalectomy results in a marked increase of POMC mRNA levels in the adenohypophysis (Nakanishi et al 1979; Birnberg et al. 1983). This appears to be due to an increased transcription rate per cell and to an increased number of POMC synthesizing cells, as revealed by in situ hybridization experiments using a probe specific for the first intron of the POMC gene (Fremeau et al. 1986). Injection of dexamethasone into adrenalectomized rats causes a rapid fall in POMC transcription (Gagner and Drouin 1985) and a more protracted decline in POMC mRNA levels (Birnberg et al. 1983). Glucocorticoids have also been shown to inhibit POMC gene expression in primary cultures of anterior pituitaries and in mouse AtT-20 tumor cells (Nakamura et al. 1978; Roberts et al. 1979; Gagner and Drouin 1985; Eberwine et al. 1987).

POMC gene expression in the anterior pituitary is under positive control by corticotropin-releasing hormone (CRH). Chronic administration of exogenous CRH for 3–7 days results in a marked increase in the POMC mRNA levels in the anterior pituitary of rats (Bruhn et al. 1984; Höllt and Haarmann 1984). An increase in POMC mRNA levels in the anterior pituitary is also seen after various *stress* treatments, such as chronic foot-shock (Höllt et al. 1986; Shiomi et al. 1986) or hypoglycemic shock induced by insulin (Tozawa et al. 1988). Repeated administration of *morphine* has also been shown to increase POMC mRNA levels in the anterior pituitary of rats (Höllt and Haarmann 1985). Naloxone (NAL)-precipitated withdrawal in morphine-tolerant rats also increased POMC mRNA levels, indicating that withdrawal stress activates POMC gene expression in the anterior pituitary (Lightman and Young 1988).

Chronic treatment of rats with ethanol in a vapor chamber decreased levels of POMC mRNA in the anterior pituitary (Dave et al. 1986). However, other experiments in which ethanol was chronically administered by liquid diet revealed an increase in the biosynthesis of POMC in the anterior pituitary (Seizinger et al. 1984a). The different findings might be related to the different modes of ethanol administration. Dehydration, commonly associated with ethanol administration in the drinking water, is also likely to affect POMC mRNA levels in the pituitary (Elkabes and Loh 1988).

In primary cultures of rat anterior pituitary cells, in AtT-20 mouse tumor cells, and in cultured human corticotrophic tumor cells, CRH has been shown to increase POMC mRNA levels (Loeffler et al. 1985; Affolter and Reisine 1985; von Dreden and Höllt 1988; Knight et al. 1987; Stalla et al. 1989a) as a result of increased gene transcription (Gagner and Drouin 1985; Jeannotte et al. 1987; Eberwine et al. 1987; Autelitano et al. 1990) The effect of CRH appears to be mediated via cAMP followed by activation of protein kinase A. In fact, insertion of a protein kinase A inhibitor in AtT-20/D16-16 tumor cells blocked the ability of CRH and 8-bromo-cAMP (8-Br-cAMP) to increase POMC mRNA levels (Reisine et al. 1985). Moreover, in primary cultures of rat anterior pituitary cells 8-Br-cAMP, forskolin (a biterpene derivative which activates adenylate cyclase), cholera toxin, and drugs which elevate cellular cAMP levels by inhibiting phosphodiesterase (e.g., Ro 20-1724) increase POMC mRNA levels (Loeffler et al. 1986b; Affolter and Reisine 1985; Simard et al. 1986; Dave et al. 1987; Stalla et al. 1988; Suda et al. 1988a). In addition, imidazole derivatives, such as ketoconazole or isoconazole which inhibit adenylate cyclase, block the CRH- or forskolin-induced increase in POMC message levels in rat anterior pituitary cultures (Stalla et al. 1988, 1989b). The inducing effect of CRH on POMC mRNA levels in cultured anterior lobe cells was partially inhibited by voltage-dependent Ca^{2+} channel blockers, such as verapamil and nifedipine (Loeffler et al. 1986b; Dave et al. 1987). Similar findings were reported with mouse AtT-20/D-16-v cell lines (von Dreden and Höllt 1988). This indicates that CRH exerts its effect on POMC gene expression via entry of *Ca^{2+}ions*.

Although arginine vasopressin (*AVP*) releases POMC peptides from anterior pituitary cells and potentiates the secretory effect of CRH in vitro, it does not have a major effect on POMC biosynthesis (von Dreden et al. 1988; Stalla et al. 1989a; Suda et al. 1989). Moreover, AVP does not potentiate the stimulation of POMC gene expression by CRH. Tumor-promoting phorbol esters, such as phorbol 12-myristate 13-acetate (TPA), which activate protein kinase C, increase POMC mRNA levels in AtT-20/D16-16 tumor cells (Affolter and Reisine 1985). However, in primary cultures of rat anterior pituitary and in the AtT-20/D16-v cell line, TPA had no effect on POMC mRNA levels (Suda et al. 1989; JP. Loeffler and V. Höllt, unpublished). This indicates that the protein kinase C-dependent signal transduction system contributes to the release, but not to the biosynthesis of POMC-peptides. Similarly, interleukin-1 (and -1β) releases POMC-peptides by a direct action on anterior pituitary cells; however, its effect on POMC gene expression is minimal, if any (Suda et al. 1988b).

2.2.2 Intermediate Pituitary

POMC gene expression in the melanotrophic cells of the intermediate pituitary is differently regulated from that in the corticotrophic cells of the anterior pituitary. Adrenalectomy or treatment with glucocorticoids affect POMC mRNA levels in the intermediate pituitary only slightly (Schachter et al. 1982). Interestingly, glucocorticoids cause an elevation of POMC mRNA levels in the intermediate pituitary of rats after denervation of the intermediate pituitary by hypothalamic lesions (Seger et al. 1988; Autelitano et al. 1987, 1989).

POMC gene expression in the intermediate pituitary is also influenced by CRH. Thus, CRH administration (Loeffler et al. 1985, 1988) and 8-Br-cAMP, forskolin, cholera toxin, and the phosphodiesterase inhibitor RO 20-1724 increased POMC mRNA levels in intermediate lobe cultures (Loeffler et al. 1986b). The effect of forskolin was partially blocked by the Ca^{2+} channel antagonist nifedipine and potentiated by Bay K 8644, a dihydropyridine which increases the opening probability of voltage-dependent calcium channels (Loeffler et al. 1986b). There is increasing evidence that long-term administration of CRH to rats in vivo decreases POMC mRNA levels (Höllt and Haarmann 1984; Lundblad and Roberts 1988). The differential effect of CRH on intermediate pituitary POMC gene expression in vivo, versus in vitro, indicates that CRH affects POMC gene expression by both direct and indirect actions.

Various stressors, such as insulin shock or chronic foot-shock for up to 7 days, increase POMC mRNA levels in the anterior pituitary, but do not affect the levels of POMC mRNA in the intermediate pituitary (Höllt et al. 1986; Tozawa et al. 1988). However, prolonged electrical foot-shock stress (administered for 2 weeks) has been reported to cause an increase in POMC mRNA levels also in the intermediate lobe (Shiomi et al. 1986).

The release of POMC-derived peptides from the intermediate pituitary is tonically inhibited by dopamine (DA) via D_2-receptors (Munemura et al. 1980). Blockade of these receptors by haloperidol causes an increase in the levels of POMC mRNA (Höllt et al. 1982; Chen et al. 1983; Meador-Woodruff et al. 1990) and POMC gene transcription (Pritchett and Roberts 1987) in the intermediate pituitary lobe of rats. On the contrary, chronic injection of bromocriptine, a DA receptor agonist, substantially decreased POMC mRNA levels in the rat intermediate pituitary (Chen et al. 1983; Levy and Lightman 1988). DA and bromocriptine inhibited POMC gene expression in cultured intermediate lobes (Cote et al. 1986) and intermediate lobe cells (Loeffler et al. 1988) in vitro. Conversely, compounds that activate the cAMP pathway (8-Br-cAMP, cholera toxin, forskolin) counteracted the DAergic inhibition of POMC mRNA biosynthesis (Loeffler et al. 1986b, c).

In addition to DA, γ-aminobutyric acid (GABA) has been shown to inhibit the release of POMC-derived peptides from the intermediate pituitary (Tomiko et al. 1983). When endogenous GABA levels in the hypothalamus and pituitary of rats were elevated by chronic administration of inhibitors of the GABA metabolizing enzyme GABAtransaminase, a time-dependent decrease in the levels of POMC mRNA was found (Loeffler et al. 1986a). GABA also caused a marked reduction of POMC mRNA levels when applied to primary cultures of intermediate lobe cells (Loeffler et al. 1986a).

Chronic application of morphine decreases POMC mRNA levels in the intermediate pituitary lobe, as revealed by RNA blot analysis (Höllt and Haarmann 1985). In contrast, in situ hybridization techniques revealed no effect of morphine upon POMC levels in this tissue (Lightman and Young 1988). It is possible, however, that high doses of morphine are required to cause a decrease of POMC in the intermediate pituitary. In addition, chronic treatment of rats with NAL, or NAL-induced withdrawal in morphine-tolerant rats, failed to alter POMC mRNA levels in the intermediate pituitary (Lightman and Young 1988).

Chronic ethanol treatment decreases POMC biosynthesis in the intermediate pituitary lobes of rats (Seizinger et al. 1984b; Dave et al. 1986). As discussed previously, this effect may result from the dehydrating action of ethanol, since salt loading has been shown to decrease POMC mRNA levels in the intermediate pituitary of mice (Elkabes and Loh 1988). This effect was antagonized by D_2-receptor antagonists indicating that salt loading increased DA release in this pituitary lobe. On the other hand, ethanol administered by a controlled drinking schedule (liquid diet) increases POMC biosynthesis in the intermediate lobe of rats (Seizinger et al. 1984a; V. Höllt, unpublished data).

2.2.3 Hypothalamus

Within the brain, the cells synthesizing POMC are almost exclusively localized in the periarcuate region of the hypothalamus as revealed by in situ hybridization (Gee et al. 1983; Civelli et al. 1982).

It was initially reported that POMC mRNA levels in the hypothalamus were unchanged 2 weeks following adrenalectomy (Birnberg et al. 1983). However, a recent paper reported that POMC mRNA levels in the hypothalamus increase after adrenalectomy and that glucocorticoids reverse this response (Beaulieu et al. 1988). Such findings indicate that glucocorticoid regulation of POMC gene expression is similar in the anterior pituitary and hypothalamus.

There is also increasing evidence that hypothalamic POMC mRNA levels are controlled by gonadal steroids. Thus, estrogen treatment in ovarectomized rats decreases hypothalamic levels of POMC mRNA (Wilcox and Roberts 1985), and POMC mRNA levels are decreased in the rat arcuate nucleus following castration, an effect that is testosterone reversible (Chowen-Breed et al. 1989a,b). The effect of castration and testosterone treatment was confined to the most rostral area of the arcuate nucleus indicating that there is a heterogeneous population of POMC neurons in the nucleus and that testosterone regulates POMC gene expression in a selected group of cells. These observations indicate that testosterone may regulate gonadotropin-releasing hormone (*GnRH*) secretion by increasing the synthesis of POMC in the arcuate nucleus. Other studies, however, showed that castration of male rats results in an increase in POMC mRNA levels in the medial basal hypothalamus (MBH) and that testosterone replacement reverses this effect (Blum et al. 1989). Similarly, ovariectomy induced an increase in POMC mRNA levels in the rostral region of the arcuate nucleus – an effect which was reversed by estradiol (Tong et al. 1990). These findings indicate that androgens and estrogens have an inhibitory action on POMC gene expression in the MBH. Evidence for an involvement of POMC in the control of GnRH came from studies showing that a marked increase in the levels of hypothalamic POMC mRNA occurs contemperaneously with the onset of puberty and that this increase was confined to the rostral portion of the arcuate nucleus (Wiemann et al. 1989).

Chronic morphine treatment has been shown to decrease POMC mRNA levels in the rat hypothalamus (Mocchetti et al. 1989; Bronstein et al. 1990). In our hands, however, POMC mRNA levels in the hypothalamus were unchanged following the chronic administration of morphine according to various treatment schedules (Höllt et al. 1989a)

2.2.4 Peripheral Tissues

Northern blot analysis revealed that the adrenal medulla, testis, spleen, kidney, thyroid gland, duodenum, and lung contain a POMC mRNA species which is 200-300 bases smaller than the species found in the pituitary and hypothalamus (Jingami et al. 1984; DeBold et al. 1988b). Primer extension and S1 nuclease mapping studies showed that this small RNA lacked exon 1 and exon 2 of the gene and that it corresponded to a set of molecules 41-162 nucleotides downstream from the 5' end of exon 3. The ratio of POMC-like mRNA to POMC-derived peptide concentrations was 1000 times greater in nonpituitary tissues than in pituitary, indicating that the POMC mRNA is much less efficiently translated and/or that the POMC-derived peptides are more rapidly released from nonpituitary, than pituitary tissues.

In the rat testis, POMC-like mRNA was predominantly found in Leydig cells (Pintar et al. 1984; Chen et al. 1984; Gizang-Ginsberg and Wolgemuth 1985). In addition, POMC mRNA was shown to be present in testicular germ cells (Kilpatrick et al. 1987; Gizang-Ginsberg and Wolgemuth 1987). Recent studies, however, revealed that administration of ethane dimethane sulfonate (EDS), a drug which selectively destroys *Leydig cells*, did not alter POMC mRNA levels in the testis of rats (Li et al. 1989). These data suggest that the predominant site of rat POMC gene expression is in testicular interstitial cells rather than Leydig cells. Moreover, some authors were unable to detect POMC mRNA in Leydig cells (Garrett and Douglass 1989).

A POMC-like transcript was also found in the rat ovary and placenta (Chen et al. 1986).

Gonadotropins markedly increase the ovarian levels of POMC mRNA in immature rats and adult rats (Melner et al. 1986; Chen and Madigan 1987; Chen et al. 1986). Similarly, androgens have a potent stimulatory effect on ovarian POMC mRNA levels (Melner et al. 1986). In addition, the levels of ovarian POMC mRNA are higher in pregnant than in nonpregnant rats (Chen and Madigan 1987; Jin et al. 1988).

Abnormal POMC transcripts have frequently been found in nonpituitary tumors. Thus, a larger, 1450-kb-POMC mRNA species has been found in a thymic carcinoid tumor (de Keyzer et al. 1985), in human pheochromocytomas (DeBold et al. 1988b), and in other nonpituitary tumors (de Keyzer et al. 1989b). S1 mapping studies revealed that these longer POMC mRNA species result from a variable mode of transcription induced by promoters located at upstream start sites of transcription located between −400 and −100 (DeBold et al. 1988a; de Keyzer et al. 1989a). In some tumors the size heterogeneity observed was due to longer poly(A) tails (Clark et al. 1989).

2.3 Regulatory Elements

The specificity of POMC promoter utilization was assessed in gene transfer studies in which a rat POMC – pRSV neomycin fusion gene was expressed in pituitary (AtT-20) or fibroblastic cells (L) (Jeannotte et al. 1987). The rat POMC promoter was only efficiently utilized and correctly transcribed in AtT-20 cells. Sequences conferring tissue specificity have been localized 480-34 bp upstream of the capping site. The tissue specificity of the rat POMC promoter was shown in transgenic mice in which a chimeric rat POMC neomycin gene was introduced into the germ line. In these mice, high levels of the fusion gene transcripts were detected in the intermediate and the anterior pituitary. They were not detected in any other tissue except for very low levels in the testes (Tremblay et al. 1988). On the other hand, in vitro experiments demonstrated that the human POMC gene could be expressed in the nonpituitary rat glial cell line C_6, although the transcripts and translation products in C_6 cells differ from those in the human pituitary (Usui et al. 1990).

Transfection of a human POMC – thymidine kinase fusion gene into mouse fibroblasts revealed that the sequences responsible for the negative regulation by glucocorticoids are located within a DNA segment that extends 670 bp upstream from the cap site for POMC mRNA (Israel and Cohen 1985). Similarly, gene transfer studies with the rat POMC gene indicated that DNA sequences responsible for the direct negative regulation of transcription reside within 706 nucleotides upstream of the transcription start site (Charron and Drouin 1986). Moreover, expression of a rat POMC-neomycin fusion gene in transgenic mice revealed that no more than 769 bp of the rat POMC gene promoter sequences are required for specific glucocorticoid inhibition of the POMC gene in the anterior pituitary (Tremblay et al. 1988). A negative glucocorticoid recognition element (nGRE) that binds to purified glucocorticoid receptors in vitro has been identified at position −63 of the rat POMC gene (Drouin et al. 1989). This nGRE differs significantly from the known glucocorticoid responsive element (GRE) consensus sequence. The nGRE was also shown to contain a binding site for a nuclear protein of the chicken ovalbumin upstream promoter (COUP) family of transcription factors. Since the binding sites for COUP and the glucocorticoid receptor overlap, glucocorticoid-dependent repression of POMC transcription was been suggested to result from mutually exclusive binding of these nuclear transcription factors (Drouin et al. 1989).

A fragment of the rat POMC gene containing DNA sequences 794-38 nucleotides upstream of the capping site confers both elevated basal activity and CRH-inducibility on the tk-CAT fusion gene when transiently expressed in AtT-20 cells. DNA sequences extending from 478 to 320 bp upstream of the capping site were identified which confer elevated basal activity and a moderate inducibility by CRH. Another fragment extending from 320 to 133 nucleotides upstream of the start of transcription was shown to be required for the strong CRH inducibility. This fragment did not confer elevated basal activity (Roberts et al. 1987). Since this DNA fragment does not have significant homology with the cAMP responsive elements of other cAMP-regulated genes, it appears that there are other elements which are responsible for mediating the cAMP-responsiveness of the rat POMC gene. The precise structures of these elements and the factors which interact with them have still to be elucidated.

3 Proenkephalin

Bovine, human, and rat PENK (also termed "PENK A"; Noda et al. 1982a) contain four copies of met-enkephalin (met-ENK) and one copy each of leu-enkephalin (leu-ENK), the heptapeptide met-ENK-Arg^6-Phe^7, and the octopeptide met-ENK-Arg^6-Gly^7-Leu^8 (Fig. 2; Noda et al. 1982a; Comb et al. 1982; Gubler et al. 1982; Legon et al. 1982). In the mouse, a different octapeptide (met-ENK-Arg^6-Ser^7-Leu^8) has been found (Zurawski et al. 1986). The PENK sequence in the frog contains an additional met-ENK and no leu-ENK (Martens and Herbert 1984). All these peptides are bounded by pairs of basic amino acids which serve as signals for proteolytic processing. PENK can also be processed into larger ENK-containing peptides, such as peptide F, peptide E, bovine adrenal medulla (BAM)-peptides, metorphamide (or adrenorphin), and amidorphin (for a review see Höllt 1986).

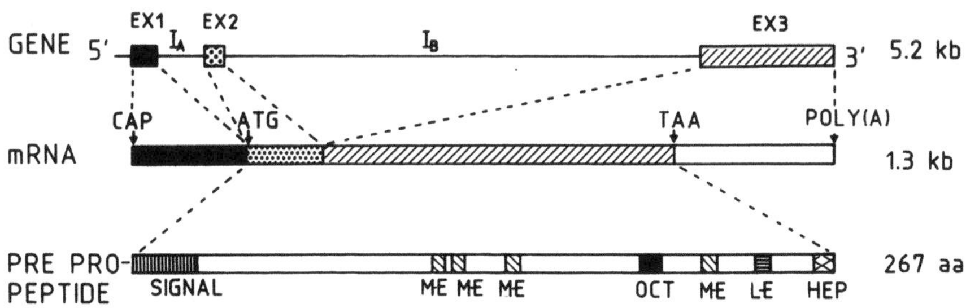

Fig. 2. Structural organization of the rat PENK gene, mRNA, and propeptide. *CAP*, transcription initiation site; *ATG*, start of translation, *TAA*, end of translation; *POLY(A)*, polydenylation site. *EX*, exon; *I*, intron; *ME*, met-enkephalin; *LE*, leu-enkephalin; *OCT*, octapeptide (met-enkephalin-Arg^6-Gly^7-Leu^8); *HEP*, heptapeptide (met-enkephalin-Arg^6-Phe^7)

3.1 Structure

The human PENK gene contains 4 exons separated by two large and one short intron (Noda et al. 1982b). The gene is about 5.2 kb long and consists of exon 1 (70 bp), intron A (87 bp), exon 2 (56 bp), intron B (469 bp), exon 3 (141 bp), intron C (about 3400 bp), and exon 4 (980 bp). In contrast, the rat PENK gene contains 3 exons only (Fig. 2; Rosen et al. 1984). The portion of PENK mRNA that is contained between exon 1 and 2 in the human PENK gene is located in a single exon (exon 1) in the rat gene. All of the species investigated appear to have only a single PENK gene. In the human, the PENK gene has been localized to chromosome 8 (Litt and Buder 1989). The size of the PENK mRNA in the brain and adrenal medulla is about 1450 nucleotides (Jingami et al. 1984; Pittius et al. 1985). A second, slightly larger, mRNA species which results from alternative splicing and contains the 87 nucleotides of intron A has been detected as a minor component in bovine hypothalamus using S1 mapping (Noda et al. 1982b). In addition to these mRNA species, larger forms of PENK mRNA have been found in some peripheral tissues. Thus, rat spermatogenic cells express a PENK mRNA species with a molecular size of 1700-1900 nucleotides (Kilpatrick and Milette 1986; Kew and Kilpatrick 1989; Kilpatrick et al. 1985, 1990). The larger PENK mRNA in the haploid sperm cells has recently been cloned and found to contain an additional 5'-flanking sequence which is derived from an alternative splicing of the PENK gene within intron A (Yoshikawa et al. 1989b; Garrett et al. 1989).

3.2 Regulation

3.2.1 Striatum

Penk gene expression in striatal neurons is under negative control of DA. Thus, chronic treatment of rats with haloperidol increases the levels of PENK mRNA in the striatum (Tang et al. 1983; Blanc et al. 1985; Sivam et al. 1986a, b; Romano et al. 1987; Morris et al. 1988b, 1989; Angulo et al. 1987, 1990b). Conversely, 6-hydroxy-DA lesion of the substantia nigra and/or the ventral tegmental area of rats caused a marked ipsilateral increase in the levels of PENK mRNA (Tang et al. 1983; Normand et al. 1988; Vernier et al. 1988; Morris et al. 1989).

There appears to be a different regulation of the PENK mRNA via D_1 and D_2 receptors. However, there is some discrepancy in the reports: Whereas Mocchetti et al. (Mocchetti et al. 1987) reported that chronic treatment of rats with the selective D_1 antagonist SCH 23390 increases PENK mRNA levels in the striatum, a clear decrease in the striatal PENK mRNA was found by Morris et al. (Morris et al. 1988b). Moreover, a clear enhancement of striatal POMC mRNA levels were seen in rats chronically treated with the D_2 antagonist sulpiride (Angulo et al. 1990b). In monkeys, bilateral or unilateral destruction of the DAergic neurons in the substantia nigra by systemic or intracarotid 1-methyl-4-phenyl-1,2,3,6-tetrahydropyridine (MPTP) injection induces parkinsonism associated with increased PENK mRNA levels in the denervated striatal neurons and with an increase in PENK producing striatal neurons (Augood et al. 1989). However, levels of PENK mRNA are decreased in the basal ganglia of patients parkinsonism with – a finding which suggests that the lesion-induced increase in PENK mRNA levels seen in animals is not a good model for the disease process in patients with parkinsonism (Mengod et al. 1990).

DA also causes a slight decrease in PENK mRNA levels in primary cultures of rat striatal cells (Kowalski et al. 1989), indicating that DA directly inhibits PENK gene expression. In these cultured striatal cells, cAMP analogues and the phorbol ester TPA increase PENK mRNA levels, suggesting that cAMP- and protein kinase C-dependent pathways are involved in the induction of PENK gene expression.

An increase in PENK mRNA levels in the striatum was also observed after chronic treatment of rats with reserpine, indicating that reserpine increases ENK synthesis by eliminating the DAergic inhibition (Mocchetti et al. 1985).

In contrast to the DAergic system, application of serotonergic drugs and/or lesioning of the raphe nuclei with 5,7-dihydroxytryptamine failed to alter striatal PENK mRNA content (Mocchetti et al. 1984; Morris et al. 1988d).

Specific alterations in striatal levels of PENK mRNA have been found after increasing endogenous GABA levels. Chronic treatment of rats with the GABA transaminase inhibitor amino-oxyacetic acid (AOAA) causes an increase in PENK mRNA levels in the striatum (Sivam et al. 1986b). On the other hand, our group observed a decrease in the level of PENK mRNA in rats chronically treated with the GABA transaminase inhibitors AOAA and ethanolamine ortho-sulfate (EOS) (Reimer and Höllt 1991). In mice, the combined application of the GABA A agonist muscimol and diazepam for 3 days causes a decrease in the striatal PENK mRNA levels (Llorens-Cortes et al. 1990).

Discrepant results have been reported with regard to the effect of opiates on striatal PENK mRNA levels. Chronic treatment with morphine was reported to cause a slight decrease in the levels of PENK mRNA in the striatum (Uhl et al. 1988), whereas we found no effect of chronic morphine treatment on striatal PENK mRNA levels (Höllt et al. 1989b). Inactivation of μ-opiate receptors by local injection of the irreversible μ-antagonist β-funaltrexamine into the striatum did not change the levels of the PENK mRNA in this structure. In contrast, local injection of the nonselective irreversible antagonist β-chlornaltrexamine causes a marked increase in PENK mRNA levels at the site of injection (Morris et al. 1988c). This indicates that the activation of non-μ, possibly δ and/or κ-opioid receptors, tonically supress striatal PENK gene expression. In addition, chronic treatment of rats with the σ-receptor antagonist BMY 14 802 increased striatal PENK mRNA levels in rats indicating that σ-receptor activity affects striatal PENK expression (Angulo et al. 1990a).

3.2.2 Hypothalamus

Cells expressing the PENK gene are concentrated in the ventromedial hypothalamus. Estrogen treatment of ovariectomized rats markedly increases PENK mRNA levels in the ventrolateral aspect of the ventromedial nucleus. This is due to both an increase in the number of PENK-expressing neurons and an increase in the amount of PENK mRNA per neuron (Romano et al. 1988). Progesterone treatment attenuated the decline of the PENK mRNA after estrogen removal. This finding suggests that the ENKergic system in the hypothalamus may be regulated by both estrogen and progesterone during the estrous cycle.

Chronic morphine treatment has been reported to decrease PENK mRNA levels in the hypothalamus of rats (Uhl et al. 1988). Other groups failed to see any effect of morphine and/or NAL treatment (Lightman and Young 1988; Höllt et al. 1989a). However, NAL-precipitated withdrawal in morphine-tolerant rats resulted in a dramatic increase in PENK mRNA levels in cells of the paraventricular nucleus (Lightman and Young 1987). Similarly, daily application of electroconvulsive shocks increases PENK mRNA in the hypothalamus of rats (Yoshikawa et al. 1985). It is tempting to speculate that an enhancement of PENK biosynthesis in areas belonging to the limbic system might contribute to the antidepressive action of electroconvulsive shock treatment in humans.

3.2.3 Hippocampus and Cortex

Within the hippocampus, cells producing PENK-derived peptides have been localized to the granule cell layer of the dentate gyrus and to various interneurons within the hippocampal formation (Gall et al. 1981). Repetitive stimulation of the amygdala (kindling) of rats, which resulted in a progressive development of generalized seizures, causes a marked increase in hippocampal levels of PENK mRNA (Naranjo et al. 1986a). These data suggest that the PENK system might participate in the development of kindling phenomena. A similar increase in hippocampal PENK mRNA levels was found in rats with recurrent limbic seizures induced by contralateral lesions of the dentate gyrus hilus (Gall et al. 1981). Moreover, direct electrical stimulation of the dentate gyrus granule cells markedly increased the levels of PENK mRNA at the site of stimulation (Morris et al. 1988a). In addition, repetitive electrical stimulation of the perforant pathway resulted in an increase in PENK mRNA levels in the ipsilateral dentate gyrus, indicating that PENK gene expression in granule cells can be induced transsynaptically (Moneta and Höllt 1990; Xie et al. 1990).

Electrical stimulation of the amygdala also increases PENK mRNA levels in the entorhinal cortex, frontal cortex, nucleus accumbens, and the amygdala itself (Naranjo et al. 1986a). Similarly, repeated electroconvulsive shock increases levels of PENK mRNA in the hippocampus and the entorhinal cortex of rats (Xie et al. 1989). The degree of increase was highest in the entorhinal cortex which contains ENKergic neurons projecting to the dentate gyrus granule cells as part of the tractus perforans. It, thus, appears that several ENKergic neurons within the entorhinal-hippocampal formation are sensitive to seizure activity.

3.2.4 Spinal Cord and Lower Brain Stem

Inflammatory stimuli have been found to be associated with an increase in the level of PENK mRNA in the spinal cord of rats (Iadarola et al. 1986, 1988b), although the effects were much smaller than those seen for PDYN mRNA. Thus, in arthritic rats suffering from chronic inflammation, no major change in the level of PENK mRNA has been found (Weihe et al. 1989). In addition, no major increase was found in the levels of PENK mRNA in the spinal cord after spinal injury or after dissection (Przewlocki et al. 1988). However, substantial but localized alterations in the PENK gene expression in the spinal cord and lower brain stem were found after manipulation of primary afferent input (Harlan et al. 1987; Nishimori et al. 1988). Levels of PENK mRNA were decreased in nucleus caudalis and spinal cord dorsal horn neurons following lesions of the primary afferents (Nishimori et al. 1988). Moreover, electrical stimulation of the trigeminus nerve elicited a rapid and dramatic induction of PENK mRNA in lamina I and lamina II neurons (Nishimori et al. 1989). Primary afferent stimulation can also enhance the expression of c-fos in subsets of lamina I and II neurons; this precedes the expression of PENK and PDYN (Hunt et al. 1987; Draisci and Iadarola 1989). The 5'-flanking region of the PENK gene contains a recognition sequence for the c-Fos/c-Jun heterodimeric complex (Comb et al. 1986; Curran et al. 1988). This supports the hypothesis that the modulatory effects of primary afferent stimulation on PENK gene expression in the spinal cord and nucleus caudalis is mediated via c-Fos/c-Jun oncoproteins.

3.2.5 Adrenal Medulla

Incubation of primary cultures of chromaffin cells from bovine adrenal medulla with 8-Br-cAMP, forskolin, or cholera toxin resulted in an increase in PENK mRNA content (Quach et al. 1984; Eiden et al. 1984b; Kley et al. 1987b; Stachowiak et al. 1990). Membrane depolarization induced by nicotine, high K^+, or veratridine causes a marked increase in PENK mRNA levels in cultured bovine chromaffin cells (Eiden et al. 1984b; Kley et al. 1986; Naranjo et al. 1986b). The effect is inhibited by drugs which block voltage-dependent Ca^{2+} channels (e.g., verapamil) or by reduced Ca^{2+}ion concentration in the medium (Siegel et al. 1985; Kley et al. 1986; Naranjo et al. 1986b, Waschek and Eiden 1988; Waschek et al. 1987). These findings indicate that the influx of extracellular Ca^{2+} through voltage-dependent Ca^{2+} channels is a prerequisite for the induction of PENK gene expression.

Histamine, bradykinin, or angiotensin-induced activation of phospholipase C, which gives rise to the second messengers diacylglycerol and inositol-3-phosphate, caused an increase in the levels of PENK mRNA in chromaffin cells (Kley et al. 1987a; Bommer et al 1987; Wan et al. 1989; Farin et al. 1990a,b). Activation of protein kinase C appears to be an important trigger for the induction of PENK gene expression, since tumor-promoting phorbol esters increase PENK mRNA levels (Kley 1988). Although muscarine activates phospholipase C (Noble et al. 1986), it does not activate PENK gene expression. However, following coincubation of muscarine with the Ca^{2+}ionophore A 23187, at concentrations which do not increase PENK mRNA levels, a clear increase in the expression of the PENK was observed (Farin et al. 1990a). Nuclear run-off experiments indicated that the increase of PENK mRNA by histamine is mediated by an increased rate of transcription. Measurements of PENK mRNA half-life indicated that histamine does not alter the stability of the mRNA coding for PENK (Farin et al. 1990a, b).

Depletion of catecholamine stores by reserpine causes a decrease in the levels of PENK mRNA (Eiden et al. 1984a). Glucocorticoids exert a permissive action on the stimulation of PENK gene expression by depolarization agents (Naranjo et al. 1986b). Dexamethasone treatment of rats causes a concomitant increase in the levels of mRNA coding for PENK and for phenylethanolamine-*N*-methyl transferase (PNMT) in the adrenal medulla (Stachowiak et al. 1990). However, another study failed to find any glucocorticoid-induced changes in PENK mRNA levels in bovine chromaffin cells, although the steroids markedly induced PNMT mRNA (Wan and Livett 1989).

In rat adrenal medulla, regulation of PENK mRNA appears to be different from that in primary cultures of bovine chromaffin cells. Thus, denervation by sectioning of the splanchnic nerve, markedly increases the level of PENK mRNA in the adrenal gland of rats (Kilpatrick et al. 1984), indicating that splanchnic innervation exerts a tonic inhibitory influence on PENK gene expression. Such an innervation-dependent suppression is supported by experiments in which rat adrenal medullae were explanted. In such an organ culture system, levels of PENK mRNA rise 74-fold after 4 days (LaGamma et al. 1985; Inturrisi et al. 1988). Depolarization with either elevated K^+ or veratridine inhibited this increase. Inhibition of Ca^{2+} influx by verapamil prevented the effect of the depolarizing agents (LaGamma et al. 1988).

These in vitro results contrast with these obtained in rats in which increased *splanchnic nerve* activity, generated by insulin, caused a dramatic increase in the levels of adrenal medullary mRNA (Kanamatsu et al. 1986; Fischer-Colbrie et al. 1988). The effect of insulin was blocked by combined treatment of rats with chlorisondamine and atropine, or by bilateral transection of the splanchnic nerves, indicating that insulin exerts its effect on PENK gene expression via splanchnic nerve activation. In addition, treatment of rats with morphine markedly increases the levels of PENK mRNA in the adrenal medulla (Höllt et al. 1989b). Moreover, morphine can also exert its effect after intracerebroventricular application, indicating that morphine activates PENK gene expression by increasing splanchnic nerve activity (Reimer and Höllt 1990). An explanation of the increase in PENK mRNA following denervation might be that the denervation causes an induction of PENK gene expression by stimulating the release of acetylcholine from splanchnic nerve endings, rather than by the unmasking of a tonic inhibition of an intact nerve.

3.2.6 Gonads

A PENK mRNA species of 1450 nucleotides was found in the uterus, oviduct, and ovary, and in testis, vas deferens, epididymidis, seminal vesicles, and prostate of rats and hamsters (Kilpatrick et al. 1985; Kilpatrick and Milette 1986; Kilpatrick and Rosenthal 1986). Such data indicate that PENK-derived peptides are synthesized at multiple sites within the male and female reproductive tracts and may locally regulate reproductive function. In rat and mouse spermatogenic cells, an additional mRNA species of 1700-1900 nucleotides was observed. This species contains a distinct 5'-untranslated mRNA sequence derived by alternative splicing within intron A of the rat gene (Yoshikawa et al. 1989b; Garrett et al. 1989; Kilpatrick et al. 1990). PENK mRNA was present in mouse spermatocytes and in spermatids, but not in extracts of mature sperm, suggesting that developing germ cells may be a major site of PENK synthesis in the tests and that PENK-derived peptides may function as germ cell-associated autocrine and/or paracrine hormones (Kilpatrick and Millette 1986; Kilpatrick and Rosenthal 1986). In rats, PENK mRNA levels were markedly changed during the estrous cycle in both the ovary and uterus. The highest concentrations occurred at estrus in the rat ovary, and at metestrus and diestrus in the rat uterus (Jin et al. 1988). In the uterus of rats and rhesus monkeys, PENK mRNA is expressed primarily in the endometrium of the uterus (Rosen et al. 1990; Low et al. 1989). Penk gene expression in the endometrium is induced by estradiol 17β, an action antagonized by progesterone (Low et al. 1989).

The regulation of PENK gene expression has been recently studied in cultured rat testicular peritubular cells. PENK mRNA was found to be increased by forskolin and by the phorbol ester TPA. Both drugs synergistically increase PENK mRNA in these cells (Yoshikawa et al. 1989a). In addition, cAMP analogues and human chorionic gonadotropin cause an increase in PENK mRNA in cultured rat Leydig cells (Garrett and Douglass 1989).

3.2.7 Cells

Activation of mouse T-helper cells in vitro by concavalin A results in a dramatic induction of PENK mRNA (Zurawski et al. 1986). In activated mouse T-helper cells as much as 0,4% of the total mRNA coded for PENK, allowing for the cloning of mouse PENK mRNA. Interestingly, the predicated amino acid sequence of mouse PENK contains a different sequence for the octapeptide. The PENK gene is expressed in normal B cells and can be stimulated with lipopolysaccharide (LPS) or *Salmonella typhimurium* (Rosen et al. 1989).

Low levels of PENK mRNA have been found in PC12 cells, a cell line derived from a rat pheochromocytoma (Byrd et al. 1987). PENK mRNA levels in PC12 cells can be increased by treatment with sodium butyrate. In C_6 glioma cells activation of β-adrenergic receptors by norepinephrine increases cAMP levels and stimulates PENK mRNA. Glucocorticoids have no effect, but potentiate the inducing effect of norepinephrine on PENK gene expression (Yoshikawa and Sabol 1986). In addition, PENK mRNA is also present in mouse neuroblastoma-rat glioma hybrid cells (NG108CC15). Treatment of the cells with opiates (etorphine, D-Ala^2-D-Met^5-ENK) increases PENK mRNA levels by an effect not involving adenylate cyclase (Schwartz 1988). PENK mRNA has also been found in several human neuroblastoma cell lines (SK-N-MC; SH-SY5Y; Folkesson et al. 1988). In SK-N-MC cells PENK mRNA levels can be induced by cAMP analogues and by β-adrenergic agonists (Monstein et al. 1990). Moreover, the PENK gene is down regulated during differentiation of these cells (Verbeeck et al. 1990). In neonatal rats the PENK gene is expressed in several tissues (kidney, liver, skin, skeletal muscle) that have undetectable levels in adults (Kew and Kilpatrick 1990). These findings indicate a regulatory role of PENK-derived peptides in cell differentiation and immune function.

3.3 Regulatory Elements

The sequence requirements for transcription of the human PENK gene were analyzed after transfection of COS monkey cells with a fusion vector of 946 bp of the 5'-flanking region and 63 bp of the 5'-noncoding region with an SV40 vector (Terao et al. 1983). Deletion up to 172 bp upstream of the capping site exerted essentially no effect on the expression of the fusion gene. in contrast, deletions up to 145, 11, 81, and 67 bp upstream of the capping site resulted in a gradual decrease in the transcriptional efficiency, indicating that these sequences contain functional enhancer and/or promoter sites.

These sequences also confer cAMP and phorbol ester inducibility. Transfection of a PENK-CAT fusion gene (pENKAT-12) into monkey CV1 cells revealed that DNA sequences required for regulation by both cAMP and phorbol ester map to the same 37-bp region located 107-71 bp 5' to the mRNA cap site of the PENK gene and exert properties of transcriptional enhancers (Comb et al. 1986; Fig. 3). Further studies provided evidence that 2 DNA elements (ENKCRE-1 and ENKCRE-2) are located within this enhancer region; these are responsible for the transcriptional response to cAMP and phorbol ester (Fig. 3). The proximal promoter element, ENKCRE-2, is essential for both basal and regulated enhancer function. The distal promoter element, ENKCRE-1, has no inherent capacity to activate transcription, but synergistically augments cAMP- and phorbol ester-inducible transcription in the presence of ENKCRE-2 (Comb et al. 1988; Hyman et al. 1988). Several transcription factors bind to the enhancer regions in vitro: a transcription factor, termed ENKTF-1, binds to the DNA region encompassing ENKCRE-1; *AP-1* and *AP-4* bind to overlapping sites spanning ENKCRE-2; and Fos and Jun polypeptides have shown to bind directly to ENKCRE-2 (AP-1 site) in a cooperative manner (Sonnenberg et al. 1989). Moreover, a synergistic transactivation of the human PENK enhancer-promoter was found by cotransfection of plasmids expressing c-*fos* and c-*jun* in F9 teratocarcinoma cells (Sonnenberg et al. 1989). A fourth transcription factor AP-2 binds to a site immediately downstream of ENKCRE-2 (Comb et al. 1988). This AP-2 DNA binding element acts synergistically with the enhancer elements to confer maximal response to cAMP and phorbol esters (Hyman et al. 1989). Further upstream from the ENKCRE elements, a novel regulatory element was recently identified that appears to bind the known transcription factor NF-KB and another brain-specific transcription activator, distinct from NF-KB, called BETA (Korner et al. 1989; Rattner et al. 1989). At a position 90 bp 5' from the TATA box the sequence GGGGACGTCCCC is found in the PENK promoter. This 12-mer

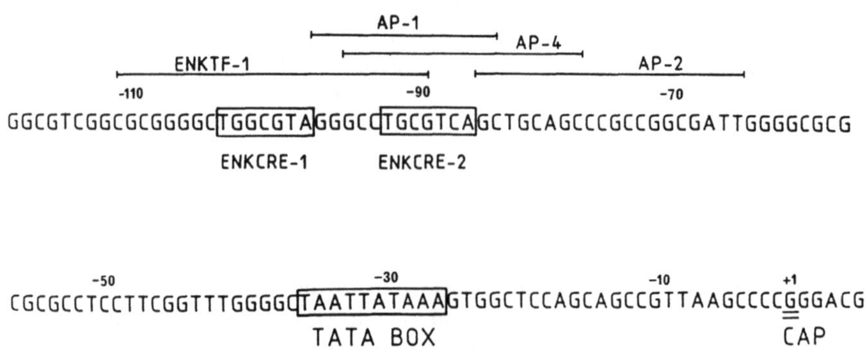

Fig. 3. Promoter and enhancer elements within the 5'-flanking region of the human PENK gene. *CAP*, transcription initiation site. (Modified according to Comb et al. 1986)

matches the classical NF-KB sequence in 9 of 11 nucleotides. A similar sequence (GGGGAGCCTCCG), 40 bp further upstream from NF-KB, appears to play a part in lymphocyte-specific expression of the gene. In fact, T helper cells have been shown to increase PENK mRNA synthesis (Zurawski et al. 1986). The role of the brain-specific transcription factor BETA remains to be established. Since multiple factors can bind to the 5' control region is likely that the PENK gene is controlled by a combinatorial interaction of several transcription factors which may be constitutive or inducible.

4 Prodynorphin

PDYN (PENK B) is a precursor of leu-ENK and of several opioid peptides which contain leu-ENK at the N-terminus, such as dynorphin A (DYN), leumorphin, dynorphin B (= rimorphin) and neoendorphins (Fig. 4 and Höllt 1986).

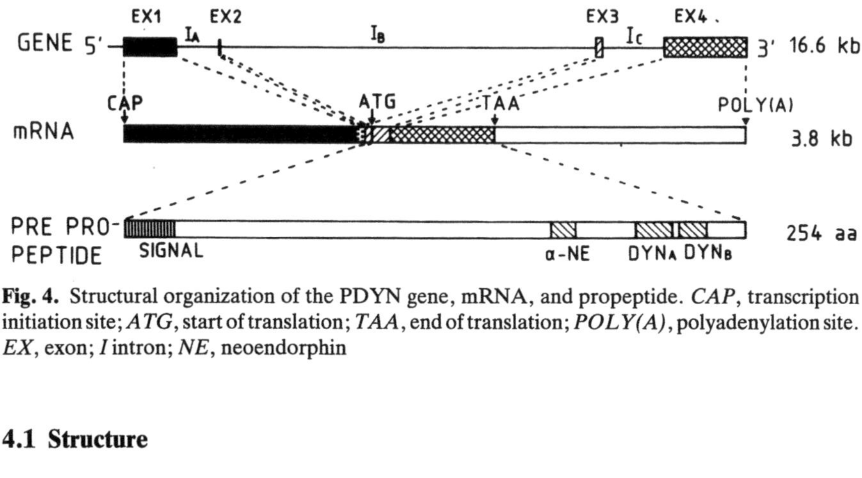

Fig. 4. Structural organization of the PDYN gene, mRNA, and propeptide. *CAP*, transcription initiation site; *ATG*, start of translation; *TAA*, end of translation; *POLY(A)*, polyadenylation site. *EX*, exon; *I* intron; *NE*, neoendorphin

4.1 Structure

The size of the mRNA in pig hypothalamus is about 3000 nucleotides (Jingami et al. 1984; Pittius et al. 1987). The PDYN mRNA possesses a very long untranslated 3'-terminal end of about 1500 nucleotides. In rats, a slightly smaller PDYN mRNA species of 2400-2600 nucleotides was observed in the striatum (Civelli et al. 1985). PDYN mRNA has been shown to have a widespread distribution in the CNS and periphery. mRNA species which are about 100- nucleotides smaller have been found in the adrenal cortex (Day et al. 1990). These differences are due to alternative splicing resulting in the deletion of exon 2 of the PDYN gene (Day et al. 1990).

Complete structural organization has been identified for the human PDYN gene (Horikawa et al. 1983). The gene contains 4 exons separated by three introns (see Fig. 1): exon 1 (1.4 kb), intron A (1.2 kb), exon 2 (60 bp), intron B (9,9 kb), exon 3 (145 bp), intron C (1.7 kb), and exon 4 (2.2 kb). As compared to the other opioid peptide genes, exons 1 and 4, which contain untranslated sequences, are very large. It is possible that the 5'-terminal region of this gene contains an additional intron (Horikawa et al. 1983). A similar structural organization has recently been reported for the rat gene (Douglass et al. 1989). Exon 1 and exon 2 encode the majority of the 5'-untranslated sequences of the mRNA while exons 3 and 4 contain the translated regions. Only a single PDYN gene appears to exist in man, pigs, and rats.

4.2 Regulation

4.2.1 Hypothalamus

In the rat hypothalamus, PDYN mRNA has been localized in the magnocellular divisions of supraoptic and paraventricular nuclei (Sherman et al. 1986; Morris et al. 1986). With chronic osmotic challenge, PDYN mRNA levels in the *supraoptic* and paraventricular nuclei are increased, in parallel with those for provasopressin (Sherman et al. 1986). These findings indicate a coordinate regulation of mRNA expression for coexisting peptide systems. A coordinate increase in the levels of PDYN and of provasopressin mRNA in the hypothalamus has also been found in rats treated with nicotine (Höllt and Horn 1989, 1991). The regulation of these mRNAs can also differ, however, since repeated electroconvulsive shock concomitantly activates PDYN and provasopressin gene expression in the supraoptic, but not in the paraventricular, hypothalamic nuclei (Schafer et al. 1989).

4.2.2 Striatum

The striatum contains many GABAergic neurons projecting to the substantia nigra and the external segment of the pallidum which express PDYN (Graybiel 1986). PDYN mRNA in the striatum, the nucleus accumbens, and the olfactory tubercle has been demonstrated by in situ hybridization (Young et al. 1986; Morris et al. 1986; Morris et al. 1989).

Destruction of the substantia nigra with 6-hydroxy-DA resulted in a decrease (Li et al. 1990) or had no effect on PDYN mRNA levels in the striatum (Young et al. 1986, 1989), indicating that the mesostriatal DA system may exert an augmenting influence on PDYN synthesis, if any, in this brain region. Chronic treatment with the DA antagonist haloperidol did not change the PDYN levels in the striatum (Morris et al. 1989). On the other hand, chronic treatment with the DA agonist apomorphine has been shown to increase the levels of PDYN mRNA in the striatum (Li et al. 1988, 1990). Moreover, a tendency of decreased PDYN levels in the striatum and nucleus accumbens was observed in rats after chronic administration of the D_1-antagonist SCH 23 390 (Morris et al. 1988b). Our recent results showed that after lesion of the medial forebrain bundle the PDYN mRNA levels in the nucleus accumbens – an area which contains a high proportion of D_1-receptors – was decreased (Reimer et al. 1992). These findings suggest that there appears to exist a tonic enhancement of PDYN synthesis in the basal ganglia via D_1-receptors.

The raphe-striatal serotonergic pathway tonically enhances PDYN gene expression in striatal target cells, since destruction of the dorsal raphe nucleus by microinjection of 5,7-dihydroxytryptamine caused significant reductions in PDYN mRNA levels in the medial nucleus accumbens and the caudomedial striatum regions which contain a particularly dense serotonergic innervation (Morris et al. 1988d).

Enhancemant of GABAergic transmission by chronic administration of GABA transaminase inhibitors (AOAA; EOS; γ-vinyl-GABA) causes a decrease in striatal levels of PDYN (Reimer and Höllt 1991). This might reflect a type of autoinhibition, since the dynorphinergic neurons in the striatum contain GABA as neurotransmitter.

In addition, irreversible inactivation of striatal opioid receptors by local application of β-chlornaltrexamine caused an increase in PDYN mRNA levels at the site of injection (Morris et al. 1988c). This indicates that endogenous opioids tonically inhibit striatal PDYN biosynthesis. This effect, however, appears to be mediated via δ- and κ-receptors rather than μ-opioid receptors, since the local injection of the irreversible antagonist β-funaltrexamine, as well as the administration of the μ-agonist morphine or the antagonist NAL have no effect on the striatal levels of PDYN (Morris et al. 1988c; V. Höllt, unpublished).

PDYN biosynthesis in the striatum was also slightly increased after electrical kindling of the deep prepyriform cortex, indicating that the striatum participates in the neuronal circuits activated by cortical kindling (Lee et al. 1989).

4.2.3 Hippocampus

Within the hippocampus, DYN peptides are synthesized in the granule cells of the dentate gyrus and transported within the mossy fiber pathways which innervate the pyramidal cells (McGinty et al. 1983; Morris et al. 1986). Following brief trains of high frequency electrical stimulation to the dentate gyrus of rats, the levels of PDYN mRNA were markedly decreased on the stimulated, but not the unstimulated, side (Morris et al. 1988a). Similarly, chronic electrical stimulation of the dentate gyrus, which resulted in stage 4 kindling seizures in rats, caused a marked decrease in PDYN mRNA levels in the granule cells of the dentate gyrus. This decrease was seen in the stimulated as well as in the unstimulated hemispheres (Morris et al. 1987). The contralateral decrease in PDYN mRNA levels might be due to the activation of commissural projections. A bilateral decrease in PDYN levels can also be induced by unilateral electrical stimulation of the perforant pathway which results in kindling seizures in rats (Moneta and Höllt 1990; Xie et al. 1990), indicating that PDYN gene expression in the hippocampus can be altered transsynaptically. PDYN mRNA levels in the hippocampus are also decreased during the development of deep prepyriform cortex kindling (Lee et al. 1989) and after repeated electroconvulsive shocks (Xie et al. 1989). The altered PDYN mRNA levels in the hippocampus return to normal 1–6 weeks following cessation of stimulation (Lee et al. 1989; Moneta and Höllt 1990). However, a further single stimulus was still effective in producing kindling seizures. These findings indicate that opioid peptides may play a role in the development, but not maintenance, of kindling. There is electrophysiological evidence for an inhibitory action of PDYN peptides on hippocampal pyramidal cells (Henricksen et al. 1982). A decrease in PDYN biosynthesis might, therefore contribute to the hyperexcitability found in the kindling state. Thus, in contrast to the PENK gene, the PDYN gene is negatively regulated by neuronal activity.

In aged rats an increase in hippocampal levels of PDYN mRNA was observed (Lacaze-Masmonteil et al. 1987). An age-related loss of perforant path afferents which inhibits dynorphin biosynthesis in the granule cells might be responsible for this effect (Collier and Routtenberg 1984).

4.2.4 Spinal Cord

In the spinal cord, a prominent role for PDYN-derived peptides has been suggested by the observation that PDYN mRNA in the cord is markedly enhanced by acute or chronic inflammatory processes (Iadarola et al. 1986, 1988a, b; Höllt et al. 1987; Weihe et al. 1989). In polyarthritic rats a pronounced elevation in PDYN mRNA was found in the lumbosacral spinal cord (Höllt et al. 1987; Weihe et al. 1989). In rats with unilateral inflammation, a pronounced increase in PDYN mRNA was observed only in those spinal cord segments that received sensory inputs from the affected limb (Iadarola et al. 1988a, b). These changes were rapid in onset, being significantly elevated as early as 4 h after the injection of the inflammatory agent into the paw (Draisci and Iadarola 1989). PDYN mRNA levels peaked after about 3 days, returning to normal after about 2 weeks (Iadorola et al. 1988a). In situ hybridization revealed that the increase in the PDYN mRNA occurs in the superficial dorsal horn laminae I and II, and in the deep dorsal horn laminae V and VI (Ruda et al. 1988). There appears to be a population of neurons in laminae IV/V, but not in laminae I/II, which coexpress PDYN and PENK peptides (Weihe et al. 1988). These neurons appear to increase their biosynthetic activity in response to inflammation, a finding that might explain the small increase in PENK mRNA observed by some groups (Iadarola et al. 1988a). In addition to inflammation, traumatic injury and cord transection also increased PDYN mRNA levels in the spinal cord (Przewlocki et al. 1988). This marked increase in PDYN biosynthesis suggests a highly specific role for PDYN-derived peptides in the modulation of pain associated with inflammation and tissue injury. Recently, it was shown that increased PDYN gene expression provoked by noxious thermal stimulation or inflammation is preceded by an early induction of c-*fos* in the same neurons (Naranjo et al. 1991; Noguchi et al. 1991). In addition, an AP-1-like element within the promoter region of the rat POMC gene has been localized and shown to be a target of Fos/Jun oncoproteins (Naranjo et al. 1991). These findings suggest that Fos/Jun oncoproteins may function as third messengers in the signal transduction mechanisms in response to painful stimuli.

4.2.5 Peripheral Tissues

PDYN mRNA has been localized in the anterior pituitary of rats (Civelli et al. 1985; Schafer et al. 1990) and pigs (Pittius et al. 1987). It is still unclear which cell type produces PDYN, although the parallel release of luteinizing hormone together with PDYN peptides in response to GnRH suggests that both peptides are produced by gonadotropic cells. PDYN mRNA in the rat anterior pituitary is decreased by estrogens and increased by antiestrogens (Spampinato et al. 1990). In the rat, in situ hybridization experiments revealed that PDYN mRNA is also produced in melanotrophic cells of the intermediate pituitary (Schafer et al. 1990).

PDYN mRNA is present in porcine heart, gut, and lung – tissues which contain low, but significant, levels of PDYN-derived peptides (Pittius et al. 1987). However, the mRNA species found in the gut and lung are smaller than those found in the brain and heart ventricle and appear to derive from a different gene which possesses a high homology to the PDYN gene (Pittius et al. 1987).

In the rat adrenal gland, a unique PDYN mRNA species, which is about 350 bases shorter than the PDYN transcripts in the hypothalamus (2100 vs 2400 nucleotides), has been found (Civelli et al. 1985). The observed size difference of the PDYN mRNA in the adrenal may be due to alternative splicing, resulting in the deletion of exon 2 of the PDYN gene (Day et al. 1990). In situ hybridization studies localized PDYN mRNA in the zona fasciculata and zona reticularis of the adrenal cortex (Day et al. 1990). PDYN mRNA has been shown to be influenced by pituitary-dependent factors, since there is a dramatic loss of PDYN mRNA in the zona fasciculata following hypophysectomy in rats (Day et al. 1990).

In the gonads, PDYN mRNA has been found in the testes of rats, rabbits and guinea-pigs, and in the rat ovary and uterus (Civelli et al. 1985; Douglass et al. 1987). Within the rat testis, PDYN-peptides have been found in Leydig cells. Moreover, PDYN mRNA is found in the R2C Leydig tumor cell line (McMurray et al. 1989). and in cultured Sertoli cells. The testicular PDYN mRNA species is about 100 bases smaller than that found in the hypothalamus. This is due to alternate splicing leading to the loss of exon 2 of the PDYN gene (Garrett et al. 1989). In R2C rat Leydig cells the PDYN gene is positively regulated by cAMP analogues, whereas phorbol esters exert a slight negative regulation on PDYN mRNA levels.

4.3 Regulatory Elements

Although the complete structure of the human PDYN gene has been known for several years (Horikawa et al. 1983), no regulatory elements of this gene have as yet been characterized. However, the triple tandem-repeated sequences at the 5'-flanking region were suggested to be involved in the regulation of PDYN gene expression (Horikawa et al. 1983).

The structure of the rat PDYN gene is similar to that of the human gene (Civelli et al. 1985; Douglass et al. 1989). Recently, transfection studies were performed in which sequences of the 5'-flanking region of the rat PDYN gene were joined with a CAT reporter gene and introduced into 2RC Leydig cells (McMurray et al. 1989). The fusion genes were positively regulated by cAMP analogues in these cells. The cAMP-responsive DNA sequence has been localized to a 210-bp fragment comprising 122 bp of 5'-flanking sequences, and 88 bp of exon 1 of the rat PDYN gene. This sequence contains a DNA fragment which is 80% homologous to the cAMP consensus sequence located downstream to the cap-site. On the other hand, expression of the PDYN fusion plasmid was not induced by TPA, indicating that the 210-bp PDYN sequence may not contain any phorbolester-inducible element. On the other hand, a functional AP-1 element has been localized within that rat PDYN gene promoter region (Naranjo et al. 1991). This DNA element is constituted by the noncanonical TGACAAACA sequence which was shown to be a target of Fos/Jun transactivation (Naranjo et al. 1991).

5 Conclusions

In this review, an attempt was made to summarize features which specifically relate to the structure and regulation of each of the three opioid peptide genes. Here, some common characteristics of the structure and regulation between these genes will be emphasized.

The three genes (see also Figs. 1, 2, 4) are strikingly similar in their general organization. They all contain a main exon which codes for the vast majority of the protein sequences. The signal peptide and the translational initiation site is coded by another exon. This exon has an almost identical site in all three genes. These similarities in structural organization suggests that the three opioid peptide genes may have evolved from a common ancestor.

The regulation of all three genes is positively regulated by cAMP, although only the PENK gene possesses an identified cAMP-responsive element within the 5'-flanking region. Moreover, all three genes are modulated in response to stress in intact animals, albeit in different tissues. This indicates that all three opioid peptide genes may play an important role under stressful conditions.

References

Affolter HU, Reisine T (1985) Corticotropin releasing factor increases proopiomelanocortin messenger RNA in mouse anterior pituitary tumor cells. J Biol Chem 260:15477–15481

Angulo JA, Christoph GR, Manning RW, Burkhart BA, Davis LG (1987) Reduction of dopamine receptor activity differentially alters striatal neuropeptide mRNA levels. Adv Exp Med Biol 221:385–391

Angulo JA, Cadet JL, McEwen BS (1990a) Sigma receptor blockade by BMY 14802 affects enkephalinergic and tachykinin cells differentially in the striatum of the rat. Eur. J Pharmacol 175:225–228

Angulo JA, Cadet JL, Woolley CS, Suber F, McEwen BS (1990b) Effect of chronic typical and atypical neuroleptic treatment on proenkephalin mRNA levels in the striatum and nucleus accumbens of the rat. J Neurochem 54:1889–1894

Augood SJ, Emson PC, Mitchell IJ, Boyce S, Clarke CE, Crossman AR (1989) Cellular localisation of enkephalin gene expression in MPTP-treated cynomolgus monkeys. Brain Res Mol Brain Res 6:85–92

Autelitano DJ Clements JA, Nikolaidis I, Canny BJ, Funder JW (1987) Concomitant dopaminergic and glucocorticoid control of pituitary proopiomelanocortin messenger ribonucleic acid and beta-endorphin levels. Endocrinology 121:1689–1696
Autelitano DJ, Lundblad JR, Blum M, Roberts JL (1989) Hormonal regulation of POMC gene expression. Annu Rev Physiol 51:715–726
Autelitano DJ, Blum M, Lopingco M, Allen RG, Roberts JL (1990) Corticotropin-releasing factor differentially regulates anterior and intermediate pituitary lobe proopiomelanocortin gene transcription, nuclear precursor RNA and mature mRNA in vivo. Neuroendocrinology 51: 123–130
Beaulieu S, Gagne B, Barden N (1988) Glucocorticoid regulation of proopiomelanocortin messenger ribonucleic acid content of rat hypothalamus. Mol Endocrinol 2:727–731
Birnberg NC, Lissitzky JC, Hinman M, Herbert E (1983) Glucocorticoids regulate proopiomelanocortin gene expression in vivo at the levels of transcription and secretion. Proc Natl Acad Sci USA 80:6982–6986
Blanc D, Cupo A, Castanas E, Bourhim N, Giraud P, Bannon MJ, Eiden LE (1985) Influence of acute, subchronic and chronic treatment with neuroleptic (haloperidol) on enkephalins and their precursors in the striatum of rat brain. Neuropeptides 5:576–570
Blum M, Roberts JL, Wardlaw SL (1989) Androgen regulation of proopiomelanocortin gene expression and peptide content in the basal hypothalamus. Endocrinology 124:2283–2288
Bommer M, Liebisch D, Kley N, Herz A, Noble E (1987) Histamine affects release and biosynthesis of opioid peptides primarily via H1-receptors in bovine chromaffin cells. J Neurochem 49:1688–1696
Bronstein DM, Przewlocki R, Akil H (1990) Effects of morphine treatment on proopiomelanocortin systems in rat brain. Brain Res 519:102–111
Bruhn TO, Sutton RE, Rivier CL, Vale WW (1984) Corticotropin-releasing factor regulates proopiomelanocortin messenger ribonucleic acid levels in vivo. Neuroendocrinology 39:170–175
Byrd JC, Naranjo JR, Lindberg I (1987) Proenkephalin gene expression in the PC12 pheochromocytoma cell line: stimulation by sodium butyrate. Endocrinology 121:299–1305
Charron J, Drouin J (1986) Glucocorticoid inhibition of transcription from episomal proopiomelanocortin gene promoter. Proc Natl Acad Sci USA 83:8903–8907
Chen CL, Madigan MB (1987) Regulation of testicular proopiomelanocortin gene expression. Endocrinology 121:590–596
Chen CL, Dionne FT, Roberts JL (1983) Regulation of the pro-opiomelanocortin mRNA levels in rat pituitary by dopaminergic compounds. Proc Natl Acad Sci USA 80:2211–2215
Chen CL, Mather JP, Morris PL, Bardin CW (1984) Expression of proopiomelanocortin-like gene in the testis and epididymis. Proc Natl Acad Sci USA 81:5672–5675
Chen CL, Chang CC, Krieger DT, Bardin CW (1986) Expression and regulation of proopiomelanocortin-like gene in the ovary and placenta: comparison with the testis. Endocrinology 118:2382–2389
Chowen-Breed J, Fraser HM, Vician L, Damassa DA, Clifton DK, Steiner RA (1989b) Testosterone regulation of proopiomelanocortin messenger ribonucleic acid in the arcuate nucleus of the male rat. Endocrinology 124:1697–1702
Chowen-Breed JA, Clifton DK, Steiner RA (1989a) Regional specificity of testosterone regulation of proopiomelanocortin gene expression in the arcuate nucleus of the male rat brain. Endocrinology 124:2875–2881
Civelli O, Birnberg N, Herbert E (1982) Detection and quantitation of proopiomelanocortin mRNA in pituitary and brain tissues from different species. J Biol Chem 257:6783–6787
Civelli O, Douglass J, Goldstein A, Herbert E (1985) Sequence and expression of the rat prodynorphin gene. Proc Natl Acad Sci USA 82:4291–4295
Clark AJ, Lavender PM, Besser GM, Rees LH (1989) Pro-opiomelanocortin mRNA size heterogeneity in ACTH-dependent Cushing's syndrome. J Mol Endocrinol 2:3–9
Collier TJ, Routtenberg A (1984) Selective impairment of declarative memory following stimulation of dentate gyrus granule cells: a naloxone-sensitive effect. Brain Res 310:384–387
Comb M, Herbert E, Crea R (1982) Partial characterization of the mRNA that codes for enkephalins in bovine adrenal medulla and human pheochromocytoma. Proc Natl Acad Sci USA 79:360–364

Comb M, Birnberg NC, Seasholtz A, Herbert E, Goodman HM (1986) A cyclic AMP- and phorbol ester-inducible DNA element. Nature 323:353–356
Comb M, Mermod N, Hyman SE, Pearlberg J, Ross ME, Goodman HM (1988) Proteins bound at adjacent DNA elements act synergistically to regulate human proenkephalin cAMP inducible transcription. EMBO J 7:3793–3805
Cote TE, Felder R, Kebabian JW, Sekura RD, Reisine T, Affolter HU (1986) D-2 dopamine receptor-mediated inhibition of proopiomelanocortin synthesis in rat intermediate lobe. Abolition by pertussis toxin or activators of adenylate cyclase. J Biol Chem 261:4555–4561
Curran T, Rauscher FJ, Cohen DR, Franza BR Jr (1988) Beyond the second messenger: oncogenes and transcription factors. Cold Spring Harb Symp Quant Biol 53:(2)769–777
Dave JR, Eiden LE, Karanian JW, Eskay RL (1986) Ethanol exposure decreases pituitary corticotropin-releasing factor binding, adenylate cyclase activity, proopiomelanocortin biosynthesis, and plasma beta-endorphin levels in the rat. Endocrinology 118:280–286
Dave JR, Eiden LE, Lozovsky D, Waschek JA, Eskay RL (1987) Calcium-independent and calcium-dependent mechanisms regulate corticotropin-releasing factor-stimulated proopiomelanocortin peptide secretion and messenger ribonucleic acid production. Endocrinology 120:305–310
Day R, Schafer MK-H, Watson SJ, Akil H (1990) Effects of hypophysectomy on dynorphin mRNA and peptide content in the rat adrenal gland. In: Quirion R, Jhamandas K Giounalakis C (eds) The International Narcotics Conference (IRNC) '89. Liss, New York, pp 207–210
DeBold CR, Mufson EE, Menefee JK, Orth DN (1988a) Proopiomelanocortin gene expression in a pheochromocytoma using upstream transcription initiation sites. Biochem Biophys Res Commun 155:895–900
DeBold CR, Nicholson WE, Orth DN (1989b) Immunoreactive proopiomelanocortin (POMC) peptides and POMC-like messenger ribonucleic acid are present in many rat nonpituitary tissues. Endocrinology 122:2648–2657
DeKeyzer Y, Bertagna X, Lenne F, Girard F, Luton JP, Kahn A (1985) Altered proopiomelanocortin gene expression in adrenocorticotropin-producing nonpituitary tumors. Comparative studies with corticotropic adenomas and normal pituitaries. J Clin Invest 76:1892–1898
DeKeyzer Y, Bertagna X, Luton JP, Kahn A (1989a) Variable modes of proopiomelanocortin gene transcription in human tumors. Mol Endocrinol 3:215–223
DeKeyzer Y, Rousseau-Merck MF, Luton JP, Girard F, Kahn A, Bertagna X (1989b) Pro-opiomelanocortin gene expression in human phaeochromocytomas. J Mol Endocrinol 2:175–181
Douglass J, Cox B, Quinn B, Civelli O. Herbert E (1987) Expression of the prodynorphin gene in male and female mammalian reproductive tissues. Endocrinology 120:707–713
Douglass J, McMurray CT, Garrett JE, Adelman JP, Calavetta L (1989) Characterization of the rat prodynorphin gene. Mol Endocrinol 3:2070–2078
Draisci G, Iadarola MJ (1989) Temporal analysis of increases in c-fos, preprodynorphin and preproenkephalin mRNAs in rat spinal cord. Brain Res Mol Brain Res 6:31–37
Drouin J, Trifiro MA, Plante RK, Nemer M, Erikson P, Wrange Ö (1989) Glucocorticoid receptor binding to a specific DNA sequence is required for hormone-dependent repression of pro-opiomelanocortin gene transcription. Mol Cell Biol 9:5305–5314
Eberwine JH, Jonassen JA, Evinger MJ, Roberts JL (1987) Complex transcriptional regulation by glucocorticoids and corticotropin-releasing hormone of proopiomelanocortin gene expression in rat pituitary cultures. DNA 6:483–492
Eiden LE, Giraud P, Affolter HU, herbert E, Hotchkiss AJ (1984a) Alternative modes of enkephalin biosynthesis regulation by reserpine and cyclic AMP in cultured chromaffin cells. Proc Natl Acad Sci USA 81:3949–3953
Eiden LE, Giraud P, Dave JR, Hotchkiss AJ, Affolter HU (1984b) Nicotinic receptor stimulation activates enkephalin release and biosynthesis in adrenal chromaffin cells. Nature 312:661–663
Elkabes S, Loh YP (1988) Effect of salt loading on proopiomelanocortin (POMC) messenger ribonucleic acid levels, POMC biosynthesis, and secretion of POMC products in the mouse pituitary gland. Endocrinology 123:1754–1760
Farin C-J, Kley N, Höllt V (1990a) Mechanisms involved in the transcriptional activation of proenkephalin gene expression in bovine chromaffin cells. J Biol Chem 265:19116–19121

Farin C-J, Höllt V, Kley N (1990b) Proenkephalin gene expression in cultured chromaffin cell is regulated at the transcriptional level. In: Quirion R, Jhadmandas K, Giounalakis C, (eds) The International Narcotics Research Conference (IRNC) '89. Liss, New York, pp 239–242

Fischer-Colbrie R, Iacangelo A, Eiden LE (1988) Neural and humoral factors separately regulate neuropeptide Y, enkephalin, and chromagranin A and B mRNA levels in the rat adrenal medulla. Proc Natl Acad Sci USA 85:3240–3244

Folkesson R, Monstein HJ, Geijer T, Pahlman S, Nilsson K, Terenius L (1988) Expression of the proenkephalin gene in human neuroblastoma cell lines. Brain Res 427:147–154

Fremeau RT,Jr., Lundblad JR, Pritchett DB, Wilcox JN, Roberts JL (1986) regulation of pro-opiomelanocortin gene transcription in individual cell nuclei. Science 234:1265–1269

Gagner JP, Drouin J (1985) Opposite regulation of pro-opiomelanocortin gene transcription by glucocorticoids and CRH. Mol Cell Endocrinol 40:25–32

Gall C, Brecha N, Karten HJ, Chang KJ (1981) Localization of enkephalin-like immunoreactivity to identified axonal and neuronal populations of the rat hippocampus. J. Comp Neurol 198:335–350

Garrett JE, Douglass JO (1989) Human chorionic gonadotropin regulates expression of the pro-enkephalin gene in adult rat Leydig cells. Mol Endocrinol 3:2093–2100

Garrett JE, Collard MW, Douglass JO (1989) Translational control of germ cell-expressed mRNA imposed by alternative splicing: opioid peptide gene expression in rat testis. Mol Cell Biol 9:4381– 4389

Gee CE, Chen CL, Roberts JL, Thompson R, Watson SJ (1983) Identification of proopiomelano-cortin neurons in rat hypothalamus by in situ cDNA-mRNA hybridization. Nature 306:374–376

Gizang-Ginsberg E, Wolgemuth DJ (1985) Localization of mRNAs in mouse testes by in situ hybridization: distribution of alpha-tubulin and developmental stage specificity of pro-opio-melanocortin transcripts. Dev Biol 111:293–305

Gizang-Ginsberg E, Wolgemuth DJ (1987) Expression of the proopiomelanocortin gene is developmentally regulated and affected by germ cells in the male mouse reproductive system. Proc Natl Acad Sci USA 84:1600–1604

Graybiel AM (1986) Neuropeptides in the basal ganglia. Res Publ Assoc Res Nerv Ment Dis 64: 135–161

Gubler U, Seeburg P, Hoffman BJ, Gage LP, Udenfriend S (1982) Molecular cloning establishes proenkephalin as precursor of enkephalin-containing peptides, Nature 295:206–208

Harlan RE, Shivers BD, Romano GJ, Howells RD, Pfaff DW (1987) Localization of prepro-enkephalin in the rat brain and spinal cord by in situ hybridization. J Comp Neurol 258:159–184

Henricksen SJ, Chouvet G, Bloom FE (1982) In vivo cellular responses to electrophoretically applied dynorphin in the rat hippocampus. Life Sci 31:1785–1788

Höllt V (1986) Opioid peptide processing and receptor selectivity. Annu Rev Pharmacol Toxicol 26:59–77

Höllt V, Haarmann I (1984) Corticotropin-releasing factor differentially regulates proopiomelano-cortin messenger ribonucleic acid levels in anterior as compared to intermediate pituitary lobes of rats. Biochem Biophys Res Commun 124:407–415

Höllt V, Haarmann I (1985) Differential alterations by chronic treatment with morphine of pro-opiomelanocortin mRNA levels in anterior as compared to intermediate pituitary lobes of rats. Neuropeptides 5:481–484

Höllt V, Horn G (1989) Nicotine and opioid peptides. In: Nordberg A, Fuxe K, Holmstedt B, Sundwall A (eds) Nicotinic receptors in the CNS: their role in synaptic transmission. Prog Brain Res 79:187–193

Höllt V, Horn G (1991) Nicotine induces opioid peptide gene expression in hypothalamus and adrenal medulla of rats. In: Adlkofer F, Thurau K (eds) Effects of nicotine of biological systems. Birkhäuser, Basel, pp 273–284

Höllt V, Haarmann I, Seizinger BR, Herz A (1982) Chronic haloperidol treatment increases the level of in vitro translatable messenger ribonucleic acid coding for the beta-endorphin/adreno-corticotropin precursor proopiomelanocortin in the pars intermedia of the rat pituitary. Endo-crinology 110:1885–1891

Höllt V, Przewlocki R, Haarmann I, Almeida OF, Kley N, Millan MJ, Herz A (1986) Stress-induced alterations in the levels of messenger RNA coding for proopiomelanocortin and prolactin in rat pituitary. Neuroendocrinology 43:277–282

Höllt V, Haarmann I, Millan MJ, Herz A (1987) Prodynorphin gene expression is enhanced in the spinal cord of chronic arthritic rats. Neurosci Lett 73:90–94
Höllt V, Haarmann I, Reimer S (1989a) Opioid peptide gene expression in rats after chronic morphine treatment. Adv Biosci 75:711–714
Höllt V, Haarmann I, Reimer S (1989b) Opioid gene expression in rats after chronic morphine treatment. Adv Biosci 75:711–714
Horikawa S, Takai T, Toyosato M, Takahashi H, Noda M, Kakidani H, Kubo T, Hirose T, Inayama S, Hayashida H, Miyata T, Numa S (1983) Isolation and structural organization of the human preproenkephalin B gene. Nature 306:611–614
Hunt SP, Pini A, Evan G (1987) Induction of c-fos-like protein in spinal cord neurons following sensory stimulation. Nature 328:632–634
Hyman SE, Comb M, Lin YS, Pearlberg J, green MR, Goodman HM (1988) A common trans-acting factor is involved in transcriptional regulation of neurotransmitter genes by cyclic AMP. Mol Cell Biol 8:4225–4233
Hyman SE, Comb M, Pearlberg J, Goodman HM (1989) An AP-2 element acts synergistically with the cyclic AMP-and phorbol ester-inducible enhancer of the human proenkephalin gene. Mol Cell Biol 9:321–324
Iadarola MJ, Douglass J, Civelli O. Naranjo JR (1986) Increased spinal cord dynorphin mRNA during peripheral inflammation. Natl Inst Drug Abuse Res Monogr Ser 75:406–409
Iadarola MJ, Brady LS, Draisci G, Dubner R (1988a) Enhancement of dynorphin gene expression in spinal cord following experimental inflammation: stimulus specificity, behavorial parameters and opioid receptor binding. Pain 35:313–326
Iadarola MJ, Douglass J, Civelli O, Naranjo JR (1988b) Differential activation of spinal cord dynorphin and enkephalin neurons during hyperalgesia: evidence using cDNA hybridization, Brain Res 455:205–212
Inturrisi CE, LaGamma EF, Franklin SO, Huang T, Nip TJ, Yoburn BC (1988) Characterization of enkephalins in rat adrenal medullary explants. Brain Res 448:230–236
Israel A, Cohen SN (1985) Hormonally mediated negative regulation of human pro-opio-melanocortin gene expression after transfection into mouse L cells. Mol Cell Biol 5:2443–2453
Jeanotte L, Trifiro MA, Plante RK, Chamberland M, Drouin J (1987) Tissue-specific activity of the pro-opiomelanocortin gene promoter. Mol Cell Biol 7:4058–4064
Jin DF, Muffly KE, Okulicz WC, Kilpatrick DL (1988) Estrous cycle- and pregnancy-related differences in expression of the proenkephalin and proopiomelanocortin genes in the ovary and uterus. Endocrinology 122:1466–1471
Jingami H, Nakanishi S, Imura H, Numa S (1984) Tissue distribution of messenger RNAs coding for opioid peptide precursors and related RNA. Eur J Biochem 142:441–447
Kanamatsu T, Unsworth CD, Diliberto EJ, Jr., Viveros OH, Hong JS (1986) Reflex splanchnic nerve stimulation increases levels of proenkephalin A mRNA and proenkephalin A-related peptides in the rat adrenal medulla. Proc Natl Acad Sci USA 83:9245–9249
Kew D, Kilpatrick DL (1989) Expression and regulation of the proenkephalin gene in rat Sertoli cells. Mol Endocrinol 3:179–184
Kew D, Kilpatrick DL (1990) Widespread organ expression of the rat proenkephalin gene during early postnatal development. Mol Endocrinol 4:337–340
Kilpatrick DL, Millette CF (1986) Expression of proenkephalin messenger RNA by mouse spermatogenic cells. Proc Natl Acad Sci USA 83:5015–5018
Kilpatrick DL, Rosenthal JL (1986) The proenkephalin gene is widely expressed within the male and female reproductive systems of the rat and hamster. Endocrinology 119:370–374
Kilpatrick DL, Howells RD, Fleminger G, Udenfriend S (1984) Denervation of rat adrenal glands markedly increases preproenkephalin mRNA. Proc Natl Acad Sci USA 81:7221–7223
Kilpatrick DL, Howells RD, Noe M, Bailey LC, Udenfriend S (1985) Expression of prepro-enkephalin-like mRNA and its peptide products in mammalian testis and ovary. Proc Natl Acad Sci USA 82:7467–7469
Kilpatrick DL, Borland K, Jin DF (1987) Differential expression of opioid peptide genes by testicular germ cells and somatic cells. Proc Natl Acad Sci USA 84:5695–5699
Kilpatrick DL, Zinn SA, Fitzgerald M, Higuchi H, Sabol SL, Meyerhardt J (1990) Transcription

of the rat and mouse proenkephalin genes is initiated at distinct sites in spermatogenic and somatic cells. Mol Cell Biol 10:3717–3726

Kley N (1988) Multiple regulation of proenkephalin gene expression by protein kinase C. J Biol Chem 263:2003–2008

Kley N, Loeffler JP, Pittius CW, Höllt V (1986) Proenkephalin A gene expression in bovine adrenal chromaffin cells is regulated by changes in eletrical activity. EMBO J 5:967–970

Kley N, Loeffler JP, Höllt V (1987a) Ca2+-dependent histaminergic regulation of proenkephalin mRNA levels in cultured adrenal chromaffin cells. Neuroendocrinology 46:89–92

Kley N, Loeffler JP, Pittius CW, Höllt V (1987b) Involvement of ion channels in the induction of proenkephalin A gene expression by nicotine and cAMP in bovine chromaffin cells. J Biol Chem 262:4083–4089

Knight RM, Farah JM, Bishop JF, O'Donohue TL (1987) CRF and cAMP regulation of POMC gene expression in corticotrophic tumor cells. Peptides 8:927–934

Korner M, Rattner A, Mauxion F, Sen R, Citri Y (1989) A brain-specific transcription activator. Neuron 3: 563–572

Kowalski C, Giraud P, Boudouresque F, Lissitzky JC, Cupo A, Renard M, Saura RM, Oliver C (1989) Enkephalins expression in striatal cell cultures. Adv Biosci 75:225–228

Lacaze-Masmonteil T, de Keyzer Y, Luton JP, Kahn A, Bertagna X (1987) Characterization of proopiomelanocortin transcripts in human nonpituitary tissues. Proc Natl Acad Sci USA 84: 7261–7265

LaGamma EF, White JD, Adler JE, Krause JE, McKelvy JF, Black IB (1985) Depolarization regulates adrenal preproenkephalin mRNA. Proc Natl Acad Sci USA 82:8252–8255

LaGamma EF, White JD, McKelvy JF, Black IB (1988) Second messenger mechanisms governing opiate peptide transmitter regulation in the rat adrenal medulla. Brain Res 441:292–298

Lee PHK, Zhao D, Xie CW, McGinty JF, Mitchell CL, Hong JS (1989) Changes of proenkephalin and prodynorphin mRNAs and related peptides in rat brain during the development of deep prepyriform cortex kindling. Mol Brain Res 6:263–273

Legon S, Glover DM, Hughes J, Lowry PJ, Rigby PW, Watson CJ (1982) The structure and expression of the preproenkephalin gene. Nucleic Acids Res 10:7905–7918

Levy A, Lightman SL (1988) Quantitative in-situ hybridization histochemistry in the rat pituitary gland: effect of bromocriptine on prolactin and pro-opiomelanocortin gene expression. J Endocrinol 118:205–210

Li H, Risbridger GP, Funder JW, Clements JA (1989) Effect of ethane dimethane sulphonate on proopiomelanocortin (POMC) mRNA and POMC-derived peptides in the rat testis. Mol Cell Endocrinol 65:203–207

Li SJ, Sivam SP, McGinty JF, Jiang HK, Douglass J, Calavetta L, Hong JS (1988) Regulation of the metabolism of striatal dynorphin by the dopaminergic system. J Pharmacol Exp Ther 246: 403–408

Li SJ, Jiang HK, Stachowiak MS, Hudson PM, Owyang V, Nanry K, Tilson HA, Hong JS (1990) Influence of nigrostriatal dopaminergic tone on the biosynthesis of dynorphin and enkephalin in rat striatum. Brain Res Mol Brain Res 8:219–225

Lightman SL, Young WS (1987) Changes in hypothalamic preproenkephalin A mRNA following stress and opiate withdrawal. Nature 328:643–645

Lightman SL, Young WS (1988) Corticotrophin-releasing factor, vasopressin and pro-opiomelanocortin mRNA responses to stress and opiates in the rat. J Physiol (Lond) 403:511–523

Litt M, Buder A (1989) A frequent RFLP identified by a human proenkephalin genomic clone [HGM9 symbol PENK]. Nucleic Acids Res 17:46555

Llorens-Cortes C, Van-Amsterdam JG, Giros B, Quach TT, Schwartz JC (1990) Enkephalin biosynthesis and release in mouse striatum are inhibited by GABA receptor stimulation: compared changes in preproenkephalin mRNA and Tyr-Gly-Gly levels. Brain Res Mol Brain Res 8:227–233

Loeffler JP, Kley N, Pittius CW, Höllt V (1985) Corticotropin-releasing factor and forskolin increase proopiomelanocortin messenger RNA levels in rat anterior and intermediate cells in vitro. Neurosci Lett 62:383–387

Loeffler JP, Kley N, Pittius CW, Almeida OF, Höllt V (1986a) In vivo and in vitro studies of GABAergic inhibition of prolactin biosynthesis. Neuroendocrinology 43:504–510

Loeffler JP, Kley N, Pittius CW, Höllt V (1986b) Calcium ion and cyclic adenosine 3',5'-monophosphate regulate proopiomelanocortin messenger ribonucleic acid levels in rat intermediate and anterior pituitary lobes. Endocrinology 119:2840–2847

Loeffler JP, Kley N, Pittius CW, Höllt V (1986c) Regulation of proopiomelanocortin (POMC) mRNA levels in primary pituitary cultures. Natl Inst Drug Abuse Res Monogr Ser 75:397–400

Loeffler JP, Demeneix BA, Kley NA, Höllt V (1988) Dopamine inhibition of proopiomelanocortin gene expression in the intermediate lobe of the pituitary. Interactions with cortiocotropin-releasing factor and the beta-adrenergic receptors and the adenylate cyclase system. Neuroendocrinology 47:95–101

Low KG, Nielsen CP, West NB, Douglass J, Brenner RM, Maslar IA, Melner MH (1989) Proenkephalin gene expression in the primate uterus: regulation by estradiol in the endometrium. Mol Endocrinol 3:852–857

Lundblad JR, Roberts JL (1988) Regulation of proopiomelanocortin gene expression in pituitary. Endocr Rev 9:135–158

Martens GJ (1986) Expression of two proopiomelanocortin genes in the pituitary gland of Xenopus laevis: complete structures of the two preprohormones. Nucleic Acids Res 14:3791–3798

Martens GJ, Herbert E (1984) Polymorphism and absence of Leu-enkephalin sequences in proenkephalin genes in Xenopus laevis. Nature 310:251–254

McGinty JF, Henriksen SJ, Goldstein A, Terenius L, Bloom FE (1983) Dynorphin is contained within hippocampal mossy fibers: immunochemical alterations after kainic acid administration and colchicine-induced neurotoxicity. Proc Natl Acad Sci USA 80:589–593

McMurray CT, Devi L, Calavetta L, Douglass JO (1989) Regulated expression of the prodynorphin gene in the R2C Leydig tumor cell line. Endocrinology 124:49–59

Meador-Woodruff JH, Pellerito B, Bronstein D, Lin HL, Ling N, Akil H (1990) Differential effects of haloperidol on the rat pituitary: decreased biosynthesis, processing and release of anterior lobe pro-opiomelanocortin. Neuroendocrinology 51:294–303

Melner MH, Young SL, Czerwiec FS, Lyn D, Puett D, Roberts JL, Koos RD (1986) The regulation of granulosa cell proopiomelanocortin messenger ribonucleic acid by androgens and gonadotropins. Endocrinology 119:2082–2088

Mengod G, Charli J-L, Palacios JM (1990) The use of in situ hybridization histochemistry for the study of neuropeptide gene expression in the human brain. Cell Mol Neurobiol 10:113–126

Mocchetti I, Giorgi O, Schwartz JP, Costa E (1984) A reduction of the tone of 5-hydroxytryptamine neurons decreases utilization rates of striatal and hypothalamic enkephalins. Eur J Pharmacol 106:427–430

Mocchetti I, Guidotti A, Schwartz JP, Costa E (1985) Reserpine changes the dynamic state of enkephalin stores in rat striatum and adrenal medulla by different mechanisms. J Neurosci 5: 3379–3385

Mocchetti I, Naranjo JR, Costa E (1987) Regulation of striatal enkephalin turnover in rats receiving antagonists of specific dopamine receptor subtypes. J Pharmacol Exp Ther 241:1120–1124

Mocchetti I, Ritter A, Costa E (1989) Down-regulation of proopiomelanocortin synthesis and betaendorphin utilization in hypothalamus of morphine-tolerant rats. J Mol Neurosci 1:33–38

Moneta ME, Höllt V (1990) Perforant path kindling induces differential alterations in the mRNA levels coding for prodynorphin and proenkephalin in the rat hippocampus. Neurosci Lett 110: 273–278

Monstein HJ, Folkesson R, Geijer T (1990) Procholecystokinin and proenkephalin A mRNA expression is modulated by cyclic AMP and noradrenaline. J Mol Endocrinol 4:37–41

Morris BJ, Haarman I, Kempter B, Höllt V, Herz A (1986) Localization of prodynorphin messenger RNA in rat brain by in situ hybridization using a synthetic oligonucleotide probe. Neurosci Lett 69:104–108

Morris BJ, Moneta ME, ten-Bruggencate G, Höllt V (1987) Levels of prodynorphin mRNA in rat dentate gyrus are decreased during hippocampal kindling. Neurosci Lett 80:298–302

Morris BJ, Feasey KJ, ten-Bruggencate G, Herz A, Höllt V (1988a) Electrical stimulation in vivo increases the expression of proenkephalin mRNA and decreases the expression of prodynorphin mRNA in rat hippocampal granule cells. Proc Natl Acad Sci USA 85:3226–3230

Morris BJ, Höllt V, Herz A (1988b) Dopaminergic regulation of striatal proenkephalin mRNA and prodynorphin mRNA. Neuroscience 25:525–532

Morris BJ, Höllt V, Herz A (1988c) Opioid gene expression in rat striatum is modulated via opioid receptors: evidence from localized receptor inactivation. Neurosci Lett 89:80–84
Morris BJ, Reimer S, Höllt V, Herz A (1988d) Regulation of striatal prodynorphin mRNA levels by the raphe-striatal pathway. Brain Res 464:15–22
Morris B, Herz A, Höllt V (1989) Location of striatal opioid gene expression, and its modulation by the mesostriatal dopamine pathway: an in situ hybridization study. J Mol Neurosci 1: 9–18
Munemura M, Cote TE, Tsuruta K, Eskay RL, Kebabian JW (1990) The dopamine receptor in the intermediate lobe of the rat pituitary gland: pharmacological characterization. Endocrinology 107:1676–1683
Nakamura M, Nakanishi S, Sueoka S, Imura H, Numa S (1978) Effects of steroid hormones on the level of corticotropin messenger RNA activity in cultured mouse-pituitary-tumor cells. Eur J Biochem 86:61–66
Nakanishi S, Inoue A, Kita T, Nakamura M, Chang AC, Cohen SN, Numa S (1979) Nucleotide sequence of cloned cDNA for bovine corticotropin-beta-lipotropin precursor. Nature 278:423–427
Naranjo JR, Iadarola MJ, Costa E (1986a) Changes in the dynamic state of brain proenkephalin-derived peptides during amygdaloid kindling. J Neurosci Res 16:75–87
Naranjo JR, Mocchetti I, Schwartz JP Costa E (1986b) Permissive effect of dexamethasone on the increase of proenkephalin mRNA induced by depolarization of chromaffin cells. Proc Natl Acad Sci USA 83:1513–1517
Naranjo JR, Mellström B, Achaval M, Sassone-Corsi P (1991) Molecular pathways of pain: Fos/jun-mediated activation of a noncanonical AP-1 site in the prodynorphin gene. Neuron 6:606–617
Nishimori T, Moskowitz MA, Uhl GR (1988) Opioid peptide gene expression in rat trigeminal nucleus caudalis neurons: normal distribution and effects of trigeminal deafferentation. J Comp Neurol 274:142–150
Nishimori T, Buzzi MG, Moskowitz MA, Uhl GR (1989) Proenkephalin mRNA expression in nucleus caudalis neurons is enhanced by trigeminal stimulation. Mol Brain Res 6:203–210
Noble EP, Bommer M, Sincini E, Costa T, Herz A (1986) H1-histaminergic activation stimulates inositol-1-phosphate accumulation in chromaffin cells. Biochem Biophys Res Commun 135: 566–573
Noda M, Furutani Y, Takahashi H, Toyosato M, Hirose T, Inayama S, Nakanishi S, Numa S (1982a) Cloning and sequence analisys of cDNA for bovine adrenal preproenkephalin. Nature 295:202–206
Noda M, Teranishi Y, Takahashi H, Toyosato M, Notake M, Nakanishi S, Numa S (1982b) Isolation and structural organization of the human proenkephalin gene. Nature 297:431–434
Noguchi K, Kowalski K, Traub R, Solodkin A, Iadarola MJ, Ruda MA (1991) Dynorphin expression and Fos-like immunoreactivity following inflammation-induced hyperalgesia are colocalized in spinal cord neurons. Mol Brain Res 10:227–233
Normand E, Popovici T, Onteniente B, Fellmann D, Piater-Tonneau D, Auffray C, Bloch B (1988) Dopaminergic neurons of the substantia nigra modulate preproenkephalin A gene expression in rat striatal neurons. Brain Res 439:39–46
Notake M, Tobimatsu T, Watanabe Y, Takahashi H, Mishina M, Numa S (1983) Isolation and characterization of the mouse corticotropin-beta-lipotropin precursor gene and a related pseudogene. FEBS Lett 156:67–71
Oates E, Herbert E (1984) 5' sequence of porcine and rat pro-opiomelanocortin mRNA. One porcine and two rat forms. J Biol Chem 259:7421–7425
Owerbach D, Rutter WJ, Roberts JL, Whitfeld P, Shine J, Seeburg PH, Shows TB (1981) The pro-opiocortin (adrenocorticotropin/beta-lipoprotein) gene is located on chromosome 2 in humans. Somatic Cell Genet 7:359–369
Pintar JE, Schachter BS, Herman AB, Durgerian S, Krieger DT (1984) Characterization and localization of proopiomelanocortin messenger RNA in the adult rat testis. Science 225:632–634
Pittius JE, Kley N, Loeffler JP, Höllt V (1985) Quantitation of proenkephalin A messenger RNA in bovine brain, pituitary and adrenal medulla: correlation between mRNA and peptide levels. EMBO J 4:1257–1260

Pittius CW, Kley N, Loeffler JP, Höllt V (1987) Proenkephalin B messenger RNA in porcine tissues: characterization, quantification, and correlation with opioid peptides. J Neurochem 48: 586–592

Pritchett DB, Roberts JL (1987) Dopamine regulates expression of the glandular-type kallikrein gene at the transcriptional level in the pituitary. Proc Natl Acad Sci USA 84:5545–5549

Przewlocki R, Haarmann I, Nikolarakis K, Herz A, Höllt V (1988) Prodynorphin gene expression in spinal cord is enhanced after traumatic injury in the rat. Brain Res 464:37–41

Quach TT, Tang F, Kageyama H, Mocchetti I, Guidotti A, Meek JL, Costa E, Schwartz JP (1984) Enkephalin biosynthesis in adrenal medulla. Modulation of proenkephalin mRNA content of cultured chromaffin cells by 8-bromo-adenosine 3',5'-monophosphate. Mol Pharmacol 26:255–260

Rattner A, Korner M, Rosen H, Raenerle PA, Citri Y (1991) Nuclear factor KB activates proenkephalin transcription in T lymphocytes. Mol Cell Biol 11:1017–1022

Reimer S, Höllt V, (1990) Morphine increases proenkephalin gene expression in the adrenal medulla by a central mechanism. In Quirion R,Jhamandas K, Giounalakis C (eds) The International narcotics Research Conference (IRNC) '89. Liss, New York, pp 215–218

Reimer S, Höllt V (1991) Gabaergic regulation of striatal opioid gene expression. Mol Brain Res 10:49–54

Reimer S, Sirinathsinghji DJS, Nikolorakis KE, Hout V (1992) Differential dopaminergic regulation of proeukephalin and prodynorphin mRNAs in the basal ganglia of rats. Mol Brain Res 12:259–266

Reisine T, Rougon G, Barbet J, Affolter HU (1985) Corticotropin-releasing factor-induced adrenocorticotropin hormone release and synthesis is blocked by incorporation of the inhibitor of cyclic AMP-dependent protein kinase into anterior pituitary tumor cells by liposomes. Proc Natl Acad Sci USA 82:8261–8265

Roberts JL Budarf ML, Baxter JD, Herbert E (1987) Selective reduction of proadrenocorticotropin/endorphin proteins and messenger ribonucleic acid activity in mouse pituitary tumor cells by glucocorticoids. Biochemistry 18:4907–4915

Roberts JL, Lundblad JR, Eberwine JH, Fremeau RT, Salton SR, Blum M (1987) Hormonal regulation of POMC gene expression in pituitary. Ann NY Acad Sci 512:275–285

Romano GJ, Shivers BD, Harlan RE, Howells RD, Pfaff DW (1987) Haloperidol increases proenkephalin mRNA levels in the caudate-putamen of the rat: a quantitive study at the cellular level using in situ hybridization. Brain Res 388:33–41

Romano GJ, Harlan RE, Shivers BD, Howells RD, Pfaff DW (1988) Estrogen increases proenkephalin messenger ribonucleic acid levels in the ventromedial hypothalamus of the rat. Mol Endocrinol 2:1320–1328

Rosen H, Douglass J, Herbert E (1984) Isolation and characterization of the rat proenkephalin gene. J Biol Chem 259:14309–14313

Rosen H, Behar O, Abramsky O, Ovadia H (1989) Regulated expression of proenkephalin A in normal lymphozytes. J Immunol 143:3703–3707

Rosen H, Itin A, Schiff R, Keshet E (1990) Local regulation within the female reproductive system and upon embroyonic implantation: identification of cells expressing proenkephalin A. Mol Endocrinol 4:146–154

Ruda MA, Iadarola MJ, Cohen LV, Young WS (1988) In situ hybridization histochemistry and immunocytochemistry reveal an increase in spinal dynorphin biosynthesis in a rat model of peripheral inflammation and hyperalgesia. Proc Natl Acad Sci USA 85:622–626

Schachter BS, Johnson LK, Baxter JD, Roberts JL (1982) Differential regulation by glucocorticoids of proopiomelanocortin mRNA levels in the anterior and intermediate lobes of the rat pituitary. Endocrinology 110:1442–1444

Schafer MK-H Day R, Hermann JP, Kwasiborski V, Sladek CD, Akil H, Watson SJ (1989) Effects of electroconvulsive shock on dynorphin in the hypothalamic-neurophysial system of the rat. Adv Bioscience 75:599–602

Schafer MK-H, Day R, Akil H, Watson SJ (1990) Identification of prodynorphin and proenkephalin cells in the neurointermediate lobe of the rat pituitary gland. In: Quirion R, Jhamadas K, Giounalakis C (eds) The International Narcotics Research Conference (IRNC) '89. Liss, New York, pp 231–234

Schwartz JP (1988) Chronic exposure to opiate agonists increases proenkephalin biosynthesis in NG108 cells. Brain Res 427:141–146
Seger MA, van-Eekelen JA, Kiss JZ, Burbach JP, de-Kloet ER (1988) Stimulation of pro-opiomelanocortin gene expression by glucocorticoids in the denervated rat intermediate pituitary gland. Neuroendocrinology 47:350–357
Seizinger BR, Bovermann K, Höllt V, Herz A (1984a) Enhanced activity of the beta-endorphinergic system in the anterior and neurointermediate lobe of the rat pituitary after chronic treatment with ethanol liquid diet. J Pharmacol Exp Ther 230:455–461
Seizinger BR, Höllt V, Herz A (1984b) Effects of chronic ethanol treatment on the in vitro biosynthesis of pro-opiomelanocortin and its posttranslational processing to beta-endorphin in the intermediate lobe of the rat pituitary. J Neurochem 43:607–613
Sherman TG Civelli O, Douglass J, Herbert E, Burke S, Watson SJ (1986) Hypothalamic dynorphin and vasopressin mRNA expression in normal and Brattleboro rats. Fed Proc 45:2323–2327
Shiomi H, Watson SJ, Kelsey JE, Akil H (1986) Pretranslational and posttranslational mechanisms for regulating beta-endorphin-adrenocorticotropin of the anterior pituitary lobe. Endocrinology 119:1793–1799
Siegel RE, Eiden LE, Affolter HU (1985) Elevated potassium stimulates enkephalin biosynthesis in bovine chromaffin cells. Neuropeptides 6:543–552
Simard J, Labrie F, Gossard F (1986) Regulation of growth hormone mRNA and proopiomelanocortin mRNA levels by cyclic AMP in rat anterior pituitary cells in culture.DNA:5:263–270
Sivam SP, Hong JS (1986) GABAergic regulation of enkephaling in rat striatum: alterations in Met5-enkephalin level, precursor content and preproenkephalin messenger RNA abundance. J Pharmacol Exp Ther 237:326–331
Sivam SP Breese GR, Napier TC, Mueller RA, Hong JS (1986a) Dopaminergic regulation of proenkephalin-A gene expression in the basal ganglia. Natl Inst Drug Abuse Res Monogr Ser 75:389–392
Sivam SP, Strunk C, Smith DR, Hong JS (1986b) Proenkephalin-A gene regulation in the rat striatum: influence of lithium and haloperidol. Mol pharmacol 30:186–191
Sonnenberg JL, Rauscher FJ, Morgan JI, Curran T (1989) Regulation of proenkephalin by Fos and Jun. Science 246:1622–1625
Spampinato S, Bachetti T, Canossa M, Ferri S (1990) Prodynorphin messenger RNA expression in the rat anterior pituitary is regulated by estrogen. In Quirion R, Jhamandas K, Giounalakis C (eds) Liss, New York 211–214
Stachowiak MH, Hong JS, Viveros OH (1990) Coordinate and differential regulation of phenylethanolamine N-methyltransferase, tyrosine hydroxylase and proenkephalin mRNAs by neural and hormonal mechanisms in cultured bovine adrenal medullary cells. Brain Res 510:277–288
Stalla GK Stalla J, Huber M, Loeffler JP, Höllt V, von-Werder K, Muller OA (1988) Ketoconazole inhibits corticotropic cell function in vitro. Endocrinology 122:618–623
Stalla GK, Stalla J, Mojto J, Oeckler R, Buchfelder M, Muller OA (1989a) Regulation of corticotrophic adenoma cells in vitro. Acta Endocrinol (Copenh) 120 [Suppl 1]:209
Stalla GK, Stalla J, von-Werder K, Muller OA, Gerzer R, Höllt V, Jakobs KH (1989b) Nitroimidazole derivates inhibit anterior pituitary cell function apparently by a direct effect on the catalytic subunit of the adenylate cyclase holoenzyme. Endocrinonology 125:699–706
Suda T, Tozawa F, Yamada M, Ushiyama T, Tomori N, Sumitomo T, Nakagami Y, Demura H, Shizume K (1988a) Effects of corticotropin-releasing hormone and dexamethasone on proopiomelanocortin messenger RNA level in human corticotroph adenoma cells in vitro. J Clin Invest 82:110–114
Suda T, Tozawa F, Yamada M, Ushiyama T, Tomori N, Sumitomo T, Nakagami Y, Shizume K (1988b) In vitro study on proopiomelanocortin messenger RNA levels in cultured rat anterior pituitary cells. Life Sci 42:1147–1152
Suda T, Tozawa F, Ushiyama T, Tomori N, Sumitomo T, Nakagami Y, Yamada M, Demura H, Shizume K (1989) Effects of protein kinase C-related adrenocorticotropin secretagogues and interleukin-1 on proopiomelanocortin gene expression in rat anterior pituitary cells. Endocrinology 124:1444–1449
Takahashi H, Hakamata Y, Watanabe Y, Kikuno R, Miyata T, Numa S (1983) Complete nucleotide sequence of the human corticotropin-beta-lipotropin precursor gene. Nucleic Acids Res 11: 6847–6858

Tang F, Costa E, Schwartz JP (1983) Increase of proenkephalin mRNA and enkephalin content of rat striatum after daily injection of haloperidol for 2 to 3 weeks. Proc Natl Acad Sci USA 80:3841–3844
Terao M, Watanabe Y, Mishina M, Numa S (1983) Sequence requirement for transcription in vivo of the human proenkephalin A gene. EMBO J 2:2223–2228
Tomiko SA, Taraskevich PS, Douglas WW (1983) GABA acts directly on cells of pituitary pars intermedia to alter hormone output. Nature 301: 706–707
Tong Y, Zhao HF, Labrie F, Peletier G (1990) Regulation of proopiomelanocortin messenger ribonucleic acid content by sex steroids in the arcuate nucleus of the female rat brain. Neurosci Lett 112:104–108
Tozawa F, Suda T, Yamada M, Ushiyama T, Tomori N, Sumitomo T, Nakagami Y, Demura H, Shizume K (1988) Insulin-induced hypoglycemia increases proopiomelanocortin messenger ribonucleic acid levels in rat anterior pituitary gland. Endocrinology 122:1231–1235
Tremblay Y, Tretjakoff I, Peterson A, Antakly T, Zhang CX, Drouin J (1988) Pituitary-specific expression and glucocorticoid regulatian of a proopiomelanocortin fusion gene in transgenic mice. Proc Natl Acad Sci USA 85:8890–8894
Uhl GR, Ryan JP, Schwartz JP (1988) Morphine alters preproenkephalin gene expression. Brain Res 459:391–397
Uhler M, Herbert E, D'Eustachio P, Ruddle FD (1983) The mouse genome contains two nonallelic pro-opiomelanocortin genes. J Biol Chem 258:9444–9453
Usui T, Nakai Y, Tsukada T, Takahashi H, Fukata J, Naito Y, Nakaishi S, Tominaga T, Murakami N, Imura H (1990) Expression of the human proopiomelonocortin gene introduced into a rat glial cell line. J Mol Endocrinol 4:169–175
Verbeeck MA, Draaijer M, Burbach JP (1990) Selective down-regulation of the proenkephalin gene during differentiation of a multiple neuropeptide-co-expressing cell line. J Biol Chem 265: 18087–18090
Vernier P, Julien JF, Rataboul P, Fourrier O, Feuerstein C, Malet J (1988) Similar time course changes in striatal levels of glutamic acid decarboxylase and proenkephalin mRNA following dopaminergic deafferentation in the rat. J Neurochem 51:1375–1380
Von Dreden G, Höllt V (1988) Vasopressin potentiates β-endorphin release but not the increase in the mRNA for proopiomelanocortin induced by corticotropin releasing factor in rat pituitary cells. Acta Endocrinol (Copenh) 117 [Suppl 287]:124
Von Dreden G, Loeffler JP, Grimm C, Höllt V (1988) Influence of calcium ions on proopiomelanocortin mRNA levels in clonal anterior pituitary cells. Neuroendocrinology 47:32–37
Wan DC, Livett BG (1989) Induction of phenylethanolamine N-methyltransferase mRNA expression by glucocorticoids in cultured bovine adrenal chromaffin cells. Eur J Pharmacol 172: 107–115
Wan DC, Marley PD, Livett BG (1989) Histamine activates proenkephalin A mRNA but not phenylethanolamine N-methyltransferase mRNA expression in cultured bovine adrenal chromaffin cells. Eur J Pharmacol 172:117–129
Waschek JA, Eiden LE (1988) Calcium requirements for barium stimulation of enkephalin and vasoactive intestinal peptide biosynthesis in adrenomedullary chromaffin cells. Neuropeptides 11:39–45
Waschek JA, Dave JR, Eskay RL, Eiden LE (1987) Barium distinguishes separate calcium targets for synthesis and secretion of peptides in neuroendocrine cells. Biochem Biophys Res Commun 146:495–501
Weihe E, Millan MJ, Leibold A, Nohr D, Herz A (1988) Co-localization of proenkephalin- and prodynorphin-derived opioid peptides in laminae IV/V spinal neurons revealed in arthritic rats. Neurosci Lett 85:187–192
Weihe E, Millan MJ, Höllt V, Nohr D, Herz A (1989) Induction of the gene encoding pro-dynorphin by experimentally induced arthritis enhances staining for dynorphin in the spinal cord of rats. Neuroscience 31:77–95
Wiemann JN, Clifton DK, Steiner RA (1989) Pubertal changes in gonadotropin-releasing hormone and proopiomelanocortin gene expression in the brain of the male rat. Endocrinology 124:1760–1767
Wilcox JN, Roberts JL (1985) Estrogen decreases rat hypothalamic proopiomeloncortin messenger ribonucleic acid levels. Endocrinology 177:2392–2396

Xie CW, Lee PH, Takeuchi K, Owyang V, Li SJ, Douglass J, Hong JS (1989) Single or repeated electroconvulsive shocks alter the levels of prodynorphin and proenkephalin mRNAs in rat brain. Brain Res Mol Brain Res 6:11–19
Xie CW, Mitchell CL, Hong JS (1990) Perforant path stimulation differentially alters prodynorphin mRNA and proenkephalin mRNA levels in the entorhinal cortex-hippocampal region. Brain Res Mol Brain Res 7:199–205
Yoshikawa K, Sabol SL (1986) expression of the enkephalin precursor gene in C6 rat glioma cells: regulation by beta-adrenergic agonists and glucocorticoids. Brain Res 387:75–83
Yoshikawa K, Hong JS, Sabol SL (1985) Electroconvulsive shock increases preproenkephalin messenger RNA abundance in rat hypothalamus. Proc Natl Acad Sci USA 82:589–593
Yoshikawa K, Aizawa T, Nozawa A (1989a) Phorbol ester regulates the abundance of enkephalin precursor mRNA but not of amyloid beta-protein precursor mRNA in rat testicular peritubular cells. Biochem Biophys Res Commun 161:568–575
Yoshikawa K, Maruyama K, Aizawa T, Yamamoto A (1989b) A new species of enkephalin precursor mRNA with a distinct 5'-untranslated region in haploid germ cells. FEBS Lett 246:193–196
Young WS, Bonner TI, Brann MR (1986) Mesencephalic dopamine neurons regulate the expression of neuropeptide mRNAs in the rat forebrain. Proc Natl Acad Sci USA 83:9827–9831
Zurawski G, Benedik M, Kamb BJ, Abrams JS, Zurawski SM, Lee FD (1986) Activation of mouse T-helper cells induces abundant preproenkephalin mRNA synthesis. Science 232:772–77

Regulation of Gene Expression of Pituitary Hormones by Hypophysiotropic Hormones

Y. NAKAI and T. TSUKADA[1]

Contents

[1] Second Division, Department of Internal Medicine, Kyoto University, Faculty of Medicine, 54 Kawaharacho, Shogoin, Sakyoku, Kyoto City, 606, Japan

Current Topics in Neuroendocrinology, Vol. 11

1 Introduction

Forty years ago, Harris (1948) proposed that the secretion of anterior pituitary hormones was regulated by factors synthesized in hypothalamic neurons, released into the hypophyseal-pituitary portal circulation and carried to the anterior pituitary, where they stimulated the secretion of their specific tropic or growth-related hormones. This remarkable concept was demonstrated regarding hypothalamic regulation of pituitary hormone secretion. During the past 20 years five hypophysiotropic hormones have been isolated and definitely characterized; thyrotropin-releasing hormone (TRH); luteotropin(folliculotropin)-releasing hormone (LH(FSH)RH) or gonadotropin-releasing hormone (GnRH); growth hormone (GH) release inhibiting hormone (GHIH) or somatotropin release-inhibiting factor (SRIF; somatostatin); corticotropin-releasing hormone (CRH); and GH-releasing hormone (GHRH). The factors which stimulate or inhibit prolactin (PRL) release have not yet been definitely identified. Although these hypophysiotropic hormones are called releasing hormones because of the rapidity with which they increase plasma levels of pituitary hormones, there is evidence that they also stimulate synthesis of pituitary hormones.

The development of molecular cloning techniques has enabled the sequencing of cloned DNA complementary to the mRNA coding for hypophysiotropic as well as pituitary hormones, and the subsequent identification of the genes ultimately directing their synthesis. The molecular biology of both the hypophysiotropic and pituitary hormones opens new vistas for studying tissue-specific expression and their complex neurohormonal regulation. In this chapter, we shall focus on recent findings on the hypophysiotropic regulation of the gene expression of anterior and intermediate pituitary hormones.

1.1 Anterior and Intermediate Pituitary Hormones

The pituitary is divided into the anterior lobe (adenohypophysis), the intermediate lobe (pars intermedia), and the posterior pituitary (neurohypophysis). The intermediate lobe is virtually missing from the human pituitary after birth, although present in most vertebrates. There are six well-recognized anterior pituitary hormones whose structures and functions have been characterized. The hormones are polypeptide in nature and can be divided into three general categories, each of which exhibits unique characteristics: ACTH and related peptides, the somatomammotropic hormones (GH, PRL), and the glycoprotein hormones (LH, FSH, and thyrotropin, TSH).

1.1.1 ACTH and Related Peptides

Corticotropin (adrenocorticotropic hormone, ACTH) is a 39-amino-acid, single-chain peptide. ACTH is one of a number of peptides derived from a common precursor molecule, proopiomelanocortin (POMC), which is a glycosylated protein of approximately 30 000 Da (Nakanishi et al. 1979). While there is only a single copy of functional POMC gene per haploid genome, the mode of gene expression varies among different tissues (Imura et al. 1982). Differential processing of POMC occurs in various cell types, resulting in different spectra of peptides; ACTH, β-lipotropin (LPH), β-endorphin, and amino-terminal peptide (big γ-melanotropin, γ-MSH) in anterior pituitary; and α-MSH, corticotropin-like intermediate lobe peptide (CLIP), acetyl-β-endorphin, and γ-MSH in intermediate lobe pituitary.

1.1.2 Somatomammotropic Hormones

The somatomammotropic hormones include two pituitary hormones, GH and PRL, and one hormone of placental origin, placental lactogen (PL) or chorionic somatomammotropin (CS), each consisting of a single peptide chain with intrachain disulfide linkages. The somatomammotropic hormones are believed to have evolved by gene duplication (Moore et al. 1982). While GH and PRL have only limited homology in amino acid sequences, there is extensive interspecies similarity of both GH and PRL, suggesting relatively small changes after gene duplication during vertebrate evolution.

1.1.3 Glycoprotein Hormones

The glycoprotein hormones of the pituitary include TSH, LH, and FSH. Also included in this family of hormones is placental chorionic gonadotropin (CG), which has similarities to LH both in structure and biological activities. The glycoprotein hormones are composed of two subunits, α and β, each consisting of a peptide core with branched carbohydrate side chains which contribute from 15% to 30% of the molecular weight (Pierce and Parsons 1981).

1.2 Hypophysiotropic Hormones

The term releasing/inhibitory factor is applied to hypothalamic substances of unknown chemical nature, whereas substances with established chemical identity are referred to as releasing/inhibitory hormones. In the past 20 years, five hypophysiotropic hormones, TRH, GnRH, SRIF, CRH, and GHRH, have been isolated and definitely characterized.

Studies with each of these hormones have revealed the rather complex actions of hypophysiotropic hormones. The actions of hypophysiotropic hormones are not limited strictly to the regulation of a single pituitary hormone. For example, TRH is a potent releaser of PRL as well as TSH. The question has therefore been raised as to whether TRH is the physiological PRL-releasing hormone or whether other factors which stimulate PRL release deserve this designation. GnRH releases both LH and FSH. SRIF inhibits the secretion of GH and TSH. The principal inhibitor of PRL secretion, dopamine, also inhibits TSH and gonadotropin secretion. Dopamine also inhibits α-MSH release from the intermediate lobe. Under certain conditions, TRH releases ACTH and GH, SRIF inhibits ACTH release, and dopamine stimulates GH release. The actions of CRH and of GHRH are relatively specific.

Hypophysiotropic hormones also exhibit behavioral effects. These behavioral effects on the central nervous system can be demonstrated even in the hypophysectomized animals, indicating that hypophysiotropic hormones play discrete extrapituitary roles. Additionally, with the development of new techniques for immunochemical identification of hypophysiotropic hormones, materials with immunoreactivities similar to these hormones have been detected not only in the hypophysiotropic area of the hypothalamus but also in other areas throughout the central nervous system, changing the terminology from "hypothalamic hormones" to "hypophysiotropic hormones."

1.2.1 Thyrotropin-Releasing Hormone

Thyrotropin-releasing hormone, a tripeptide, was the first of the hypophysiotropic hormones to be chemically identified, while its mRNA and gene structures were the last to be characterized (Lechan et al. 1986). There is evidence that almost all the TRH in the median eminence (ME) and much of that in the arcuate nucleus originate from neurons in the paraventricular nuclei, although TRH-containing cell bodies are also present in other hypothalamic areas. Extrahypothalamic TRH accounts for as much as 70% of total brain TRH.

1.2.2 Gonadotropin-Releasing Hormone

Gonadotropin-releasing hormone provides the primary hormonal link between the brain and the pituitary-gonadal axis. Immunoreactive GnRH, a decapeptide, has been demonstrated in the hypothalamus of several species. GnRH is present almost exclusively in the ME and organum vasculosum, with lesser concentrations in the arcuate and ventromedial nuclei.

There is still lingering controversy as to the existence of a distinct FSH-releasing factor. Most available evidence, however, indicates that the differential effect of GnRH on FSH and LH release can be explained by variation in dose, time course, and pulsatile frequencies of GnRH administration, and also by interactions of gonadal steroid hormones and gonadal peptide hormones at the pituitary level in response to a single LH and FSH release.

1.2.3 Somatostatin

Somatostatin was first isolated from the hypothalamus and characterized as a tetradecapeptide (somatostatin-14). It exhibits profund inhibitory effects on GH release. It also plays a physiological role as a TSH release inhibitory factor. It is now recognized that the originally described somatostatin-14 is a part of an amino-terminal extended somatostatin (somatostatin-28) which, in turn, is contained within a larger molecule, prosomatostatin. Somatostatin-28 is biologically more potent than somatostatin-14 in certain assay systems and is secreted along with somatostatin-14; thus, somatostatin-28 appears to serve as a true hormone and not merely, as was originally suggested, as a prohormone for the smaller form.

The highest levels of SRIF are found in the ME, and appreciable concentrations are found in other hypothalamic areas, as well as in the thalamus, cortex, preoptic area, midbrain, and spinal cord. SRIF also exists in the gastrointestinal tract and the pancreas.

1.2.4 Corticotropin-Releasing Hormone

Although CRH was the first hypophysiotropic factor whose activity was demonstrated in hypothalamic extracts, it was not until 1981 that the chemical structure of the ovine CRH was established (Vale et al. 1981). CRH thus characterized was larger than the other previously reported hypophysiotropic factors, being a straight-chain peptide of 41 amino acids. The majority of CRH-containing cell bodies have been identified in the paraventricular nucleus of several species. This nucleus contains subpopulations of cells, magnocellular neurons, in which immunoreactive CRH and vasopressin are colocalized.

1.2.5 GH-Releasing Hormone

Isolation of a peptide with GH-releasing activity has recently been reported. Unlike the other hypophysiotropic peptides, this peptide was first isolated and characterized not from the hypothalamus, but from pancreatic tumors associated with acromegaly in two patients, both of whom were successfully treated by removal of the pancreatic tumor.

Immunochemical studies indicate that antisera against GHRH stain neuronal cell bodies in the arcuate nucleus of the human hypothalamus, with fibers projecting to the ME. Cell bodies are also seen in the ventromedial nucleus, which had previously been reported as a center for GHRH secretion. Thus GHRH, in contrast to SRIF, exhibits a rather restricted localization in the hypophysiotropic area of the hypothalamus.

1.3 Biosynthesis of Peptide Hormones

Because pituitary hormones are peptide hormones which are secretory proteins, their biosynthesis and secretion occur via the same processes as nonhormonal secretory proteins. Transcription of genetic information takes place within the nucleus to form mRNA precursors (heterogeneous nuclear RNA) followed by RNA splicing, which includes RNA cleavage, excision of introns, and rejoining of exons. The RNA is also modified by "capping" at the 5' prime end and by polyadenylation at the 3' prime end. The initial translation products have an amino-terminal signal peptide consisting of about 20 hydrophobic amino acid residues, which is believed to direct the growing polypeptide chain to the endoplasmic reticulum. Peptide hormones are often synthesized as prohormones and are subsequently processed by cleavage and/or chemical modification to form the active molecules. Variations at these biosynthetic steps, such as alternative RNA splicing and differential posttranslational processing, account for biological diversity of the expression of a gene.

2 Molecular Biology of Pituitary POMC

2.1 Structure of POMC

The primary structure of the ACTH-β-LPH precursor, POMC, was elucidated by determining the nucleotide sequence of cloned DNA complementary to the mRNA coding for the precursor (Nakanishi et al. 1979). The amino terminus of β-LPH, located at the carboxyl end of POMC, is connected by the paired basic residues Lys-Arg with the carboxyl terminus of ACTH. In addition to α-MSH present in ACTH, and β-MSH present in β-LPH, a third MSH-like sequence, named γ-MSH, is found in the amino-terminal half of the precursor molecule. Each of the component peptides known or assumed to be biologically active is bounded by paired basic amino acid residues which are considered to be the sites of proteolytic processing (Fig. 1). The processing of POMC has been studied extensively by using a pulse-chase technique applied to mouse ACTH-producing tumor cell line AtT-20 cells (Eipper and Mains 1980).

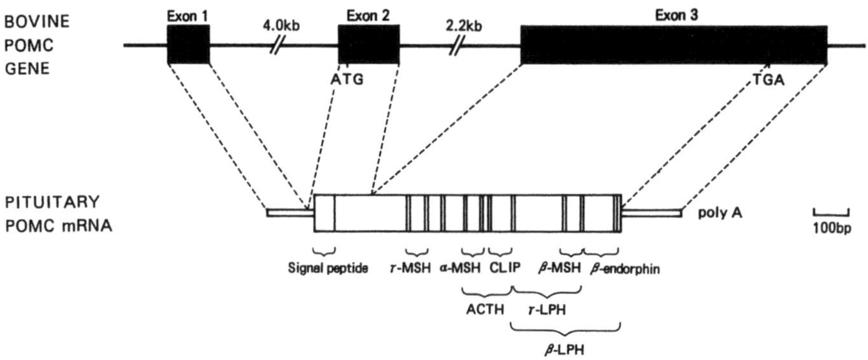

Fig. 1. The structure and expression of the bovine POMC gene is schematically shown (*top*) together with the structure of POMC mRNA present in the pituitary (*bottom*). The initiation codon is indicated by *ATG*, the termination codon by *TGA*. *Thin open bars* indicate 5'- and 3'-untranslated sequences of the POMC mRNA. The position of the major POMC-derived peptides is shown on the pituitary mRNA structure. Pairs of basic amino acid residues cleaved during processing are indicated by *double vertical lines*

2.2 Structure of the POMC Gene

Genomic DNA fragments containing POMC-encoding sequences have been isolated and sequenced from several mammalian species including human (Takahashi et al. 1983), rat (Drouin et al. 1985), cow (Nakanishi et al. 1981), pig (Boileau et al. 1983), and mouse (Notake et al. 1983b; Uhler et al. 1983), and only one functional gene has been identified per haploid genome. The human POMC gene is located on chromosome 2.

The bovine gene is composed of approximately 7.3 kb, with three exons separated by two relatively large introns of approximately 4.0 and 2.2 kb (Fig. 1). The first exon is approximately 100 nucleotides in length and comprises most of the 5'-untranslated portion of the pituitary POMC mRNA. The second exon of about 150 nucleotides contains a small portion of 5'-untranslated sequence and the protein coding portion of the mRNA, including the signal peptide and a portion of the amino-terminal peptide. The third exon encodes all the peptides with known biological activity, including ACTH, β-endorphin, and the melanotropins.

The POMC gene is expressed in a variety of tissues in the mammal, but its main site of expression is the pituitary anterior lobe corticotrophs and intermediate lobe melanotrophs. The POMC molecule is posttranslationally processed to different biologically active peptides in these two cell types, with ACTH, β-LPH, β-endorphin, and big γ-MSH being major products in the corticotroph; and α-MSH, CLIP, and acetyl-β-endorphin, and γ-MSH in the melanotroph (Imura et al. 1982). This array of different biological activities derived from the POMC molecule possibly necessitates the observed complex regulation of POMC peptide secretion and POMC gene expression in the pituitary.

POMC-like mRNAs have been detected in a number of nonpituitary tissues, including the hypothalamus and other portions of the brain, adrenal medulla, spleen macrophages, ovary, and testis (Imura et al. 1982; Lacaze-Masmonteil et al. 1987). The POMC mRNAs of the reproductive tissues as well as the brain and adrenal medulla are smaller (approximately 850 – 1000 bases) than the pituitary mRNA (~1100 bases). These smaller mRNAs are the products of transcripts initiated from the 3' end of the second intron rather than alternative processing of RNA transcripts initiated at the pituitary promoter (Lacaze-Masmonteil et al. 1987; Nakai et al. 1988). Whereas DNA sequences upstream of the transcription initiation site in the pituitary contain the usual promoter elements such as TATA box, no such element can be found upstream of the putative site of transcription initiation in nonpituitary tissues. This region only contains GC-rich elements, which are reminiscent of the promoters of house-keeping genes.

2.3 Regulation of POMC Gene Expression

The secretion of POMC-derived peptides is under hormonal control. In particular, secretion of pituitary POMC peptides is subject to lobe-specific multihormonal regulation (Table 1). The levels of POMC mRNA are usually regulated in the same manner as secretion. In the anterior pituitary, CRH increases both ACTH release and POMC mRNA levels, whereas glucocorticoids inhibit both processes. Another class of neurohormones act on the anterior pituitary corticotroph, including vasopressin, angiotensin II, cholecystokinin C-terminal octapeptide (CCK-8), and α-adrenergic agonists (Fig. 2).

By contrast, in the intermediate lobe, glucocorticoids have minimum effect on POMC transcription and on the secretion of POMC-derived peptides. The intermediate lobe melanotroph is under the negative control of hypothalamic dopamine which acts through the D_2 dopamine receptor. In addition to the inhibitory effects of dopamine on intermediate lobe cells, γ-aminobutyric acidadrenergic (GABAergic) innervation and GABA receptor activation also negatively influence both ACTH release and POMC mRNA levels. β-Adrenergic agonists stimulate POMC gene transcription in the melanotroph.

2.3.1 Regulation of POMC mRNA Levels by CRH

Dallman et al. (1985) have shown the importance of hypothalamic input in maintaining high levels of POMC gene expression in the anterior pituitary. Using Halasz knife lesions to destroy CRH and vasopressin input to the anterior pituitary, they showed that POMC mRNA levels were decreased in lesioned rats relative to those of sham-lesioned rats. Furthermore, by lesioning the para-

Table 1. Summary of inputs to POMC-producing cells

Cell type	Stimulation	Inhibition
Corticotroph	CRH AVP CCK Angiotensin II β-Adrenergic α-Adrenergic	Glucocorticoids
Melanotroph	β-Adrenergic (CRH)	Dopamine GABA
AtT-20 cell	CRH AVP TRH α-Adrenergic β-Adrenergic	Glucocorticoids Dopamine Somatostatin

CRH, corticotropin-releasing hormone; AVP, arginine-8-vasopressin; CCK, cholecystokinin; TRH, thyrotropin-releasing hormone; GABA, γ-aminobutyric acid

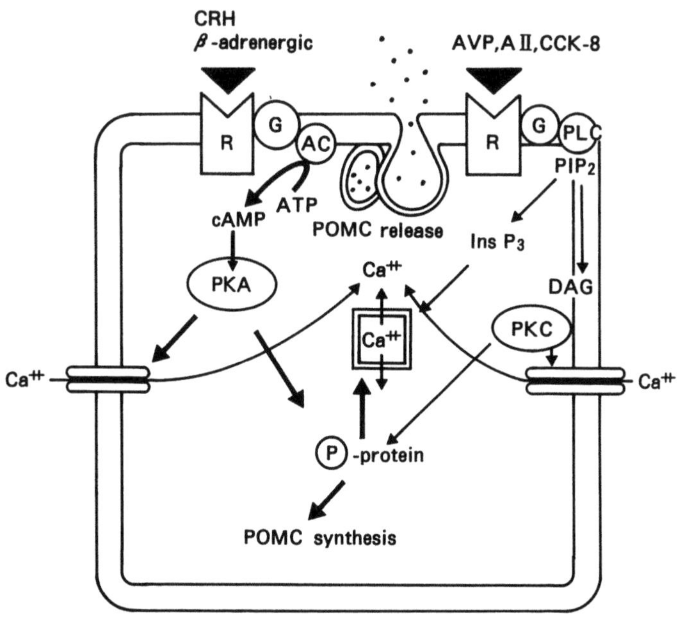

Fig. 2. Schematic representation of the anterior pituitary corticotroph indicating different plasma membrane receptors and intracellular signal transduction. *R*, receptor; *G*, GTP-binding protein; *PKA*, protein kinase A; Ca^{++}, Calcium ion; PIP_2, phosphatidylinositol diphosphate; $InsP_3$, inositol triphosphate; *DAG* diacylglycerol; *PLC*, phospholipase C

ventricular nucleus, a major site of CRH neurons which project to the ME, Bruhn et al. (1984) showed a decrease in the levels of POMC mRNA in the anterior pituitary in adrenalectomized rats.

Several groups have examined the effects of CRH administration on POMC mRNA levels, pharmacologically mimicking chronic hypothalamic input to the pituitary. Bruhn et al. (1984) showed that administration of CRH by long-term intravenous infusion to intact rats caused two- to three fold increase in the levels of POMC mRNA in the anterior pituitary but not intermediate pituitary. Chronic CRH administration to intact animals by subcutaneous implantation of osmotic minipumps results in stimulation of POMC gene expression in the anterior pituitary relative to that in non-CRH-treated animals (Hollt and Haarmann 1984). However, intermediate pituitary POMC mRNA levels in these chronically CRH-treated animals were decreased to about 30% of control animals (Hollt and Haarmann 1984; Lundblad and Roberts 1988), in contrast to the finding of Bruhn et al. (1984). These observations are in sharp contrast to the acute stimulatory effects of CRH on intermediate lobe POMC peptide secretion observed in vivo, and peptide secretion and gene transcription in vitro (Loeffler et al. 1985b).

In addition to its in vivo effects, CRH has been demonstrated to have direct stimulatory effects on ACTH release and POMC mRNA accumulation both in AtT-20 cells (Affolter and Reisine 1985) and on anterior and intermediate pituitary cells in dispersed primary cell cultures (Loeffler et al. 1985b; Dave et al. 1987).

CRH acts through a cytoplasmic membrane receptor coupled through a stimulatory guanine nucleotide binding protein (Gs) to adenylate cyclase and increases intracellular cAMP levels in corticotrophs (Aguilera et al. 1983; Perrin et al. 1986). The addition of 8-bromo-cAMP and phorbol ester stimulated POMC mRNA levels in AtT-20 cells in culture (Affolter and Reisine 1985). The cAMP analogue 8-bromo-cAMP increased POMC mRNA in primary cultures of anterior lobe cells (Simard et al. 1986; Loeffler et al. 1986b) as well as in intermediate lobe cells (Loeffler et al. 1986b). Forskolin, an adenylate cyclase-activating agent, also stimulated POMC mRNA accumulation in both anterior and intermediate lobe cultures (Loeffler et al. 1986b), suggesting a common role for cAMP in regulating expression of the POMC gene in both pituitary cell types in culture.

The effects on POMC mRNA of CRH and other agents that increase cAMP production (forskolin, isoproterenol, and 8-bromo-cAMP), but not phorbol ester, are blocked by fusion of AtT-20 cells with liposomes containing cAMP-dependent protein kinase inhibitor (PKI; the Walsh inhibitor), demonstrating an essential role for cAMP in mediating the CRH effects on ACTH release and POMC gene expression (Reisine et al. 1985a).

Basal levels of POMC gene expression in vitro may be determined by cAMP-dependent processes. Treatment of anterior and intermediate lobe primary cultures with phosphodiesterase inhibitor elevated POMC mRNA levels (Simard et al. 1986; Loeffler et al. 1986b). In addition, basal levels of POMC mRNA were reduced in AtT-20 cells fused with liposomes containing PKI (Reisine et al. 1985a).

2.3.2 Regulation of POMC mRNAs by Another Class of Neuropeptides

Another class of neuropeptides acting on the anterior pituitary corticotroph, including vasopressin, angiotensin II, CCK-8, and α-adrenergic agonists act not through adenylate cyclase-dependent pathways but probably through phosphatidylinositol turnover, with mobilization of intracellular calcium stores and activation of protein kinase C (PKC; Raymond et al. 1985; Abou-Samra et al. 1986; Schoenenberg et al. 1987). These substances alone may not be potent secretagogues but act to potentiate the CRH-stimulated secretion of POMC-derived peptides in vitro (Abou-Samra et al. 1987). In fact, both vasopressin and angiotensin II act to facilitate CRH-stimulated cAMP formation in anterior lobe cultures, possibly through the inhibition of phosphodiesterase and indirect stimulation of adenylate cyclase activity (Abou-Samra et al. 1987), although they do not significantly increase cAMP formation by themselves (Vale et al. 1983).

In cells responsive to a number of secretagogues, the interrelationships between these pathways are probably complex. For example, CRH and other agents that increase intracellular cAMP levels elevate cytosolic calcium levels in AtT-20 cells (Guild et al. 1986; Luini et al. 1985) through the activation of protein kinase A (Guild and Reisine 1987). SRIF blocks the stimulatory effects of CRH and β-adrenergic agonists in AtT-20 cells by preventing cAMP formation (Heisler et al. 1982; Reisine et al. 1985b); but in addition, SRIF can also block POMC peptide release in response to 8-bromo-cAMP; suggesting a mechanism beyond the inhibition of adenylate cyclase. SRIF has been shown to decrease cytosolic calcium current in AtT-20 cells independent of its ability to block cAMP formation (Luini et al. 1986; Reisine and Guild 1985). This inhibition is dependent on the participation of an inhibitory G protein since it is blocked by the addition of pertussis toxin (Reisine et al. 1985b).

Therefore, in a specialized cell like the corticotroph or melanotroph, pathways mediating secretion of peptides and probably expression of the POMC gene operate not in isolation but in a complex coordinated fashion to give an integrated response to complex physiological inputs.

2.3.3 Regulation of POMC mRNAs by Amines and Amino Acids

The β-adrenergic agonist, isoproterenol, was shown to stimulate POMC gene expression in both anterior and intermediate pituitary cultures and was antagonized by propranolol (Lundblad and Roberts 1988). The effect of adrenergic agonists are mediated through activation of cAMP-dependent protein kinase.

Dopamine has been shown to be inhibitory to POMC mRNA in the intermediate lobe, working through an inhibition of the adenylate cyclase system. Treatment of rats with haloperidol, a dopamine antagonist, caused a time-dependent increase in POMC mRNA levels (Hollt et al. 1982; Chen et al. 1983). In several studies, dopamine agonists such as 8-bromo-ergocryptine or bromocryptine (CB-154) were shown to have an inhibitory effect on POMC mRNA levels in the melanotrophs (Chen et al. 1983; Cote et al. 1986). There were essentially no changes in anterior lobe POMC mRNA levels with either haloperidol or bromocriptine treatment (Chen et al. 1983).

Simultaneous addition of bromocriptine and forskolin to rat intermediate lobe elevated POMC mRNA but not to the same extent as forskolin alone (Cote et al. 1986). Cholera toxin stimulated accumulation of POMC mRNA in intermediate lobe cells (Loeffler et al. 1986b). These results, with the findings that pertussis toxin blocks the inhibition by dopamine agonists of the biosynthesis of POMC peptides in intermediate lobe cells, suggest that dopamine and CRH oppositely regulate POMC mRNA in the melanotroph via actions mediated through G proteins on a basal adenylate cyclase activity.

In addition to the inhibitory effects of dopamine on intermediate lobe POMC mRNA, GABAergic innervation and GABA receptor activation have been implicated in negative influence on POMC mRNA levels in vivo in the intermediate but not anterior pituitary (Loeffler et al. 1986a). The decrease in POMC gene expression is also observed in primary cultures of the intermediate pituitary treated with GABA or GABAergic agonists (Loeffler et al. 1986a).

2.3.4 Regulation of POMC mRNAs by Glucocorticoids

Glucocorticoids have major effects on POMC mRNA in the anterior lobe with minimal effects on the neurointermediate lobe. Adrenalectomy increases POMC gene transcription in the anterior pituitary (Nakanishi et al. 1977; Schachter et al. 1982; Birnberg et al. 1983). The treatment of primary cultures of the anterior pituitary or AtT-20 cells with glucocorticoids causes a decrease in the level of POMC mRNA (Nakamura et al. 1978; Roberts et al. 1979; Eberwine et al. 1987). Thus, it appears that at least part of the effect of glucocorticoid treatment in vivo on anterior pituitary mRNA is mediated by direct effects on the corticotrophs. However, several studies suggest that the inhibitory effects of glucocorticoids on anterior pituitary POMC mRNA levels observed in the intact animal are also mediated through some regions at or above the level of the hypothalamus (Bruhn et al. 1984; Dallman et al. 1985; Jingami et al. 1985).

Eberwine et al. (1987) evaluated transcription of the POMC gene in primary culture in response to combinations of the two hormones, CRH and dexamethasone (DEX), under different orders of addition. Pretreatment of corticotroph cultures with DEX for varying periods of time inhibited subsequent CRH-stimulated POMC gene transcription. However, if CRH was given first to the cultures, subsequent glucocorticoid addition had no effect at a time when CRH maximally stimulated POMC gene transcription.

In contrast, glucocorticoids were shown to have essentially no effect on intermediate lobe POMC gene transcription in intact animals. This can be attributed to the fact that under normal physiological conditions the melanotrophs do not express glucocorticoid receptor (GR; Antakly et al. 1985). It has been shown that the GR may be expressed in melanotrophs in vitro after long-term culture (Antakly et al. 1985). In concordance with this finding, Eberwine et al. (1987) demonstrated an inhibitory effect of glucocorticoids on basal POMC gene transcription as well as the inhibition of subsequent CRH-stimulated POMC gene expression in intermediate lobe cells after 3–4 days in cultures.

2.4 Transcriptional Regulatory Sequences of the POMC Gene

Identification of the genomic sequences responsible for regulated transcription of the POMC gene is necessary for the ultimate biochemical characterization of the mediators of hormonal regulation of POMC gene transcription. The regulation of gene transcription is currently considered to be exerted by the specific interaction of regulatory proteins (*trans* factors) with given DNA sequences (*cis* elements). Different types of *cis/trans* coupling are thought to allow both tissue-specific expression and hormonal induction or inhibition of gene transcription. The *cis* regulatory sequences can be located within promoter regions, or may be part of an enhancer that can regulate transcription at any distance inside or outside the gene, upstream or downstream from its promoter. Changes in the three-dimensional structure of DNA allow the DNA-bound *trans* factors to interact with the transcription initiation complex (Mitchell and Tjian 1989).

2.4.1 Promoter Sequences

The first characterization of the POMC promoter by transcription systems in vivo utilized transient expression of the human POMC gene linked to an SV40 vector in COS cells (Mishina et al. 1982). The cloned POMC gene, connected with an SV40 vector and introduced into COS monkey cells, was transcribed from its own promoter. The DNA sequences required for promoter function were identified by using 5'-deletion mutants of the fusion gene that contained the 5'-flanking sequence and capping site of the human POMC gene and the structural sequence of the herpes simplex virus (HSV) thymidine kinase gene. A characteristic feature was that the deletion of the sequence located between 53 and 59 bp upstream of the capping site stimulated the transcription about threefold; the deletion of the TATA box region abolished the transcription. A similar stimulatory effect was observed in the cell-free transcription system derived from AtT-20 cells but hardly at all in the system derived from HeLa cells (Notake et al. 1983a).

2.4.2 Tissue-Specific Elements of the POMC Promoter

Since the POMC gene is expressed at much higher levels in the pituitary than in any other tissues, it is likely that unique DNA sequences of the POMC gene are responsible for this specificity.

The introduction of cloned eukaryotic genes into the cultured cells provides a useful tool to investigate the molecular mechanisms of the regulation of gene expression. A fragment of human genomic DNA containing the entire POMC gene, together with the neomycin-resistant gene (*neo*), was introduced by transfection into the AtT-20 cells and the mouse fibroblast L cells or rat glial C6 cells. In the transformed AtT-20 cells human POMC gene was transcribed correctly and the transcript was spliced faithfully. Deletion analysis demonstrated that no more than 1.6 kb in the 5'-flanking region of the human POMC gene was required for transcription. In the transformed L cells or the transformed C6 cells, however, most of the transcripts of human POMC gene were not correctly initiated. These results suggest the presence of tissue-specific transcription factors or repressors in these tissues (Nakai et al. 1988; Usui et al. 1990).

The in vivo specificity of POMC 5'-flanking sequences was assessed in transgenic mice bearing a POMC-*neo* chimeric gene (Tremblay et al. 1988). The pattern of transgene expression mimicked very closely the expression of endogenous POMC, i.e., the transgene was expressed at a high level in the pituitary and the expression was undetectable in other tissues, except in the testes of some transgenic lines. Testicular expression was not observed in all transgenic lines and, when present, the transgene mRNAs were at least 100 times less abundant than in the pituitary; in addition, these mRNAs had aberrant 5' ends in some lines. It was demonstrated that no more than –760 bp of the POMC gene was required to direct cell-specific expression in both the anterior and intermediate pituitary lobes.

Jeannotte et al. (1987) described the identification of the POMC promoter portions which were necessary for expression of the POMC gene in AtT-20 cells by an antibiotic resistance focus-forming assay using progressive 5' deletions of the rat POMC promoter linked to *neo*. Deletion of sequences between −480 and −323 pb and between −166 and −34 bp decreased activity of the POMC promoter when transfected into AtT-20 cells. Roberts et al. (1987) also found the region of the rat POMC promoter between −478 and −320 bp partially responsible for elevated basal activity of the POMC promoter in AtT-20 cells by transient expression assay. Both groups showed fragments of the POMC 5'-flanking region conferred elevated gene expression in AtT-20 cells when linked to heterologous promoters, such as the HSV thymidine kinase promoter and the Rous sarcoma virus (RSV) promoter. By DNA mobility-shift binding assay and subsequent footprinting analysis, Roberts et al. (1987) showed the region of the POMC promoter between −478 and −320 bp bound a factor present in the nuclear extracts of AtT-20 cells. Purification of this factor and reconstitution of the tissue-specific expression in a cell-free transcription system will confirm the role of this factor in mediating the pituitary-specific regulation of the POMC gene.

2.4.3 Sequences Mediating Responses to Second Messengers

POMC gene transcription is under the control of neurohormones that act at the cell surface, as described above. CRH and other agents that increase intracellular cAMP levels stimulate transcription of the POMC gene in primary culture of anterior lobe cells and in AtT-20 cells.

Roberts et al. (1987) have identified a region within the 5'-flanking region of the rat POMC gene that is responsible for CRH inducibility in AtT-20 cells, using progressive 5' deletions of the rat POMC promoter linked to the chloramphenicol acetyltransferase (CAT) gene. A discrete fragment of the flanking sequence between −320 and −133 bp confers both CRH and forskolin inducibility on the HSV thymidine kinase promoter in AtT-20 cells when placed upstream of this normally nonresponsive promoter. However, this sequence does not share similarity with the cAMP-responsive elements (CREs) described for other genes (Montminy et al. 1986; Tsukada et al. 1987).

To identify the region in the human POMC gene responsible for this regulation, we constructed chimeric genes containing different portions of the 5'-flanking region of the human POMC gene fused to the CAT gene. The transcriptional activity of the fusion genes introduced into the rat glial cell line C6 was assayed by measuring CAT activity in the cell lysate. Forskolin stimulated the expression of POMC-CAT fusion genes containing 2.9 kb of the 5'-flanking sequence. Deletion analysis demonstrated that the region between −417 and −97 bp from the transcriptional start site of the human POMC gene was responsible for regulation by cAMP (Fig. 3). This region was shown to confer cAMP-responsiveness when placed upstream of a heterologous viral promoter (Usui et al. 1989).

The DNA enhancer elements responsible for regulation through the cAMP/protein kinase A signaling pathway, CREs, consist of core octamer motifs, 5'-TGACGTCA-3' (Montminy et al. 1986; Tsukada et al. 1987). Two regions with 75% sequence homologies to this motif were found around −340 and −320 bp upstream from the transcriptional start site of the human POMC gene (Fig. 3).

A

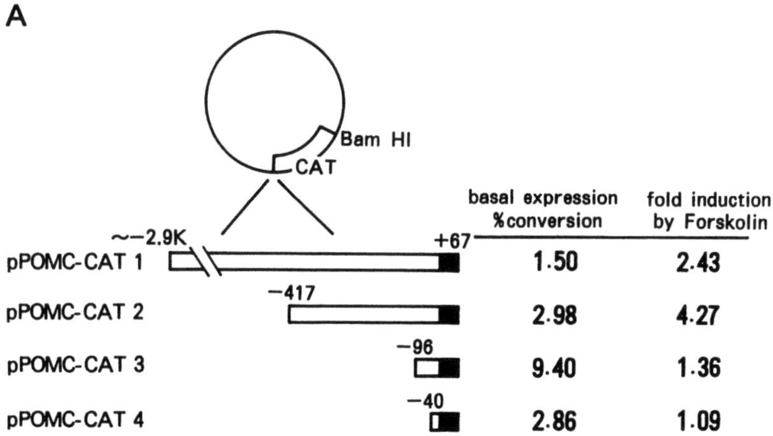

B

```
                                        CCCACGACTTCCACATCACAGTTTCCAAACAATGGGGAAATCGGAGGCCTCCCCGTGTGCAGAC -401
GGTGATATTTACCGCCAAATGCGAACCAGGCAGATGCCAGCCCCAGCACGCACGCAGG[TAACTTCA]CCCTCGCCTCAA[CGACCTCA]GAGGCTGCCCGGCC -301
TGCCCCACACGGGGGTGCTAAGCCTCCCGCCCGTTCTAAGCGGAGACCCAACGCCATCCATAATTAAGTTCTTCCTGAGGGCGAGCGGCCAGGTGCGCCT -201
TCGGCAGGACAGTGCTAATTCCAGCCCCTTTCCAGCGCGTCTCCCCGCGCTCGTCCCCCGTCTGGAAGCCCCCCTCCCACGCCCCGCGGCCCCCCTTCCC -101
CTGGCCCGGGGAGCTGCTCCTTGTGCTGCCGGGAAGGTCAAAGTCCCGCGCCCACCAGGAGAGCTCGGCAAGTATATAAGGACAGAGGAGCGCGGGACCA   -1
                                                                         ------
```

Fig. 3 A, B Effect of forskolin on POMC-CAT fusion genes in C6 cells. **A** Schematic representation of the pPOMC-CAT1–4 and calculated basal and stimulated CAT expression. **B** Nucleotide sequence of the 5'-flanking region involved in human POMC gene responsible for cAMP regulation. The TATA box is indicated by *short dashes*. Two 75% homologous sequences to the 8 bp motif of the cAMP-regulatory element are indicated by *boxes*

2.4.4 Glucocorticoid Regulatory Sequences

Glucocorticoid induction of transcription is mediated by direct GR contact with DNA within glucocorticoid response elements (GREs; Beato 1989). On the other hand, glucocorticoids have a negative influence on mRNA levels for POMC as well as PRL, the α subunit of glycoprotein hormones, and a number of protooncogenes (Beato 1989). The cellular mechanisms mediating this inhibition have not yet been elucidated. The POMC gene offers an interesting model for negative regulation of gene transcription by glucocorticoids.

Israel and Cohen (1985) demonstrated glucocorticoid hormone regulation of a transfected POMC promoter using a fragment of human gene extending from −670 to −38 bp fused to the HSV promoter; the glucocorticoid-responsive sequene lies between −670 and −38 bp in the human POMC 5'-flanking region. Drouin et al. (1989b) demonstrated, using DNA-mediated gene transfer into transgenic mice and tissue culture cells, glucocorticoid regulatory elements were located between positions −706 and +63 bp in the 5'-flanking region of the rat POMC gene. Multiple regulatory elements that bind nuclear proteins are present within this region. In particular, a sequence that binds the GR and behaves as a negative GRE (nGRE) also binds nuclear proteins of the chicken ovalbumin upstream promoter (COUP) family of transcription factors. Thus glucocorticoid repression of POMC transcription may result from the mutually exclusive binding of the GR and the COUP transcription factor to the POMC nGRE (Drouin et al. 1989a).

In order to localize the POMC promoter sequences responsible for glucocorticoid inhibition of transcription, different portions of the 5'-flanking region of the human POMC gene were fused to the CAT gene and stably transfected into AtT-20 cells and L cells, together with the *neo* gene. Deletion analysis demonstrated that no more than 417 bp in the 5'-flanking region of the human POMC gene are required for transcriptional repression by glucocorticoid. This region was also responsible for the transcription induction of human POMC gene by cAMP (Usui et al. 1989). The addition of DEX to the transformed L cells did not significantly affect the contents of human POMC mRNA, although these cells expressed GR. However, the increase of the transcripts induced by forskolin was partially but significantly suppressed by DEX in the transformed L cells. These results suggest that binding of GR to the nGRE could lead to steric occlusion of positive transcription factors, such as cAMP response element binding (CREB) protein and tissue-specific factors, or that GR which is bound to nGRE could interact with DNA-bound positive factors in such a way as to prevent their transcriptional stimulatory activity.

3 Molecular Biology of the Pituitary Somatomammotropic Hormones

3.1 Structure of Somatomammotropic Hormones

The somatomammotropic hormones include two pituitary hormones, GH and PRL, and one hormone of placental origin, placental lactogen (PL) or chorionic somatomammotropin (CS), each consisting of a single peptide chain with intrachain disulfide linkages (Niall et al. 1973). There is extensive interspecies similarity of both GH and PRL, suggesting relatively limited changes after gene duplication during vertebrate evolution. Despite this similarity, subprimate GHs are biologically inactive in humans. There is considerable homology between the primary structures of GH and CS. In contrast, there is only 16% identity of amino acid residues between GH and PRL and 13% identity between PRL and CS. GH and CS each contain two disulfide bonds, whereas PRL, which is seven amino acids longer than the other two, contains three. Despite the differences in their structures each of the three hormones has intrinsic lactogenic and growth-promoting activity.

3.2 Structure of the Somatomammotropic Hormone Gene Family

The genes coding for GH and CS are related and form part of a family which also comprises PRL (Miller and Eberhardt 1983). In rat (r) there is a single rGH gene (Barta et al. 1981), a single rPRL gene (Cooke and Baxter 1982), and several rCS genes (Duckworth et al. 1986). In man (h), while the single hPRL gene (Truong et al. 1984) is located on chromosome 6 (Owerbach et al. 1981), there is a cluster of related hGH and hCS genes grouped together on chromosome 17 (Owerbach et al. 1980; Hirt et al. 1987).

Only one of the hGH genes (hGH-N) is expressed in the pituitary and is responsible for directing hGH synthesis (Miller and Eberhardt 1983). Alternative splicings of the hGH-N nuclear pre-mRNA have been shown to generate multiple mRNAs, one of which leads to the production of 20-kDa variant protein which constitutes 5%–10% of pituitary GH (DeNoto et al. 1981). The other (hGH-V), while present in hGH-deficient patients who lack the hGH-N gene, is apparently incapable of substituting for the hGH-N gene even though it can be expressed under certain experimental conditions and also in the placenta (Frankenne et al. 1987).

Two of the five hCS genes appear to be expressed in syncytiotrophoblast tissue, where they lead to the production of hCS with identical amino acid sequences (Barrera-Saldana et al. 1983).

PRL gene is expressed mainly in the pituitary, but PRL mRNA has been detected in human decidual cells. The human decidual PRL cDNA has been cloned and the sequence found to be almost identical to that of the pituitary cDNA, confirming that there is only a single copy of the hPRL gene per haploid genome (Takahashi et al. 1984).

3.3 Regulation of Somatomammotropic Hormone Gene Expression

The regulation of GH and PRL gene transcription involves two aspects: tissue-specific expression and hormonal regulation. A general approach to understanding these mechanisms should involve the identification of *cis/trans* interactions and purification and cloning of *trans*-acting factors and assessment of their regulation. Such an approach has been used with great success in the case of GH and PRL genes, with the identification of tissue-specific factor Pit-1/GHF-1, and also the definition of steroid and thyroid hormone receptor structure and DNA binding characteristics.

The cell-type-specific characteristics of GH and PRL gene regulation appear to be very similar, and the mechanisms may be partly mediated by a common factor for both genes. However, within the pituitary their hormonal regulation differs markedly. A large number of hormonal signals may act on given cell types, and in principle these might act either independently or by modulating tissue-specific control mechanisms.

3.3.1 Tissue-Specific Expression

A pituitary-specific *trans*-acting factor(s) has recently been identified which appears to regulate tissue-specific expression of the GH and PRL genes and possibly also the TSH β-subunit gene (Nelson et al. 1988). This factor, GHF-1 or Pit-1, has subsequently been cloned and identified by two independent groups (Bodner et al. 1988; Ingraham et al. 1988). Whether this factor regulates both GH and PRL has been disputed (Castrillo et al. 1989), but evidence that some similar tissue-specific targeting mechanism operates in vivo is provided by the finding that upstream elements of the rGH gene can direct expression of a linked hGH structural gene in lactotrophs and thyrotrophs as well as in somatotrophs of transgenic mice (Lira et al. 1988).

3.3.1.1 GH Gene

Transfection studies using rGH promoter-CAT fusion genes have shown that as little as 235 bp of rat GH 5'-flanking sequence can confer cell-type specificity (Nelson et al. 1986). Transfection studies using much longer stretches of rGH promoter and 5'-flanking sequences (1800 bp) confirmed that CAT expression was restricted to pituitary cells, but this restriction of specificity was lost when 5' sequences were deleted down to the −309 or −183 bp positions, suggesting the presence of upstream repressor elements which normally prevented gene activation in nonpituitary cell types (Larsen et al. 1986b). These data conflict with those of Nelson et al. (1986), suggesting that tissue specificity is not conferred solely by 235 bp of 5'-flanking sequences.

Localization of protein-binding sites within rGH 5'-flanking DNA has been undertaken with footprinting studies. The exact details vary slightly between different reports, but two consensus footprints may be deduced between positions −98 and −62 bp (GC1), and −140 and −106 bp (GC2) (Catanzaro et al. 1987;

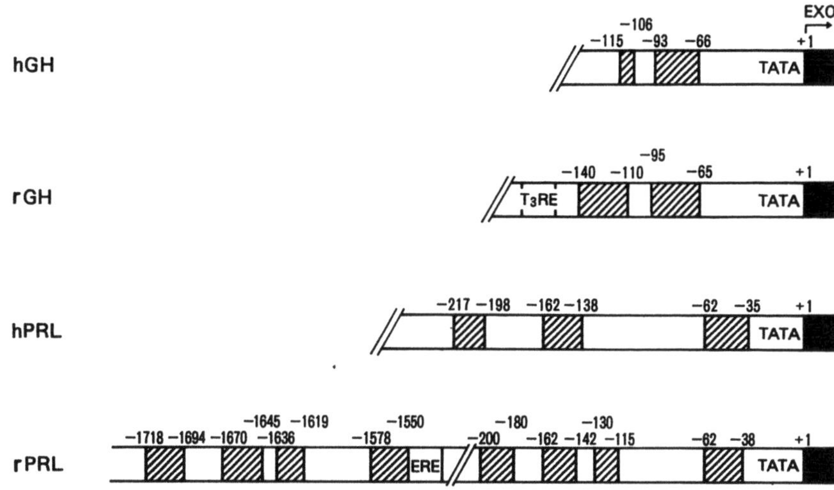

Fig. 4. Tissue-specific *cis* elements of GH and PRL genes. Schematic diagram of the cell-specific elements in the human and rat PRL and GH genes required for cell-specific gene expression. Proposed binding sites for Pit-1 are shown as *hatched boxes*, and ERE and thyroid hormone response element (T_3RE) are shown as *open boxes*

West et al. 1987; Guerin and Moore 1988; Ye et al. 1988) (Fig. 4). A site at −220/−241 (GC3) has recently been shown to be pituitary-specific but binds a different protein factor from the GC1 and GC2 sites (Guerin and Moore 1988).

Transfection studies of the hGH gene showed the presence of two pituitary-specific protein-binding sites, which mutually competed for binding of a single protein factor, GHF-1 (Lefevre et al. 1987). They contain regions highly homologous to the GC1 and GC2 sites described for the rGH gene promoter and have similar locations (Catanzaro et al. 1987; West et al. 1987; Fig. 4).

3.3.1.2 PRL Gene

The function of 5'-flanking DNA from the rPRL gene has been assessed by transfection studies. There are two distinct upstream enhancer regions, the distal region located between −1831 and −1530 bp upstream of the transcription initiation site accounted for 98%–99% of basal tissue-specific expression and the proximal one between −422 and −36 bp upstream accounted for 1%–2% of activity (Nelson et al. 1986; Fig. 4).

However, in vitro transcription experiments in GH_3 cell extracts suggest that only the proximal regulatory elements were necessary for tissue-specific expression of the rPRL gene (Gutierrez-Hartmann et al. 1987). Different studies and different techniques have yielded slightly different results, but up to four protein-binding sites have been identified within the first 210 bp 5' to the cap site

(Gutierrez-Hartmann et al. 1987; Nelson et al. 1988; Schuster et al. 1988; Fig. 4). At least three of these proximal sites are pituitary specific, as the protein binding is seen only with pituitary cell extracts and not with proteins from other tissues. Using DNase I footprinting, Nelson et al. (1988) found that, in addition to the four proximal protein-binding sites described above, there are four distal sites between positions −1579 and −1718 bp in the rPRL gene (Fig. 4).

The tissue-specific regulation of the hPRL gene has been studied recently, but the organization of at least the proximal element of the 5'-flanking region appears to be similar to the rPRL gene. DNase I footprinting studies have indicated pituitary-specific *trans* factor binding to three sites, corresponding to the highly homologous sites 1, 3, and 4 in the rPRL gene (Lemaigre et al. 1989; Fig. 4).

3.3.1.3 Trans-Acting Factor Pit-1/GHF-1

Two sites within the rat and human GH promoter DNA are homologous and appear to bind the same protein factor GHF-1 (Fig. 4). Eight sites upstream of the rPRL gene and at least three sites for the hPRL gene also appear to bind a single *trans*-acting factor Pit-1 (Fig. 4). Thus it appears that a single protein Pit-1/GHF-1 may be responsible for directing expression of both GH and PRL genes and may also bind to the promoter of the TSH-β gene (Nelson et al. 1988). The cDNA encoding Pit-1/GHF-1 has recently been cloned from the two groups (Bodner et al. 1988; Ingraham et al. 1988). However, the two reports disagree on the crucial point as to whether Pit-1/GHF-1 protein can bind to the rPRL gene (Castrillo et al. 1989; Ingraham et al. 1988, 1990).

Pit-1/GHF-1 contains a DNA-binding region closely similar to the homeodomain first described in a class of proteins involved in embryonic development in *Drosophila* (Ingraham et al. 1990; Fig. 4). Recently four new members of the POU-domain gene family were identified; three from brain cDNA (Brn-1, Brn-2, and Brn-3) and one from rat testes cDNA (Tst-1; Ingraham et al. 1990).

3.3.2 Hormonal and Intracellular Regulation of GH Gene Expression

The cell-type-specific characteristics of GH and PRL gene regulation appear to be very similar, and the mechanisms may indeed be partly mediated by a common factor for both genes, as described above. However, within the pituitary their hormonal regulation differs markedly. A large number of hormonal signals may act on given cell types and these might act either independently or by modulating tissue-specific control mechanisms. Steroid hormones and thyroid hormones exert their effects by binding to specific nuclear receptors which have been shown to be DNA-binding proteins, and much of our present understanding of DNA-receptor interactions derives from recent work on nuclear receptor structure (Evans 1988; O'Malley 1990). However, peptide hormones and amines, which bind to membrane receptors, must act through an intermediate intracellular signaling system via second messengers (Roesler et al. 1988; Kikkawa et al. 1989).

GH gene transcription is stimulated or inhibited by a variety of hormones (Table 2). The best studied of these regulators is thyroid hormone (Samuels et al.

1988), which is well known to be a major regulator of GH secretion in vivo in the rat. The hypophysiotropic peptide hormones, GHRH and SRIF, have so far received less attention.

Table 2. Summary of known inputs to GH- or PRL-producing cells

	PRL	rGH	hGH
GHRH		+	+
TRH	+	−	
VIP	+		
SRIF		−	−
IGFI		−	−
Dopamine	−		
GABA	−		
Estrogen	+		
Thyroid hormone	−	+	−
Glucocorticoids	−	+	+
Vitamin D	+		

GHRH, growth hormone-releasing hormone; TRH, thyrotropin releasing hormone; VIP, vasoactive intestinal peptide; IGFI, insulin-like growth factor I; SRIF, somatostatin; GABA, γ-aminobutyric acid; +, stimulation; −, inhibition

3.3.2.1 Peptide Hormone Regulation and Intracellular Messenger Systems

Hypothalamic GHRH stimulates GH gene transcription in primary cell cultures (Barinaga et al. 1983, 1985; Gick et al. 1984) and also in transgenic mice overexpressing the human GHRH precursor gene (Hammer et al. 1985). GHRH binds to membrane receptors on the pituitary somatotrophs. Most evidence suggests that these are coupled directly to the stimulatory GTP binding protein (Gs), which in turn activates the catalytic subunit of adenylate cyclase to form cAMP (Fig. 5). Calcium influx also independently occurs and may be further enhanced by cAMP-mediated phosphorylation of calcium channels (Frohman and Jansson 1986). GHRH thus exerts its effects on somatotrophs through stimulation of adenylate cyclase, and its effects on rGH gene transcription may be mimicked by adenylate cyclase activation and exogenous cAMP (Clayton et al. 1986), but not by activation of protein kinase C or cell depolarization, which stimulates only secretion of GH (Barinaga et al. 1985; Morita et al. 1987).
Transcription of the rGH gene may be modulated independently of GH release. GHRH stimulation of GH release is highly dependent on extracellular calcium, whereas both basal and GHRH-stimulated GH gene transcriptions are much less so (Barinaga et al. 1985). Thus, while the release of GH may be mediated by calcium influx, cAMP-dependent protein kinase may stimulate GH gene transcription. Similarly, while SRIF inhibits GHRH-stimulated GH release via activation of the Gi protein (Fig. 5), it has minor effects on cAMP generation and GH gene transcription (Lamberts 1988). SRIF may reduce rGH mRNA accumu-

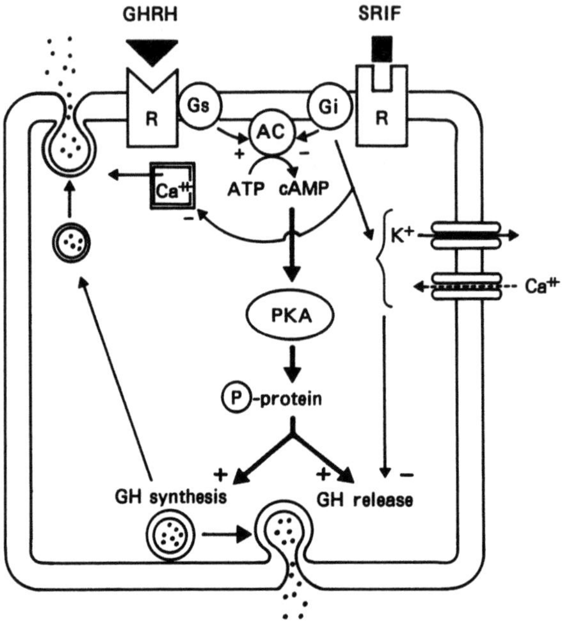

Fig. 5. Schematic representation of GHRH- and SRIF-receptor signal transduction. The receptors of both hypophysiotropic hormones are coupled to stimulatory (*Gs*, in the case of GHRH) or inhibitory (*Gi*, in the case of SRIF) G proteins, which regulate the activity of the adenylate cyclase (*AC*). Formation of cAMP and subsequent protein phosphorylation modulate GH synthesis and secretion. SRIF may have other direct inhibitory effects on calcium influx or calcium-mediated exocytosis

lation (Wood et al. 1987a), although a separate study failed to show short-term effects on transcriptional rate (Barinaga et al. 1985). Ectopic hypersecretion of GH by transplanted tumors in the rat has been shown to inhibit GH mRNA accumulation in vivo and this autoregulation probably involves negative feedback from insulin-like growth factor I (IGF-I; Yamashita et al. 1986a).

Little work has been reported on the effects of peptide hormones on the hGH gene. Davis et al. (1988) have shown the stimulation of endogenous hGH mRNA levels and GH release by GHRH and by cAMP, whereas stimulation of protein kinase C caused only hGH release without affecting hGH mRNA levels. Brent et al. (1988) showed that the hGH promoter was cAMP-responsive. Yamashita et al. (1986b) showed suppression of hGH mRNA levels by IGF-I using cultured cells from five human pituitary adenomas. IGF-I also prevented GHRH-induced stimulation of GH mRNA levels.

3.3.2.2 Regulation of GH mRNAs by Thyroid Hormone

Rat GH gene transcription is induced by triiodothyronine (T_3) in the pituitary GC cell line (Yaffe and Samuels 1984). This induction coincides with T_3 receptor occupancy and alterations in chromatin structure (Nyborg and Spindler 1986). The transfection studies with staged deletions of 5' and 3' regions in rGH fragments mapped the *cis* elements involved in this T_3 regulation to position −194/ −169 (Wight et al. 1988) corresponding closely to a T_3 receptor footprint described by Koenig et al. (1987) and which contains part of a DNA sequence that binds the c-*erbA* protooncogene translation product (Glass et al. 1987).

The data obtained in another series of studies are slightly different from the results described above (Larsen et al. 1986a; Flug et al. 1987). Larsen et al. (1986a) found that sequences between positions −183 and −202 bp were able to confer thyroid hormone responsiveness, but only in pituitary cells, indicating that a thyroid hormone response element (TRE) itself may have tissue-specific characteristics. Ye et al. (1988) proposed that two interacting tissue-specific *cis* elements (−317/−107 and −95/−65) are functionally linked with the T_3 response element to act as an enhancer-like unit which confers both cell-type-specific and T_3-regulated gene expression.

Gel retardation and DNase I footprinting studies have shown that partially purified T_3-receptor preparations show binding to upstream regions of rGH gene and that there may be multiple sites which can interact with receptor within the region −530/+7 (Ye and Sammuels 1987; Lavin et al. 1988).

Analysis of stable transformants obtained by transfection of the hGH gene with 500 bp of promoter sequences into GH_3 cells (Cattini et al. 1986) has shown that T_3 inhibits hGH gene transcription while stimulating that of the rGH gene. However, this T_3 inhibition is not clearly linked to the hGH promoter as it could not be seen in transient expression studies using the CAT reporter gene (Cattini and Eberhardt 1987; Brent et al. 1988). Although a T_3 response element has not yet been precisely identified in the human gene, T_3 receptor from lymphoblastic IM-9 cell nuclear extracts is able to bind between positions −129 and −290 bp in the 5'-flanking region of the gene (Barlow et al. 1986).

3.3.2.3 Regulation of GH mRNAs by Steroid Hormones

Transcription of the rGH gene is induced by glucocorticoid in the GH cell line (Evans et al. 1982). A rGH promoter region as short as 248 bp was able to confer glucocorticoid induction to a reporter gene in heterologous cells (Heiser and Eckhart 1985), but in transient transfection analysis in pituitary cells, Flug et al. (1987) found only minimal glucocorticoid effects. Birnbaum and Baxter (1986) found that a promoterless rGH gene was glucocorticoid inducible, with mRNA of a nearly correct size, suggesting that GREs were contained downstream, not upstream, from the transcription initiation site. Glucocorticoid stimulates expression of the transfected hGH gene (Cattini et al. 1986; Brent et al. 1988) similar to their effect on the rGH gene, and GRE has been demonstrated within the first intron of the gene, downstream of the transcription initiation site (Moor et al. 1985; Slater et al. 1985). Although such a site has not yet been shown in the rGH gene, its first intron does contain a TGTCCT sequence typical of GR binding sites. A weaker GR binding site is found within the promoter region (Slater et al. 1985).

3.3.3 Hormonal and Intracellular Regulation of PRL Gene Expression

Although many neural stimuli can induce PRL release, secretion is tonically inhibited by the hypothalamus. The number of probable hormonal regulators of the PRL gene is known. For some of these, hormone-responsive elements have been identified in the 5'-flanking region of the rPRL gene, but few studies have yet characterized the regulation of the hPRL gene.

3.3.3.1 Regulation by Peptides, Amines, and Amino Acids, and Intracellular Messenger Pathways

Thyrotropin-releasing hormone stimulates rPRL gene transcription and increases cytoplasmic mRNA accumulation (Laverriere et al. 1983; Murdoch et al. 1983; White and Bancroft 1983). In addition, TRH increases the half-life of cytoplasmic mRNA (Laverriere et al. 1983). Similar stimulation is seen with epidermal growth factor (EGF; White and Bancroft 1983; Murdoch et al. 1985b). Vasoactive intestinal peptide (VIP) also elevates PRL mRNA levels in pituitary cells (Carrillo et al. 1985). Dopamine inhibits PRL gene transcription (Mauer 1981) and GABA also has an inhibitory effect on PRL mRNA accumulation (Loeffler et al. 1985a).

These various hormones are thought to act through different intracellular mechanisms. TRH stimulates the hydrolysis of membrane phospholipids resulting in PKC activation and calcium mobilization, and leading to phosphorylation of cytoplasmic and nuclear proteins (Gershengorn 1986).

The rPRL gene is likely to be controlled by a number of interacting intracellular messenger systems (Bancroft et al. 1985; Murdoch et al. 1985a). cAMP has been shown to increase rPRL gene transcription (Maurer 1981) and forskolin increases cytoplasmic PRL mRNA accumulation (Murdoch et al. 1985b). Day and Maurer (1989) have recently shown that a distal element (−1713/−1495) conferred regulation by TRH, EGF, and cAMP to a reporter gene.

Intracellular calcium appears to be crucial for PRL gene expression (Hinkle et al. 1988; Laverriere et al. 1988). Studies with calmodulin antagonist drugs suggest that PRL mRNA accumulation is calmodulin dependent (Bancroft et al. 1985; Murdoch et al. 1985b; White 1985). Recent evidence from transfection studies has suggested that a calcium regulatory element may lie within the first 174 bases of the 5'-flanking DNA (Jackson and Bancroft 1988).

PRL gene transcription is stimulated by activation of PKC with phorbol esters (Murdoch et al. 1985b). Studies of rPRL gene 5'-flanking DNA have demonstrated an upstream phorbol ester-responsive element (Supowit et al. 1984). This coincides exactly with the EGF-responsive element, and lies between −78 and −35 bp upstream from the transcription start site (Elsholtz et al. 1986). This region of DNA forms part of a putative calcium regulatory element and a site of DNA topoisomerase action (Jackson and Bancroft 1988; White and Preston 1988). This region also represents the first proximal binding site of the tissue-specific factor Pit-1/GHF-1 (Nelson et al. 1988).

In summary, responses to various intracellular messengers seem to involve proximal 5'-flanking sequences in the rPRL gene and the relevant response elements may coincide with binding site for a tissue-specific protein factor.

3.3.3.2 Regulation of PRL mRNAs by Thyroid Hormone and Vitamin D

T_3 causes inhibition of PRL mRNA accumulation in rat pituitary cells (Maurer 1982). Vitamin D has been shown to stimulate cytoplasmic accumulation of PRL mRNA and this stimulation is antagonized by glucocorticoids (Wark and Gurtler 1986). Nuclear receptors for T_3 and vitamin D may interact with DNA or other *trans* factors in a manner similar to estrogen receptor (ER) and GR, and indeed it has been found that T_3 receptor may bind to estrogen-responsive element (ERE) in a transcriptionally inactive form, to decrease overall level of gene transcription (Glass et al. 1988).

3.3.3.3 Regulation of PRL mRNAs by Steroid Hormones

Physiological effects of estrogen on PRL mRNA accumulation in vivo have been clearly shown by in situ hybridization (Steel et al. 1988). In vitro, estradiol-17β causes increased accumulation of PRL mRNA in rat pituitary cells due to a rapid stimulation of the rate of gene transcription. The stimulatory effect of estrogen is exerted by direct binding of ER to DNA, and ERE in rPRL 5'-flanking DNA has been located more than 1500 bp distant from the transcription initiation site (Mauer and Notides 1987; Somasekhar and Gorski 1988). However, it was found that an inhibitory effect of estrogen on the proximal region of the rPRL promoter could be expressed by inducing overexpression of the ER. It is interesting to note that the ERE is immediately adjacent to one of the four distal Pit-1 binding sites described by Nelson et al. (1988), suggesting possible interactions between ER and a separate tissue-specific protein (Kim et al. 1988a).

DEX inhibits bovine PRL gene transcription by an action exerted within the first 250 bp of 5'-flanking DNA (Camper et al. 1985). DEX has a similar effect on proximal promoter elements of the rPRL gene (Adler et al. 1988; Somasekhar and Gorski 1988). Interestingly, it was shown that an inhibitory effect of the GR and the overexpressed ER was not exerted by their DNA-binding domain (Adler et al. 1988). They suggest that repression is brought about by steroid-activated protein-protein interaction between the receptor and Pit-1, which prevents Pit-1 from binding DNA in a manner similar to NF-*K* B inhibition. Thus, the stimulatory and inhibitory effects of these receptors were exerted by different parts of the molecules and the negative effects did not involve direct DNA-protein interaction but most probably indirect effects on other DNA-binding proteins. Various protein-protein interactions have recently been proposed for negative GR action on other genes (Goodbourn 1990).

4 Molecular Biology of the Pituitary Glycoprotein Hormones

4.1 Structure

The glycoprotein hormones of the pituitary include TSH, LH, and FSH. A placental CG with both structural and biological similarities to LH is included in this class of hormones. The glycoprotein hormones are composed of two subunits, α and β, each consisting of a peptide core with branched carbohydrate side chains, the presence of which is necessary for the hormones' biological activity (Pierce and Parsons 1981). Within a single species, the α subunits of the glycoprotein hormones are identical or nearly identical, whereas the β subunits vary, providing the biological specificity to each hormone. The linear peptides of the glycoprotein subunits are synthesized individually, and the addition of the carbohydrates appears to occur following completion of the peptide chain synthesis. Since the concentration of the free α subunit in the pituitary greatly exceeds that of the free β subunit, it is believed that glycoprotein hormone synthesis is regulated primarily at the level of β subunit synthesis.

4.2 Structure of the Glycoprotein Hormone Subunits Gene

In recent years, cDNAs and genes of all pituitary glycoprotein hormone subunits have been isolated and characterized in several species (Fig. 6). The structures of the β subunit genes are very similar, whereas α subunit gene, although it undoubtedly evolved from a common ancestor to the β subunit genes, has a somewhat different organization.

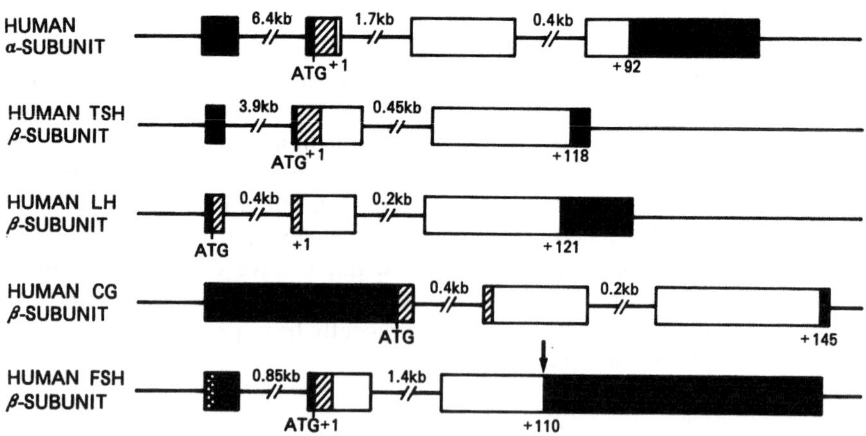

Fig. 6. Structure of the human glycoprotein hormone genes. Schematic diagram of the human α, TSH β, LH β, CG β, and FSH β genes are shown. Untranslated regions are depicted by *solid box*. Signal peptides and mature apoproteins are represented by *hatched* and *unshaded regions*, respectively. Numbers below each diagram signify amino acid position with respect to the first amino acid of the mature apoprotein. In the diagram of FSH β subunit gene, the *cross-hatched region* represents the extended 5'-untranslated region produced by alternative splicing. The *solid arrow* indicates an alternative polyadenylation site

4.2.1 α Subunit Gene

The cDNAs and genes for the α subunit in the human (Fiddes and Goodman 1979, 1981), bovine (Erwin et al. 1983; Goodwin et al. 1983), mouse (Chin et al. 1981; Gordon et al. 1988b), and rat (Godine et al. 1982; Burnside et al. 1988) have been isolated and characterized. In each species, a single gene encodes the α subunit and is composed of four exons and three introns (Fig. 6). The positions of the introns are highly conserved in all four species. There is considerable variation in the size of the α subunit gene in these species, 8–16.5 kb, and this variation is attributable to differences in the size of the first intron which interrupts the 5'-untranslated region.

An α subunit mRNA species of 730–800 nucleotides is produced in all species. There is a 24-amino-acid leader peptide followed by a 96-amino-acid mature protein in all species studied to date, except in human, which encodes a 92-amino-acid mature peptide. The 4-amino-acid discrepancy can be explained by the 12-codon deletion. The gene for the human α subunit is located on chromosome 6.

4.2.2 TSH β Subunit Gene

The cDNAs and genes for the TSH β subunit in the rat (Croyle and Maurer 1984; Carr et al. 1987), bovine (Maurer et al. 1984), mouse (Gurr et al. 1983; Gordon et al. 1988a), and human (Wondisford et al. 1988) have been isolated and characterized. Rat TSH β subunit gene is 4.9 kb long and consists of three exons interrupted by introns of 3.9 and 0.4 kb. Two transcriptional start sites 43 bases apart were identified. The human TSH β subunit gene also consists of three exons interrupted by two introns of 3.9 and 0.4 kb (Fig. 6). The human gene may have only one transcriptional start site (Wondisford et al. 1988), unlike the rat and mouse genes which contain two, due to alteration of the 5' distal TATA box. The gene for the human TSH β subunit is located on chromosome 1. The murine TSH β subunit gene has similarities and differences from the TSH β subunit gene in other mammals. A unique and striking feature of the murine TSH β subunit gene is found in the 5'-untranslated region, in which three exons are found instead of a single exon. Thus, the murine TSH β subunit gene is 5 kb in size and unlike the rat and human TSH β subunit gene, it consists of five exons and four introns. Two transcription start sites were found to be present in the murine TSH β subunit gene, located 40 bp apart (Wood et al. 1987b).

4.2.3 LH β and CG β Subunit Gene Family

The human (Talmadge et al. 1984), rat (Chin et al. 1983; Tepper and Roberts 1984), and bovine (Virgin et al. 1985) cDNAs and genes encoding the LH β subunit as well as the hCG β subunit (Fiddes and Goodman 1980) have been isolated and characterized by several groups. Unlike the β subunits of the other members of the glycoprotein family, the hCG β subunit is not encoded by a single gene. Its organization is much more complex, with at least seven distinct hCG β subunit genes or hCG β like genes arranged in a cluster on human chromosome 19, one of which encodes the LH β subunit (Policastro et al. 1983; Julier et al. 1984).

Unlike the α subunit genes, the LH β subunit genes are smaller, with a total length of approximately 1.5 kb. There are three exons and two introns, which are 352 and 233 nucleotides long in the human, considerably smaller than the sizes of the introns of the α subunit genes (Fig. 6). The encoded protein has a leader peptide that is 24 amino acids long, and the mature protein is 121 amino acids long. The positions of the introns are highly conserved in all species, with the first intron interrupting the region of the gene that encodes the signal peptide and the second intron interrupting the coding sequence of the mature protein between amino acid residues +41 and +42.

Although hLH β subunit and hCG β subunit genes possess nearly identical (> 90% similarity) sequences upstream from the translational start site, they have different transcriptional start sites, indicating that the two genes utilize different promoters. Consequently, the 5'-untranslated region of the human LH β subunit gene is short, only 9 bases in length, whereas the hCG β subunit gene possesses a 5'-untranslated region that is 366 bases. Thus, in spite of the presence of a consensus TATAA sequence in the same location in the two genes, only in the hLH β subunit gene is this sequence employed as a promoter. In the hCG β subunit gene, a different promoter is used, resulting in an mRNA species that begins 366 nucleotides upstream from the translational start site (Fig. 6). Interestingly, there is no consensus TATAA sequence upstream from the hCG β subunit gene transcriptional start site. This finding indicates that tissue-specific factors may be acting upon different promoter sites in two genes of identical sequence in this region (Jameson et al. 1986).

4.2.4 FSH β Subunit Gene

The rat (Maurer 1987) and bovine (Maurer and Beck 1986) cDNAs and the rat (Gharib et al. 1989), bovine (Kim et al. 1988b), and human (Watkins et al. 1987) genes that encode the FSH β subunit have been isolated and characterized. Like the genes that encode the β subunits of LH, CG, and TSH, the FSH β subunit gene possesses three exons and two introns (Fig. 6). A single gene encodes this subunit in all species studied to date. The nucleotide and amino acid sequences of the FSH β subunit gene, respectively, are highly conserved between species.

The FSH β subunit gene differs from the other glycoprotein hormone β subunits in that it possesses an extremely long 3'-untranslated region (1 kb, 1.2 kb, and 1.5 kb in the rat, bovine, and human genes, respectively). The significance of the unusual length of the 3'-untranslated region is not yet known. However, sequences within the 3'-untranslated region of several genes have been shown to be important in determining RNA stability.

The human FSH β subunit gene differs from the bovine and rat in that four mRNA species are transcribed (Jameson et al. 1988a), whereas the other two species transcribe a single, 1.7-kb mRNA species from this gene. In the human, the first exon contains an alternative splicing donor site resulting in a 5'-untranslated region of 63 and 33 bases. Furthermore, two polyadenylation sites are utilized. One polyadenylation site incorporates the stop codon in its consensus AATAAA sequences, whereas the other site is positioned in approximately the same location as it is in the bovine and rat genes, and results in the 1.7-kb mRNA species. The gene for the human FSH β subunit is located on chromosome 11.

4.3 Regulation of TSH Subunits Gene Expression

Circulating TSH levels are modulated by neurotransmitters released from the hypothalamus to the pituitary. The most prominent and thoroughly investigated of these are TRH and dopamine. T_3 and thyroxine (T_4) directly inhibit both TSH release and TSH synthesis. They inhibit TSH secretion by decreasing the number of TRH receptors and by decreasing TSH biosynthesis; these actions correlate with the extent of nuclear T_3 binding.

4.3.1 Hypothalamic Regulation of TSH Subunits Gene Expression

Treatment of rat pituitary cells in primary culture with TRH increased levels of TSH β subunit and α subunit mRNAs (Kourides et al. 1984; Franklyn et al. 1986; Shupnik et al. 1986b), but had no effect in vivo when administered in osmotic pumps or in animals in which the pituitary was transplanted beneath the kidney (Lippman et al. 1986). TRH appears to stimulate TSH β subunit and α subunit mRNAs to a similar extent (Shupnik et al. 1986b), while T_3 suppresses the TSH β subunit mRNA to a much greater extent than the α subunit (Franklyn and Sheppard 1986).

Dopamine treatment of cultured pituitary cells from hypothyroid rats results in decreased TSH subunits mRNAs and suppressed transcription of both TSH subunits genes (Shupnik et al. 1986b). Dopamine partially reverses TRH stimulation of TSH subunits gene transcription, reducing stimulated mRNA synthesis to near control values.

Since both TRH and dopamine act via cell-surface receptors (Foord et al. 1983), their actions at the gene level must be mediated through second messengers which can modify the activity of transcription factors. TRH does not appear to act through cAMP-dependent pathways, but may require Ca^{2+}-dependent or inositol phosphate-dependent protein phosphorylation for its biological actions (Martin 1983; Tan and Tashjian 1981; Rebecchi and Gershengorn 1983; Sobel and Tashjian 1983; Drust and Martin 1984; Gershengorn et al. 1980). Dopamine may act by decreasing intracellular cAMP levels (Shupnik et al. 1986b). The addition of cAMP partially overcomes dopamine suppression of transcription, although it has little effect on basal or TRH-stimulated mRNA synthesis. While TSH transcription was not altered by exogenous cAMP, increases in intracellular cAMP by forskolin have been shown to increase TSH β subunit mRNA (Franklyn et al. 1986). Although dopamine appears to act to suppress intracellular cAMP levels (Franklyn et al. 1986), no specific nuclear or DNA-binding protein that is sensitive to dopamine has been described.

4.3.2 Transcriptional Effects of Thyroid Hormone on TSH Subunits Gene

Dramatic decreases in hybridizable TSH β subunit and α subunit mRNAs are observed after thyroid hormone treatment in rat and mouse (Gurr and Kourides 1984; Chin et al. 1985; Shupnik et al. 1986a). TSH β subunit mRNA concentrations are more sensitive to changes in thyroid status than α subunit mRNA. Franklyn and Sheppard (1986) observed that in the hypothyroid rat a tenfold increase in TSH β subunit mRNA compared with euthyroid levels and twofold increase in α subunit mRNA, changes reversed by administration of T_3.

Thyroid hormones have measureable high affinity receptor sites on the chromatin in thyrotrophs, and the binding of hormones to these sites is believed to initiate the biological response. Estimations of the rate of TSH gene transcription have demonstrated that T_3 rapidly decreases transcription of the TSH β and α subunit genes, the changes in transcription rate being directly proportional to the occupancy of nuclear receptors for T_3 (Franklyn and Sheppard 1986; Shupnik et al. 1986a).

4.4 Regulation of Gonadotropin Subunits Gene Expression

The secretion and synthesis of the gonadotropins are regulated in a complex fashion, with involvement of factors from the hypothalamus, GnRH, and from the gonads, sex steroids, and gonadal peptides.

4.4.1 Regulation of Gonadotropin Subunits mRNAs by GnRH

GnRH is the first key hormone in the reproductive system. It is synthesized and stored in neurosecretory cells of the medial basal hypothalamus. GnRH is released in a pulsatile manner into the hypophyseal portal circulation and is transported to the anterior pituitary; here it regulates the secretion of the gonadotropins, LH and FSH, into the systemic circulation.

Many investigators studied the effects of GnRH on gonadotropin biosynthesis using pulse-chase experiments, and the results of these studies were conflicting (Vogel et al. 1986; Starzec et al. 1986). Interestingly, Ramey et al. (1987) found that when GnRH alone was added to cultured cells, there was an increase in glycosylation of LH, but no change in incorporation of radiolabeled precursors into the protein. However, when estradiol was added either before or concomitantly with GnRH, an increase in the incorporation of radiolabeled amino acids into LH could be appreciated. It is possible that the presence of estradiol is necessary to increase GnRH receptor numbers on gonadotrophs before the induction of LH synthesis by GnRH.

Another evidence that GnRH increases the biosynthesis of gonadotropin is offered by experiments in which the effects of GnRH on gonadotropin subunits mRNA levels have been examined both in vivo and in vitro. Postcastration levels of all three gonadotropin subunits mRNAs decline when male or female rats are treated with adequate doses of a potent GnRH antagonist (Wierman et al. 1989).

Similarly, the postcastration rise in subunits mRNA levels are abolished when castrated animals are treated with GnRH antagonists or GnRH antisera (Lalloz et al. 1988a; Kato et al. 1989; Rodin et al. 1989). Administration of GnRH by constant infusion into castrated animals prevents the postcastration rise of LH β subunit and FSH β subunit mRNA levels but has no effect on those of α subunit (Lalloz et al. 1988b), suggesting that β subunit is more tightly coupled to GnRH action.

The pituitary is divorced of input from the hypothalamus by any one of a number of means, following which GnRH is administered and effects on subunits mRNAs are measured. Hamernik and Nett (1988) and Mercer et al. (1988) have shown that administration of GnRH in a pulsatile fashion to hypothalamic-pituitary-disconnected and ovariectomized ewes restored the lowered gonadotropin subunits mRNA levels to levels observed in ovariectomized animals.

Administration of testosterone to castrated male rats results in infrequent serum LH pulses, suggesting that GnRH secretion is markedly suppressed. This animal model was used to study the effects of pulse amplitude and frequency on gonadotropin subunits mRNAs (Papavasiliou et al. 1986; Haisenleder et al. 1988; Dalkin et al. 1989). The GnRH pulse amplitudes that produced the highest levels of α and LH β subunit mRNAs were 75 and 25 ng/pulse, respectively. Furthermore, it was found GnRH administered at 8-min pulse intervals increased α subunit levels but had little or no effect on LH β subunit and FSH β subunit mRNA levels. Thirty-minute pulse intervals increased the levels of all three gonadotropin subunits mRNA levels, whereas 120-min pulse intervals increased only those of FSH β subunit. Leung et al. (1987) have utilized a comparable experimental design to study the effects of GnRH pulse frequencies to ovariectomized ewes in which endogenous GnRH secretion had been suppressed by progesterone administration. As in the rat, the higher frequency GnRH pulses favored increases in α but not LH β subunit or FSH β subunit mRNAs. Several groups have examined the effects of pulsatile administration of GnRH (every 2 h) to hypogonadal (hpg) mice and have obtained different results. Saade et al. (1989) have shown that this treatment results in an elevation of both α and LH β subunit mRNA levels, whereas Stanley et al. (1988) showed that the same treatment regimen resulted in increases in α but not LH β subunit mRNA levels. Thus, from these studies, it appears that gonadotropin subunit mRNA levels are regulated by GnRH and that they may be regulated differentially.

The second group of studies involves experiments in which the effects of GnRH on gonadotropin subunits mRNA levels or synthesis is examined in vitro. These experiments have produced different results in different laboratories. Salton and coworkers (1988) could not demonstrate any changes in the transcription of gonadotropin subunits mRNAs in cultured pituitary cells treated for 20 min with GnRH. Hubert et al. (1988) have found that GnRH in concentrations of $10^{-7}M$ when administered to primary pituitary cell cultures for 24–72 h resulted in stimulation (two- to three fold) of α subunit but no change in LH β subunit mRNA levels. Subsequently, however, Andrews et al. (1988) demonstrated that low doses of GnRH ($10^{-10}M$) stimulated increases in LH β subunit mRNA levels in cultured anterior pituitary cells.

In summary, there is abundant evidence that GnRH regulates gonadotropin subunits biosynthesis in in vivo systems. There are conflicting data from in vitro studies, but the failure of some of these studies to demonstrate induction of gonadotropin subunits biosynthesis may be attributable to artifacts of the experimental models employed.

Although GnRH can elevate both cAMP and cGMP under certain conditions, the cyclic nucleotides are not involved in the acute exocytotic response to GnRH (Naor and Catt 1980). cAMP however might be involved in GnRH-induced gonadotropin biosynthesis (Starzec et al. 1989).

It has recently been demonstrated that after binding to its specific receptors, GnRH stimulates phosphoinositide turnover, mobilizes Ca^{2+}, activates PKC, and induces arachidonic acid (AA) release (Naor 1990). The production of multiple second messenger molecules is responsible for gonadotropin release and synthesis (Huckle and Conn 1988). The relationships between Ca^{2+}/calmodulin- and PKC-dependent pathways of GnRH action are summarized in Fig. 7.

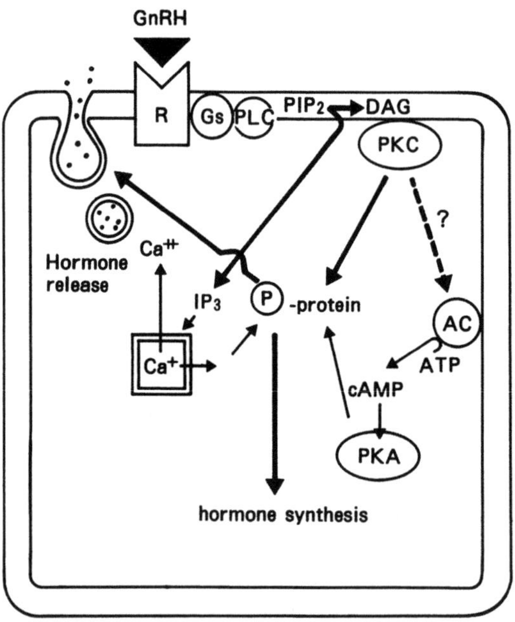

Fig. 7. Schematic representation of GnRH-receptor signal transduction. *R*, receptor; *PKA*, protein kinase A; *AC*, adenylate cyclase; *G*, GTP-binding protein; *PLC*, phospholipase C; IP_3, inositol triphosphate; *DAG*, diacylglycerol; Ca^{++}, calcium ion

It was reported that GnRH increased LH β subunit mRNA levels by a PKC-mediated mechanism (Andrews et al. 1988). Activators of PKC have been found to increase the rates of LH biosynthesis (Starzec et al. 1989). Papavasiliou et al. (1986) have found that depletion of PKC activity inhibits the ability of GnRH to increase gonadotropin subunits mRNA levels. Thus, the action of GnRH on LH mRNA levels appear to require the presence of PKC. The action of this enzyme on events at the transcriptional levels has been implicated in a variety of systems (Brostrom et al. 1987).

4.4.2 Regulation of Gonadotropin Subunits mRNAs by Sex-Steroid Hormones and Gonadal Peptides

The effects of sex-steroid hormone administration on gonadotropin subunits mRNA levels have been studied extensively. The different types of sex-steroid hormones exert different regulatory effects on gonadotropin secretion and subunits synthesis (Gharib et al. 1990). Using the cell-free translation system, Corbani et al. (1984) found that translatable α and LH β subunit mRNA levels rose gradually after castration, reaching plateaus in male and female rats approximately 3 weeks postcastration, but the changes in translatable FSH β subunit mRNAs were modest in both sexes during this time. When the cDNAs encoding the gonadotropin subunits were isolated, a more direct approach to the measurement of steady-state gonadotropin mRNAs became possible (Gharib et al. 1986, 1987). Castration increases the levels of gonadotropin subunit mRNAs in both sexes, albeit to a different extent and with different kinetics.

Another approach to study the effects of castration on gonadotropin subunits synthesis has been the examination of changes within the gonadotroph cell population of gonadotropin subunit mRNA concentrations utilizing the technique of in situ hybridization (Ibrahim et al. 1986). The number of LH β subunit mRNA-containing cells doubles in male rats 2 or 4 weeks postcastration compared with intact controls. Furthermore, the amount of LH β subunit mRNA per cell and the area of cells that contain LH β subunit mRNA also double.

The gonadal peptides, inhibin and follistatin, suppress FSH secretion, at least in part, by negatively regulating FSH β subunit mRNA levels directly at the level of the anterior pituitary gland (Gharib et al. 1990).

4.5 Transcriptional Regulatory Sequences of the Glycoprotein Hormones

TSH β subunit is produced in the pituitary thyrotrophs. Both LH and FSH β subunits are produced in the pituitary gonadotrophs. The synthesis and secretion of these subunits is under complex hormonal control. Since the α subunit gene is expressed not only in gonadotrophs and thyrotrophs of the pituitary but also in placental trophoblast cells, it may be anticipated that it contains multiple regulatory elements for various hormonal and physiologic effectors. However, relatively little information is known about potential regulatory sequences.

4.5.1 Transcriptional Regulatory Sequences of the α Subunit Gene

cAMP regulates transcription of the human α subunit gene. In the 5'-flanking region of the human α subunit gene, a directly repeated 18-bp palindromic sequence has been identified that is capable of conferring cAMP responsiveness to reporter genes. In the bovine, mouse, and rat genes there is a single copy of this palindrome sequence, differing by a single base pair in the 8-bp core of the palindrome (Delegeane et al. 1987; Deutsch et al. 1987). In contrast to the human, there is no evidence for this region conferring cAMP responsiveness in the mouse (Gordon et al. 1988b), rat, or cow (Gharib et al. 1990).

Many studies have suggested that *trans*-activating factors might bind to the 5'-flanking region of the human α subunit gene. DNase I protection experiments showed that DNA-binding factors bound to this element in both placental and nonplacental cells, but that an upstream tissue-specific factor bound factors only in placental cells, suggesting that the CRE acts in concert with a cAMP-responsive enhancer to confer tissue specificity to this gene (Delegeane et al. 1987). Furthermore, other studies revealed that the 18-bp region could successfully compete for limiting cellular transcription factors, lending further support to the notion that *trans*-acting factors interacted with the CRE (Jameson et al. 1987, 1988b).

Subsequently, CREB protein has been characterized and has been shown to bind as a dimer to CREs (Montminy and Bilezikjian 1987). The cDNAs encoding CREB protein were isolated and characterized recently from human placenta (Hoeffler et al. 1988) and rat cortex. The CREB proteins possess a carboxyl-terminal basic region, similar to the leucine zipper sequence that is present in many DNA-binding proteins, and structural similarities to the yeast transcription factor GCN4 and the phorbol ester-responsive, c-Jun protein.

In an intriguing set of experiments, Akerblom and coworkers (1988) have shown that glucocorticoids negatively regulate human α subunit gene expression in placental cells by binding to a site in the 5'-flanking region of the α subunit gene that overlaps the CRE. They have hypothesized that glucocorticoids inhibit α subunit gene expression by interfering with the binding of the CRE-binding transcription factor.

Thyroid hormone regulation of the rat α subunit has been studied by transient expression analysis of α subunit promoter-reporter gene constructs, and by binding of labeled receptor to specific α subunit gene regions (Burnside et al. 1988). T_3 suppressed α subunit promoter activity in transfected GH_3 cells, with the T_3-responsive region between bases −80 and −4 relative to the transcriptional start site. An oligodeoxynucleotide representing bases −74 to −38 was found to bind $[^{125}I]T_3$-receptor complex derived from pituitary cell nuclear extracts as well as in vitro synthesized human c-*erbA* β subunit.

4.5.2 Transcriptional Regulatory Sequences of TSH β Subunit Gene

Stimulation by TRH of transcription of the rTSH β subunit gene has been examined by gene deletion analysis and transient expression assays (Carr et al. 1989). TSH β subunit promoter activity in these studies was stimulated in response to TRH, and the TRH-responsive region of the gene lies in a region approximately 1 kb upstream.

The avidin-biotin complex DNA binding assay was utilized to examine DNA sequence-specific binding of T_3 receptor to portions of the rat TSH β subunit gene (Darling et al. 1989). Both nuclear extracts from GH_3 cells and in vitro synthesized T_3 receptors bound to the gene at two sites, one within the first intron and another to a region immediately adjacent to the second transcriptional start site. Inhibition of binding with various oligonucleotides and mutational analysis indicates that the sequences −22 to −5 and +3 to +31 bases relative to the second transcriptional start site are most important for T_3 receptor binding. These regions are also important for transcriptional regulation. The placement of these binding sites suggest that binding of the T_3 receptor to the gene may block binding of other transcription factors and thus inhibit transcription. Wondisford et al. (1988) have noted that the human TSH β subunit gene contains a region from +23 to +37 bases in the first exon, which bears a high degree of sequence homology to other TREs.

Recent work has indicated that several nuclear proteins bind distally to the murine TSH β subunit transcriptional start sites (Alexander et al. 1989). DNase I protection assays demonstrate 5 separate binding sites for thyrotropic tumor (TtT97) nuclear extracts in the mTSH β subunit promoter region. These sites have been termed distal, D1 (−253 to −227 bp), and proximal, P1 (−76 to −68 bp), P2 (−106 to −98 bp), P3 (−126 to −112 bp), and P4 (−142 to −131 bp) footprints. The D1 site is completely specific for TSH β-subunit-expressing cells, since nuclear proteins from nonexpressing cells (L cells) or thyrotrophic tumors (MGH 101A) do not footprint in this region; minimal footprinting activity is noted in the P1–P4 regions. At this time, it appears that these proteins and DNA regions may be required for optimal TSH β subunit gene expression, but their role in basal and regulated transcription is yet to be determined.

4.5.3 Transcriptional Regulatory Sequences of LH β Subunit Gene

The 5'-flanking region of the rat LH β subunit gene contains a putative ERE that has been detected by filter binding studies and gene transfer experiments (Shupnik et al. 1989). From these studies, it has been learned that the portion of the rat LH β subunit gene that is able to confer estrogen responsiveness to reporter genes and that also binds to purified ER is located between −1388 bp and −1105 bp upstream from the transcriptional start site. This region contains a 15-bp imperfect palindrome sequence that displays a high degree of sequence similarity to EREs in the PRL gene (Maurer and Notides 1987) and frog vitellogenin gene.

4.5.4 Transcriptional Regulatory Sequences of FSH β Subunit Gene

The 5'-flanking regions of the rat, bovine, and human genes do not display a high degree of similarity. However, within the 5'-flanking regions of all three genes, there are three regions of greater than 80% similarity that may contain *cis*-DNA elements that may be important in the regulation of transcription of the FSH β subunit gene. Thus, they might be the sites of association of regulatory factors that uniquely affect the expression of the FSH β subunit gene (Gharib et al. 1990).

5 Conclusions

In this article, we have reviewed the structures and regulatory mechanisms of the anterior and intermediate pituitary hormone genes. Thanks to the recent progress in molecular biology of eukaryotes, mechanisms for the tissue-specific expression and hormonal regulation are rapidly being elucidated.

A wide variety of hormones appear to have significant transcriptional effects on pituitary hormone genes. The function of nuclear hormone receptors with DNA-binding activity (e.g., the steroid hormone receptors and the thyroid hormone receptor) is increasingly well understood with evidence both for protein-DNA interaction and for protein-protein interaction (Evans 1988; Beato 1989; Goodbourn 1990; O'Malley 1990).

The picture is much less clear for the mechanisms of tissue-specific expression and the signaling mechanisms after the hormone binding to the membrane receptors. Each of the membrane receptors is coupled to intermediary G proteins, with SRIF binding uniquely to the inhibitory subclass of G proteins. The hypophysiotropic hormones could be classified into two groups by the postreceptor mechanisms in their target cells: one group includes CRH and GHRH, which work through the cAMP-protein kinase A system; the other consists of GnRH and TRH, which utilize the phosphoinositide-PKC cascade. The recent identification and molecular cloning of *trans*-acting factors, such as Pit-1/GHF-1 for tissue-specific expression, CREB for the cAMP-protein kinase A system, and Jun and Fos for the phosphoinositide-PKC system, are the major achievements in the field.

Advances in our understanding of mechanisms of the pituitary hormone gene expression and signal transduction of the hypophysiotropic hormone stimuli will provide a basis for more sophisticated clinical application of these hormones. Whether specific defects in gene regulation of hormone and/or their receptors are involved in pituitary tumor development is another exciting issue.

Acknowledgements. We are indebted to Drs. Hiroo Imura, Shigetada Nakanishi, and Shosaku Numa for their continuing support and critical comments. Ms. Rie Shiraishi and Ms. Hitomi Kawamura expertly typed the manuscript.

References

Abou-Samra AB, Catt KJ, Aguilera G (1986) Involvement of protein kinase C in the regulation of adrenocorticotropin release from rat anterior pituitary cells. Endocrinology 118:212–217

Abou-Samra AB, Harwood JP, Manganiello VC, Catt, KJ, Aguilera G (1987) Phorbol 12-myristate 13-acetate and vasopressin potentiate the effect of corticotropin-releasing factor on cyclic AMP production in rat anterior pituitary cells: Mechanism of action. J Biol Chem 262:1129–1136

Adler S, Waterman ML, He X, Rosenfeld MG (1988) Steroid receptor-mediated inhibition of rat prolactin gene expression does not require the receptor DNA-binding domain. Cell 52:685–695

Affolter HU, Reisine T (1985) Corticotropin releasing factor increases proopiomelanocortin messenger RNA in mouse anterior pituitary tumor cells. J Biol Chem 260:15477–15481

Aguilera G, Harwood JP, Wilson JX, Morell J, Brown JH, Catt KJ (1983) Mechanisms of action of corticotropin-releasing factor and other regulators of corticotropin release in rat pituitary cells. J Biol Chem 258:8039–8045

Akerblom IE, Slater EP, Beato M, Baxter JD, Mellon PL (1988) Negative regulation by glucocorticoids through interference with a cAMP responsive enhancer. Science 241:350–353

Alexander LM, Gordon DF, Wood WM, Kao MY, Ridgway EC, Gutierrez-Hartmann A (1989) Identification of thyrotroph-specific factors and *cis*-acting sequences of the murine thyrotropin β subunit gene. Mol Endocrinol 3:1037–1045

Andrews WV, Maurer RA, Conn PM (1988) Stimulation of rat luteinizing hormone-β messenger RNA levels by gonadotropin releasing hormone: apparent role for protein kinase C. J Biol Chem 263:13755–13761

Antakly T, Sasaki A, Liotta AS, Palkovits M, Krieger DT (1985): Induced expression of the glucocorticoid receptor in the rat intermediate pituitary lobe. Science 229:277–279

Bancroft C, Gick GG, Johnson ME, White BA (1985) Regulation of growth hormone and prolactin gene expression by hormones and calcium. Biochem Actions Horm 12:173–213

Barinaga M, Yamonoto G, Rivier C, Vale W, Evans R, Rosenfeld MG (1983) Transcriptional regulation of growth hormone gene expression by growth hormone-releasing factor. Nature 306:84–85

Barinaga M, Bilezikjian LM, Vale WW, Rosenfeld MG, Evans RM (1985) Independent effects of growth hormone releasing factor on growth hormone release and gene transcription. Nature 314:279–281

Barlow JW, Voz MLJ, Eliard PH, Mathy-Hartert M, de Nayer P, Economidis IV, Belayew A, Martial JA, Rousseau GG (1986) Thyroid hormone receptors bind to defined regions of the growth hormone and placental lactogen genes. Proc Natl Acad Sci USA 83:9021–9025

Barrera-Saldana HA, Seeburg PH, Saunders GF (1983) Two structurally different genes produce the same secreted human placental lactogen hormone. J Biol Chem 258:3787–3793

Barta A, Richards RI, Baxter JD, Shine J (1981) Primary structure and evolution of rat growth hormone gene. Proc Natl Acad Sci USA 78:4867–4871

Beato M (1989) Gene regulation by steroid hormones. Cell 56:335–344

Birnbaum MJ, Baxter JD (1986) Glucocorticoids regulate the expression of a rat growth hormone gene lacking 5'-flanking sequences. J Biol Chem 261:291–297

Birnberg NC, Lissitzky JC, Hinman M, Herbert E (1983) Glucocorticoids regulate proopiomelanocortin gene expression *in vivo* at the levels of transcription and secretion. Proc Natl Acad Sci USA 80:6982–6986

Bodner M, Castrillo JL, Theill LE, Deerinck T, Ellisman M, Karin M (1988) The pituitary-specific transcription factor GHF-1 is a homeobox-containing protein. Cell 55:505–518

Boileau G, Barbeau C, Jeannotte L, Chretien M, Drouin J (1983) Complete structure of the porcine pro-opiomelanocortin mRNA derived from the nucleotide sequence of cloned cDNA. Nucleic Acids Res 11:8063–8071

Brent GA, Harney JW, Moore DD, Larsen PR (1988) Multihormonal regulation of the human, rat, and bovine growth hormone promoters: differential effects of 3',5'-cyclic adenosine monophosphate, thyroid hormone, and glucocorticoids. Mol Endocrinol 2:792–798

Brostrom MA, Chin KV, Cade C, Gmitter D, Brostrom CO (1987) Stimulation of protein synthesis in pituitary cells by phorbol esters and cyclic AMP: evidence for rapid induction of a component of translational initiation. J Biol Chem 262:16515–16523

Bruhn TO, Sutton RE, Rivier CL, Vale WW (1984) Corticotropin-releasing factor regulates proopiomelanocortin messenger ribonucleic acid levels in vivo. Neuroendocrinology 39:170–175

Burnside J, Buckland PR, Chin WW (1988) Isolation and characterization of the gene encoding the α subunit of the rat pituitary glycoprotein hormones. Gene 70:67–74

Camper Sa, Yao YAS, Rottman FM (1985) Hormonal regulation of the bovine prolactin promoter in rat pituitary tumor cells. J Biol Chem 260:12246–12251

Carr FE, Need LR, Chin WW (1987) Isolation and characterization of the rat thyrotropin β subunit gene: differential regulation of two transcriptional start sites by thyroid hormone. J Biol Chem 262:981–987

Carr FE, Shupnik MA, Burnside J, Chin WW (1989) Thyrotropin-releasing hormone stimulates the activity of the rat thyrotropin β subunit gene promoter transfected into pituitary cells. Mol Endocrinol 3:717–724

Carrillo AJ, Pool TB, Sharp ZD (1985) Vasoactive intestinal peptide increases prolactin messenger ribonucleic acid content in GH3 cells. Endocrinology 116:202–206

Castrillo JL, Bodner M, Karin M (1989) Purification of growth hormone-specific transcription factor GHF-1 containing homeobox. Science 243:814–817

Catanzaro DF, West BL, Baxter JD, Reudelhuber TL (1987) A pituitary-specific factor interacts with an upstream promoter element in the rat growth hormone gene. Mol Endocrinol 1:90–96

Cattini PA, Eberhardt NL (1987) Regulated expression of chimaeric genes containing the 5'-flanking regions of human growth hormone-related genes in transiently transfected rat anterior pituitary tumor cells. Nucleic Acids Res 15:1297–1309

Cattini PA, Anderson TR, Baxter JD, Mellon P, Eberhardt NL (1986) The human growth hormone gene is negatively regulated by triiodothyronine when transfected into rat pituitary tumor cells. J Biol Chem 261:13367–13372

Chen CLC, Dionne FT, Roberts J (1983) Regulation of the proopiomelanocortin mRNA levels in rat pituitary by dopaminergic compounds. Proc Natl Acad Sci USA 80:2211–2215

Chin WW, Kronenberg HM, Dee PC, Maloof F, Habener JF (1981) Nucleotide sequence of the mRNA encoding the pre-α-subunit of mouse thyrotropin. Proc Natl Acad Sci USA 78:5329–5333

Chin WW, Godine JE, Klein DR, Chang AS, Tan LK, Habener JF (1983) Nucleotide sequence of the cDNA encoding the precursor of the β subunit of rat luteotropin. Proc Natl Acad Sci USA 80:4649–4653

Chin WW, Shupnik MA, Ross DS, Habener JF, Ridgway EC (1985) Regulation of the α and thyrotropin β subunit messenger ribonucleic acids by thyroid hormones. Endocrinology 116:873–887

Clayton RN, Bailey LC, Abbot SD, Detta A, Docherty K (1986) Cyclic adenosine nucleotides and growth hormone-releasing factor increase cytosolic growth hormone messenger RNA levels in cultured rat pituitary cells. J Endocrinol 110:51–57

Corbani M, Counis R, Starzec A, Jutisz M (1984) Effect of gonadectomy on pituitary levels of mRNA encoding gonadotropin subunits and secretion of luteinizing hormone. Mol Cell Endocrinol 35:83–87

Cooke NE, Baxter JD (1982) Structural analysis of the prolactin gene suggests a separate origin for its 5' end. Nature 297:603–606

Cote TE, Felder R, Kebabian JW, Sekura RD, Reisine T, Affolter HU (1986) D-2 dopamine receptor-mediated inhibition of pro-opiomelanocortin synthesis in rat intermediate lobe: abolition by pertussis toxin or activators of adenylate cyclase. J Biol Chem 261:4555–4561

Croyle ML, Maurer RA (1984) Thyroid hormone decreases thyrotropin subunit mRNA levels in rat anterior pituitary. DNA 3:231–236

Dalkin AC, Haisenleder DJ, Ortolano GA, Ellis TR, Marshall JC (1989) The frequency of gonadotropin-releasing-hormone stimulation differentially regulates gonadotropin subunit messenger ribonucleic acid expression. Endocrinology 125:917–924

Dallman MF, Makara GB, Roberts JL, Levin N, Blum M (1985) Corticotrope response to removal of releasing factors and corticosteroids *in vivo*. Endocrinology 117:2190–2197

Darling DS, Burnside J, Chin WW (1989) Binding of thyroid hormone receptors to the rat thyrotropin-β gene. Mol Endocrinol 3:1359–1368

Dave JR, Eiden LE, Lozovsky D, Waschek JA, Eskay RL (1987) Calcium-independent and calcium-dependent mechanisms regulate corticotropin-releasing factor-stimulated proopiomelanocortin peptide secretion and messenger ribonucleic acid production. Endocrinology 120:305–310

Davis JRE, Wilson EM, Sheppard MC (1988) Human pituitary tumours in cell culture: modulation of prolactin and growth hormone messenger RNA levels in vitro. Neuroendocr Perspect 6:221–228

Day RN, Maurer RA (1989) The distal enhancer region of the rat prolactin gene contains elements conferring response to multiple hormones. Mol Endocrinol 3:3–9

Delegeane AM, Ferland LH, Mellon PL (1987) Tissue-specific enhancer of the human glycoprotein hormone α subunit gene: dependence on cyclic AMP-inducible elements. Mol Cell Biol 7:3994–4002

DeNoto FM, Moore DD, Goodman HM (1981) Human growth hormone DNA sequence and mRNA structure: possible alternative splicing. Nucleic Acids Res 9:3719–3730

Deutsch PJ, Jameson JL, Habener JF (1987) Cyclic AMP responsiveness of human gonadotropin-α gene transcription is directed by a repeated 18-base pair enhancer: α-promoter receptivity to the enhancer confers cell-preferential expression. J Biol Chem 262:12169–12174

Drouin J, Chamberland M, Charron J, Jeannotte L, Nemer M (1985) Structure of the rat pro-opiomelanocortin (POMC) gene. FEBS Lett 193:54–58

Drouin J, Nemer M, Charron J, Gagner JP, Jeannotte L, Sun YL, Therrien M, Tremblay Y (1989a) Tissue-specific activity of the pro-opiomelanocortin (POMC) gene and repression by glucocorticoids. Genome 31:510–519

Drouin J, Trifiro MA, Plante RK, Nemer M, Eriksson P, Wrange O (1989b) Glucocorticoid receptor binding to a specific DNA sequence is required for hormone-dependent repression of pro-opiomelanocortin gene transcription. Mol Cell Biol 9:5305–5314

Drust DS, Martin TFJ (1984) Thyrotropin-releasing hormone rapidly activates protein phosphorylation in GH_3 pituitary cells by a lipid-linked, protein kinase C-mediated pathway. J Biol Chem 259:14520–14530

Duckworth ML, Kirk KL, Friesen HG (1986) Isolation and identification of a cDNA clone of rat placental lactogen II. J Biol Chem 261:10871–10878

Eberwine JH, Jonassen JA, Evinger MJQ, Roberts JL (1987) Complex transcriptional regulation by glucocorticoids and corticotropin-releasing hormone of proopiomelanocortin gene expression in rat pituitary cultures. DNA 6:483–492

Eipper BA, Mains RE (1980) Structure and biosynthesis of proadrenocorticotropin/endorphin and related peptides. Endocr Rev 1:1–27

Elsholtz HP, Mangalam HJ, Potter E, Albert VR, Supowit S, Evans RM, Rosenfeld MG (1986) Two different *cis*-active elements transfer the transcriptional effects of both EGF and phorbol esters. Science 234:1552–1557

Erwin CR, Croyle ML, Donelson JE, Maurer RA (1983) Nucleotide sequence of cloned complementary deoxyribonucleic acid for the α subunit of bovine pituitary glycoprotein hormones. Biochemistry 22:4856–4860

Evans RM (1988) The steroid and thyroid hormone receptor superfamily. Science 240:889–895

Evans RM, Birnberg NC, Rosenfeld MG (1982) Glucocorticoid and thyroid hormones transcriptionally regulate growth hormone gene expression. Proc Natl Acad Sci USA 79:7659–7663

Fiddes JC, Goodman HM (1979) Isolation, cloning and sequence analysis of the cDNA for the α subunit of human chorionic gonadotropin. Nature 281:351–356

Fiddes JC, Goodman HM (1980) The cDNA for the β subunit of human chorionic gonadotropin suggests evolution of a gene by readthrough into the 3'-untranslated region. Nature 286:684–687

Fiddes JC, Goodman HM (1981) The gene encoding the common alpha subunit of the four human glycoprotein hormones. J Mol Appl Genet 1:3–18

Flug F, Copp RP, Casanova J, Horowitz ZD, Janocko L, Plotnick M, Samuels HH (1987) *cis*-Acting elements of the rat growth hormone gene which mediate basal and regulated expression by thyroid hormone. J Biol Chem 262:6373–6382

Foord SM, Peters JR, Dieguez C, Scanlon MF, Hall R (1983) Dopamine receptors on intact anterior pituitary cells in culture: functional association with the inhibition of prolactin and thyrotropin. Endocrinology 112:1567–1577

Frankenne F, Rentier-Delrue F, Scippo ML, Martial J, Hennen G (1987) Expression of the growth hormone variant gene in human placenta. J Clin Endocrinol Metab 64:635–637

Franklyn JA, Sheppard MC (1986) Regulation of TSH gene transcription. Eur J Clin Invest 16:452–454

Franklyn JA, Wilson M, Davis JR, Ramsden DB, Docherty K, Sheppard MC (1986) Demonstration of thyrotropin β subunit messenger RNA in primary culture:evidence for regulation by thyrotropin-releasing hormone and forskolin. J Endocrinol 111:R1–R2

Frohman LA, Jansson JO (1986) Growth hormone-releasing hormone. Endocr Rev 7:223–253

Gershengorn MC (1986) Mechanism of thyrotropin releasing hormone stimulation of pituitary hormone secretion. Annu Rev Physiol 48:515–526

Gershengorn MC, Rebecchi MJ, Geras E, Arevalo CO (1980) Thyrotropin-releasing hormone (TRH) action in mouse thyrotropic tumor cells in culture: evidence against a role for adenosine 3',5'-monophosphate as a mediator of TRH-stimulated thyrotropin release. Endocrinology 107:665–670

Gharib SD, Bowers SM, Need LR, Chin WW (1986) Regulation of rat luteinizing hormone subunit messenger ribonucleic acids by gonadal steroid hormones. J Clin Invest 77:582–589

Gharib SD, Wierman ME, Badger TM, Chin WW (1987) Sex steroid hormone regulation of follicle-stimulating hormone subunit messenger ribonucleic acid (mRNA) levels in the rat. J Clin Invest 80:294–299

Gharib SD, Roy A, Wierman ME, Chin WW (1989) Isolation and characterization of the gene encoding the β subunit of rat follicle-stimulating hormone. DNA 8:339–349

Gharib SD, Wierman ME, Shupnik MA, Chin WW (1990) Molecular biology of the pituitary gonadotropins. Endocr Rev 11:177–198

Gick GG, Zeytin FN, Brazeau P, Ling NC, Esch FS, Bancroft C (1984) Growth hormone-releasing factor regulates growth hormone mRNA in primary cultures of rat pituitary cells. Proc Natl Acad Sci USA 81:1553–1555

Glass CK, Franco R, Weinberger C, Albert VR, Evans RM, Rosenfeld MG (1987) A c-*erb*-A binding site in rat growth hormone gene mediates *trans*-activation by thyroid hormone. Nature 329:738–741

Glass CK, Holloway JM, Devary OV, Rosenfeld MG (1988) The thyroid hormone receptor binds with opposite transcriptional effects to a common sequence motif in thyroid hormone and estrogen response elements. Cell 54:313–323

Godine JE, Chin WW, Habener JF (1982) α Subunit of rat pituitary glycoprotein hormones: primary structure of the precursor determined from the nucleotide sequence of cloned cDNAs. J Biol Chem 257:8368–8371

Goodbourn S (1990) Negative regulation of transcriptional initiation in eukaryotes. Biochim Biophys Acta 1032:53–77

Goodwin RG, Moncman CL, Rottman FM, Nilson JH (1983) Characterization and nucleotide sequence of the gene for the common α subunit of the bovine pituitary glycoprotein hormones. Nucleic Acids Res 11:6873–6882

Gordon DF, Wood WM, Ridgway EC (1988a) Organization and nucleotide sequence of the gene encoding the β subunit of murine thyrotropin. DNA 7:17–26

Gordon DF, Wood WM, Ridgway EC (1988b) Organization and nucleotide sequence of the mouse α subunit gene of the pituitary glycoprotein hormones. DNA 7:679–690

Guerin SL, Moore DD (1988) DNAse I footprint analysis of nuclear proteins from pituitary and nonpituitary cells that specifically bind to the rat growth hormone promoter and 5'-regulatory region. Mol Endocrinol 2:1101–1107

Guild S, Reisine T (1987) Molecular mechanisms of corticotropin-releasing factor stimulation of calcium mobilization and adrenocorticotropin release from anterior pituitary tumor cells. J Pharmacol Exp Ther 241:125–130

Guild S, Itoh Y, Kebabian JW, Luini A, Reisine T (1986) Forskolin enhances basal and potassium-evoked hormone release from normal and malignant pituitary tissue: the role of calcium. Endocrinology 118:268–279

Gurr JA, Kourides IA (1984) Ratios of α to β TSH mRNA in normal, and hypothyroid pituitaries and TSH-secreting tumors. Endocrinology 115:830–832

Gurr JA, Catterall JF, Kourides IA (1983) Cloning of cDNA encoding the pre-β subunit of mouse thyrotropin. Proc Natl Acad Sci USA 80:2122–2126

Gutierrez-Hartmann A, Siddiqui S, Loukin S (1987) Selective transcription and DNase I protection of the rat prolactin gene by GH_3 pituitary cell-free extracts. Proc Natl Acad Sci USA 84:5211–5215

Haisenleder DJ, Katt JA, Ortolano GA, El-Gewely MR, Duncan JA, Dee C, Marshall JC (1988) Influence of gonadotropin-releasing hormone pulse amplitude, frequency, and treatment duration on the regulation of luteinizing hormone (LH) subunit messenger ribonucleic acids and LH secretion. Mol Endocrinol 2:338–343

Hamernik DL, Nett TM (1988) Gonadotropin-releasing hormone increases the amount of messenger ribonucleic acid for gonadotropins in ovariectomized ewes after hypothalamic-pituitary disconnection. Endocrinology 122:959–966

Hammer RE, Brinster R, Rosenfeld MG, Evans RM, Mayo KE (1985) Expression of human growth hormone-releasing factor in transgenic mice results in increased somatic growth. Nature 315:413–416

Harris GW (1948) Neural control of the pituitary gland. Physiol Rev 28:139–179

Heiser W, Eckhart W (1985) Hormonal regulation of a polyoma virus middle-size T-antigen gene linked to growth hormone control sequences. J Gen Virol 66:2147–2160

Heisler S, Reisine TD, Hook VYH, Axelrod J (1982) Somatostatin inhibits multireceptor stimulation of cyclic AMP formation and corticotropin secretion in mouse pituitary tumor cells. Proc Natl Acad Sci USA 79:6502–6506

Hinkle PM, Jackson AE, Thompson TM, Zavacki AM, Coppola DA, Bancroft C (1988) Calcium channel agonists and antagonists: effects of chronic treatment on pituitary prolactin synthesis and intracellular calcium. Mol Endocrinol 2:1132–1138

Hirt H, Kimelman J, Birnbaum MJ, Chen EY, Seeburg PH, Eberhardt NL, Barta A (1987) The human growth hormone gene locus: structure, evolution, and allelic variation. DNA 6:59–70

Hoeffler JP, Meyer TE, Yun Y, Jameson JL, Habener JF (1988) Cyclic AMP-responsive DNA-binding protein: structure based on a cloned placental cDNA. Science 242:1430–1433

Hollt V, Haarmann I (1984) Corticotropin-releasing factor differentially regulates pro-opiomelanocortin messenger ribonucleic acid levels in anterior as compared to intermediate pituitary lobes of rats. Biochem Biophys Res Commun 124:407–415

Hollt V, Haarmann I, Seizinger BR, Herz A (1982) Chronic haloperidol treatment increases the level of *in vitro* translatable messenger ribonucleic acid coding for the β-endorphin/adrenocorticotropin precursor proopiomelanocortin in the pars intermedia of the rat pituitary. Endocrinology 110:1885–1891

Hubert JF, Simard J, Gagne B, Barden N, Labrie F (1988) Effect of luteinizing hormone releasing hormone (LHRH) and [D-Trp^6, Des Gly-NH_2^{10}]LHRH ethylamide on α subunit and LH β messenger ribonucleic acid levels in rat anterior pituitary cells in culture. Mol Endocrinol 2:521–527

Huckle WR, Conn PM (1988) Molecular mechanisms of gonadotropin releasing hormone action. II. The effector system. Endocr Rev 9:387–395

Ibrahim SN, Moussa SM, Childs GV (1986) Morphometric studies of rat anterior pituitary cells after gonadectomy: correlation of changes in gonadotropes with the serum levels of gonadotropins. Endocrinology 119:629–637

Imura H, Nakai Y, Nakao K, Oki S, Fukata J, Tanaka I, Kinoshita F, Tsukada T, Yoshimasa T (1982) Biosynthesis and regulation of secretion of adrenocorticotrophin, β-endorphin and related peptides. In: Motta M, Zanisi M, Piva F (eds) Pituitary hormones and related peptides. Academic, London, pp 243–269

Ingraham HA, Chen R, Mangalam HJ, Elsholtz HP, Flynn SE, Lin CR, Simmons DM, Swanson L, Rosenfeld MG (1988) A tissue-specific transcription factor containing a homeodomain specifies a pituitary phenotype. Cell 55:519–529

Ingraham HA, Albert VR, Chen R, Crenshaw III EB, Elsholtz HP, He X, Kapiloff MS, Mangalam HJ, Swanson LW, Treacy MN, Rosenfeld MG (1990) A family of POU-domain and Pit-1 tissue-specific transcription factors in pituitary and neuroendocrine development. Annu Rev Physiol 52:773–791

Israel A, Cohen SN (1985) Hormonally mediated negative regulation of human pro-opiomelanocortin gene expression after transfection into mouse L cells. Mol Cell Biol 5:2443–2453

Jackson AE, Bancroft C (1988) Proximal upstream flanking sequences direct calcium regulation of the rat prolactin gene. Mol Endocrinol 2:1139–1144

Jameson JL, Lindell CM, Habener JF (1986) Evolution of different transcriptional start sites in the human luteinizing hormone and chorionic gonadotropin β subunit genes. DNA 5:227–234

Jameson JL, Deutsch PJ, Gallagher GD, Jaffe RC, Habener JF (1987) *trans*-Acting factors interact with a cyclic AMP response element to modulate expression of the human gonadotropin α gene. Mol Cell Biol 7:3032–3040

Jameson JL, Becker CB, Lindell CM, Habener JF (1988a) Human follicle-stimulating hormone β subunit gene encodes multiple messenger ribonucleic acids. Mol Endocrinol 2:806–815

Jameson JL, Jaffe RC, Deutsch PJ, Albanese C, Habener JF (1988b) The gonadotropin α-gene contains multiple protein binding domains that interact to modulate basal and cAMP-responsive transcription. J Biol Chem 263:9879–9886

Jeannotte L, Trifiro MA, Plante RK, Chamberland M, Drouin J (1987) Tissue-specific activity of the pro-opiomelanocortin gene promoter. Mol Cell Biol 7:4058–4064

Jingami H, Matsukura S, Numa S, Imura H (1985) Effects of adrenalectomy and dexamethasone administration on the level of preprocorticotropin-releasing factor messenger ribonucleic acid (mRNA) in the hypothalamus and adrenocorticotropin/β-lipotropin precursor mRNA in the pituitary in rats. Endocrinology 117:1314–1320

Julier C, Weil D, Couillin P, Cote JC, Van Cong N, Foubert C, Boue A, Thirion JP, Kaplan JC, Junien C (1984) The beta chorionic gonadotropin-beta luteinizing gene cluster maps to human chromosome 19. Hum Genet 67:174–177

Kato Y, Imai K, Sakai T, Inoue K (1989) Simultaneous effect of gonadotropin-releasing hormone (GnRH) on the expression of two gonadotropin β genes by passive immunization to GnRH. Mol Cell Endocrinol 62:135–139

Kikkawa U, Kishimoto A, Nishizuka Y (1989) The protein kinase C family: heterogeneity and its implications. Annu Rev Biochem 58:31–44

Kim KE, Day RN, Maurer RA (1988a) Functional analysis of the interaction of a tissue-specific factor with an upstream enhancer element of the rat prolactin gene. Mol Endocrinol 2:1374–1381

Kim KE, Gordon DF, Maurer RA (1988b) Nucleotide sequence of the bovine gene for follicle-stimulating hormone β-subunit. DNA 7:227–233

Koenig RJ, Brent GA, Warne RL, Larsen PR, Moore DD (1987) Thyroid hormone receptor binds to a site in the rat growth hormone promoter required for induction by thyroid hormone. Proc Natl Acad Sci USA 84:5670–5674

Kourides IA, Gurr JA, Wolf O (1984) The regulation and organization of thyroid stimulating hormone genes. Recent Prog Horm Res 40:79–120

Lacaze-Masmonteil T, de Keyzer Y, Luton JP, Kahn A, Bertagna X (1987) Characterization of proopiomelanocortin transcripts in human nonpituitary tissues. Proc Natl Acad Sci USA 84:7261–7265

Lalloz MRA, Detta A, Clayton RN (1988a) Gonadotropin-releasing hormone is required for enhanced luteinizing hormone subunit gene expression *in vivo*. Endocrinology 122:1681–1688

Lalloz MRA, Detta A, Clayton RN (1988b) Gonadotropin-releasing hormone desensitization preferentially inhibits expression of the luteinizing hormone β subunit gene expression *in vivo*. Endocrinology 122:1689–1694

Lamberts SW (1988) The role of somatostatin in the regulation of anterior pituitary hormone secretion and the use of its analogs in the treatment of human pituitary tumors. Endocr Rev 9:417–436

Larsen PR, Harney JW, Moore DD (1986a) Sequences required for cell-type-specific thyroid hormone regulation of rat growth hormone promoter activity. J Biol Chem 261:14373–14376

Larsen PR, Harney JW, Moore DD (1986b) Repression mediates cell-type-specific expression of the rat growth hormone gene. Proc Natl Acad Sci USA 83:8283–8287

Laverriere JN, Morin A, Tixier-Vidal A, Truong A, Gourdji D, Martial JA (1983) Inverse control of prolactin and growth hormone gene expression: effect of thyroliberin on transcription and RNA stabilization. EMBO J 2:1493–1499

Laverriere JN, Tixier-Vidal A, Buisson N, Morin A, Martial JA, Gourdji D (1988) Preferential role of calcium in the regulation of prolactin gene transcription by thyrotropin-releasing hormone in GH_3 pituitary cells. Endocrinology 122:333–340

Lavin TN, Baxter JD, Horita S (1988) The thyroid hormone receptor binds to multiple domains of the rat growth hormone 5'-flanking sequence. J Biol Chem 263:9418–9426

Lechan RM, Wu P, Jackson IMD, Wolf H, Cooperman S, Mandel G, Goodman RH (1986) Thyrotropin-releasing hormone precursor: characterization in rat brain. Science 231:159–161

Lefevre C, Imagawa M, Dana S, Grindlay J, Bodner M, Karin M (1987) Tissue-specific expression of the human growth hormone gene is conferred in part by the binding of a specific *trans*-acting factor. EMBO J 6:971–981

Lemaigre FP, Peers B, Lafontaine DA, Mathy-Hartert M, Rousseau GG, Belayew A, Martial JA (1989) Pituitary-specific factor binding to the human prolactin, growth hormone, and placental lactogen genes. DNA 8:149–159

Leung K, Kaynard AH, Negrini BP, Kim KE, Maurer RA, Landefeld TD (1987) Differential regulation of gonadotropin subunit messenger ribonucleic acids by gonadotropin-releasing hormone pulse frequency in ewes. Mol Endocrinol 1:724–728

Lippman SS, Amr S, Weintraub BD (1986) Discordant effects of thyrotropin (TSH)-releasing hormone on pre- and posttranslational regulation of TSH biosynthesis in rat pituitary. Endocrinology 119:343–348

Lira SA, Crenshaw EB, Glass CK, Swanson LW, Rosenfeld MG (1988) Identification of rat growth hormone genomic sequences targeting pituitary expression in transgenic mice. Proc Natl Acad Sci USA 85:4755–4759

Loeffler JP, Kley N, Pittius CW, Hollt V (1985a) γ-Aminobutyric acid decreases levels of messenger ribonucleic acid encoding prolactin in the rat pituitary. Neurosci Lett 53:121–125

Loeffler JP, Kley N, Pittius CW, Hollt V (1985b) Corticotropin-releasing factor and forskolin increase proopiomelanocortin messenger RNA levels in rat anterior and intermediate cells in vitro. Neurosci Lett 62:383–387

Loeffler JP, Demeneix BA, Pittius CW, Kley N, Haegele KD, Hollt V (1986a) GABA differentially regulates the gene expression of proopiomelanocortin in rat intermediate and anterior pituitary. Peptides 7:253–258

Loeffler JP, Kley N, Pittius CW, Hollt V (1986b) Calcium ion and cyclic adenosine 3',5'-monophosphate regulate proopiomelanocortin messenger ribonucleic acid levels in rat intermediate and anterior pituitary lobes. Endocrinology 119:2840–2847

Luini A, Lewis D, Guild S, Corda D, Axelrod J (1985) Hormone secretagogues increase cytosolic calcium by increasing cAMP in corticotropin-secreting cells. Proc Natl Acad Sci USA 82:8034–8038

Luini A, Lewis D, Guild S, Schofield G, Weight F (1986) Somatostatin, an inhibitor of ACTH secretion, decreases cytosolic free calcium and voltage-dependent calcium current in a pituitary cell line. J Neurosci 6:3128–3132

Lundblad JR, Roberts JL (1988) Regulation of proopiomelanocortin gene expression in pituitary. Endocr Rev 9:135–158

Martin TFJ (1983) Thyrotropin-releasing hormone rapidly activates the phosphodiester hydrolysis of polyphosphoinositides in GH_3 pituitary cells: evidence for the role of a polyphosphoinositide-specific phospholipase C in hormone action. J Biol Chem 258:14816–14822

Maurer RA (1981) Transcriptional regulation of the prolactin gene by ergocryptine and cyclic AMP. Nature 294:94–97

Maurer RA (1982) Thyroid hormone specifically inhibits prolactin synthesis and decreases prolactin messenger ribonucleic acid levels in cultured pituitary cells. Endocrinology 110:1507–1514

Maurer RA (1987) Molecular cloning and nucleotide sequence analysis of complementary deoxyribonucleic acid for the β subunit of rat follicle stimulating hormone. Mol Endocrinol 1:717–723

Maurer Ra, Beck A (1986) Isolation and nucleotide sequence analysis of a cloned cDNA encoding the β subunit of bovine follicle-stimulating hormone. DNA 5:363–369

Maurer RA, Notides AC (1987) Identification of a estrogen-responsive element from the 5'-flanking region of the rat prolactin gene. Mol Cell Biol 7:4247–4254

Maurer RA, Croyle ML, Donelson JE (1984) The sequence of a cloned cDNA for the β subunit of bovine thyrotropin predicts a protein containing both NH_2- and COOH-terminal extensions. J Biol Chem 259:5024–5027

Mercer JE, Clements JA, Funder JW, Clarke IJ (1988) Luteinizing hormone-β mRNA levels are regulated primarily by gonadotropin-releasing hormone and not by negative estrogen feedback on the pituitary. Neuroendocrinology 47:563–566

Miller WL, Eberhardt NL (1983) Structure and evolution of the growth hormone gene family. Endocr Rev 4:97–130

Mishina M, Kurosaki T, Yamamoto T, Notake M, Masu M, Numa S (1982) DNA sequences required for transcription *in vivo* of the human corticotropin-β-lipotropin precursor gene. EMBO J 1:1533–1538

Mitchell PJ, Tjian R (1989) Transcriptional regulation in mammalian cells by sequence-specific DNA binding proteins. Science 245:371–378

Montminy MR, Bilezikjian LM (1987) Binding of a nuclear protein to the cyclic-AMP response element of the somatostatin gene. Nature 328:175–178

Montminy MR, Sevarino KA, Wagner JA, Mandel G, Goodman RH (1986) Identification of a cyclic-AMP-responsive element within the rat somatostatin gene. Proc Natl Acad Sci USA 83:6682–6686

Moore DD, Walker MD, Diamond DJ, Conkling MA, Goodman HM (1982) Structure, expression and evolution of growth hormone genes. Recent Prog Horm Res 38: 197–225

Moore DD, Marks AR, Buckley DI, Kapler G, Payvar F, Goodman HM (1985) The first intron of the human growth hormone gene contains a binding site for glucocorticoid receptor. Proc Natl Acad Sci USA 82:699–702

Morita S, Yamashita S, Melmed S (1987) Insulin-like growth factor I action on rat anterior pituitary cells: effects of intracellular messengers on growth hormone secretion and messenger ribonucleic acid levels. Endocrinology 121:2000–2006

Murdoch GH, Franco R, Evans RM, Rosenfeld MG (1983) Polypeptide hormone regulation of gene expression: Thyrotropin-releasing hormone rapidly stimulates both transcription of the prolactin gene and the phosphorylation of a specific nuclear protein. J Biol Chem 258:15329–15335

Murdoch GH, Evans RH, Rosenfeld MG (1985a) Polypeptide hormone regulation of prolactin gene transcription. Biochem Actions Horm 12:37–68

Murdoch GH, Waterman M, Evans RM, Rosenfeld MG (1985b) Molecular mechanisms of phorbol ester, thyrotropin-releasing hormone, and growth factor stimulation of prolactin gene transcription. J Biol Chem 260:11852–11858

Nakai Y, Tsukada T, Usui T, Takahashi H, Jingami H, Fukata J, Imura H (1988) Proopiomelanocortin (POMC) synthesis in the pituitary and non-pituitary tissues. In: Imura H, Shizume K, Yoshida S (eds) Progress in endocrinology. Elsevier, Amsterdam, pp 737–742

Nakamura M, Nakanishi S, Sueoka S, Imura H, Numa S (1978) Effects of steroid hormones on the level of corticotropin messenger RNA activity in cultured mouse-pituitary-tumor cells. Eur J Biochem 86:61–66

Nakanishi S, Kita T, Taii S, Imura H, Numa S (1977) Glucocorticoid effect on the level of corticotropin messenger RNA activity in rat pituitary. Proc Natl Acad Sci USA 74:3283–3286

Nakanishi S, Inoue A, Kita T, Nakamura M, Chang ACY, Cohen SN, Numa S (1979) Nucleotide sequence of cloned cDNA for bovine corticotropin-β-lipotropin precursor. Nature 278:423–427

Nakanishi S, Teranishi Y, Watanabe Y, Notake M, Noda M, Kakidani H, Jingami H, Numa S (1981) Isolation and characterization of the bovine corticotropin/β-lipotropin precursor gene. Eur J Biochem 115:429–438

Naor Z (1990) Signal transduction mechanisms of Ca^{2+} mobilizing hormones: the case of gonadotropin-releasing hormone. Endocr Rev 11:326–353

Naor Z, Catt KJ (1980) Independent actions of gonadotropin releasing hormone upon cyclic GMP production and luteinizing hormone release. J Biol Chem 255:342–344

Nelson C, Crenshaw III EB, Franco R, Lira SA, Albert VR, Evans RM, Rosenfeld MG (1986) Discrete *cis*-active genomic sequences dictate the pituitary cell type-specific expression of rat prolactin and growth hormone genes. Nature 322:557–562

Nelson C, Albert VR, Elsholtz HP, Lu LIW, Rosenfeld MG (1988) Activation of cell-specific expression of rat growth hormone and prolactin genes by a common transcription factor. Science 239:1400–1405

Niall HD, Hogan ML, Tregear GW, Segre GV, Hwang P, Friesen H (1973) The chemistry of growth hormone and the lactogenic hormones. Recent Prog Horm Res 29:387–416

Notake M, Kurosaki T, Yamamoto T, Handa H, Mishina M, Numa S (1983a) Sequence requirement for transcription *in vitro* of the human corticotropin/β-lipotropin precursor gene. Eur J Biochem 133:599–605

Notake M, Tobimatsu T, Watanabe Y, Takahashi H, Mishina M, Numa S (1983b) Isolation and characterization of the mouse corticotropin-β-lipotropin precursor gene and a related pseudogene. FEBS Lett 156:67–71

Nyborg JK, Spindler SR (1986) Alterations in local chromatin structure accompany thyroid hormone induction of growth hormone gene transcription. J Biol Chem 261:5685–5688

O'Malley B (1990) The steroid receptor superfamily: more excitement predicted for the future. Mol Endocrinol 4:363–369

Owerbach D, Rutter WJ, Martial JA, Baxter JD, Shows TB (1980) Genes for growth hormone, chorionic somatomammotropin, and growth hormone-like gene on chromosome 17 in humans. Science 209:289–292

Owerbach D, Rutter WJ, Cooke NE, Martial JA, Shows TB (1981) The prolactin gene is located on chromosome 6 in humans. Science 212:815–816

Papavasiliou SS, Zmeili S, Khoury S, Landefeld TD, Chin WW, Marshall JC (1986): Gonadotropin-releasing hormone differentially regulates expression of the genes for luteinizing hormone α and β subunits in male rats. Proc Natl Acad Sci USA 83:4026–4029

Perrin MH, Haas Y, Rivier JE, Vale WW (1986) Corticotropin-releasing factor binding to the anterior pituitary receptor is modulated by divalent cations and guanyl nucleotides. Endocrinology 118:1171–1179

Pierce JG, Parsons TF (1981) Glycoprotein hormones: structure and function. Annu Rev Biochem 50:465–495

Policastro P, Ovitt CE, Hoshina M, Fukuoka H, Boothby MR, Boime I (1983) The β subunit of human chorionic gonadotropin is encoded by multiple genes. J Biol Chem 258:11492–11499
Ramey JW, Highsmith RF, Wilfinger WW, Baldwin DM (1987) The effects of gonadotropin-releasing hormone and estradiol on luteinizing hormone biosynthesis in cultured rat anterior pituitary cells. Endocrinology 120:1503–1513
Raymond V, Leung PCK, Veilleux R, Labrie F (1985) Vasopressin rapidly stimulates phosphatidic acid-phosphatidylinositol turnover in rat anterior pituitary. FEBS Lett 182:196–200
Rebecchi MJ, Gershengorn MC (1983) Thyroliberin stimulates rapid hydrolysis of phosphatidylinositol 4,5-biphosphate by a phosphodiesterase in rat mammotropic pituitary cells: evidence for an early Ca^{2+}-independent action. Biochem J 216:287–294
Reisine T, Guild S (1985) Pertussis toxin blocks somatostatin inhibition of calcium mobilization and reduces the affinity of somatostatin receptors for agonists. J Pharmacol Exp Ther 235:551–557
Reisine T, Rougon G, Barbet J, Affolter HU (1985a) Corticotropin-releasing factor-induced adrenocorticotropin hormone release and synthesis is blocked by incorporation of the inhibitor of cyclic AMP-dependent protein kinase into anterior pituitary tumor cells by liposomes. Proc Natl Acad Sci USA 82:8261–8265
Reisine T, Zhang YL, Sekura R (1985b) Pertussis toxin treatment blocks the inhibition of somatostatin and increases the stimulation by forskolin of cyclic AMP accumulation and adrenocorticotropin secretion from mouse anterior pituitary tumor cells. J Pharmacol Exp Ther 232:275–282
Roberts JL, Budarf ML, Baxter JD, Herbert E (1979) Selective reduction of proadrenocorticotropin/endorphin proteins and messenger ribonucleic acid activity in mouse pituitary tumor cells by glucocorticoids. Biochemistry 18:4907–4914
Roberts JL, Lundblad JR, Eberwine JH, Fremeau RT, Salton SRJ, Blum M (1987) Hormonal regulation of POMC gene expression in pituitary. Ann NY Acad Sci 512:275–285
Rodin DA, Lalloz MRA, Clayton RN (1989) Gonadotropin-releasing hormone regulates follicle-stimulating hormone β subunit gene expression in the male rat. Endocrinology 125:1282–1289
Roesler WJ, Vandenbark GR, Hanson RW (1988) Cyclic AMP and the induction of eukaryotic gene transcription. J Biol Chem 263:9063–9066
Saade G, London DR, Clayton RN (1989) The interaction of gonadotropin-releasing hormone and estradiol on luteinizing hormone and prolactin gene expression in female hypogonadal (hpg) mice. Endocrinology 124:1744–1753
Salton SRJ, Blum M, Jonassen JA, Clayton RN, Roberts JL (1988) Stimulation of pituitary luteinizing hormone secretion by gonadotropin-releasing hormone is not coupled to β-luteinizing hormone gene transcription. Mol Endocrinol 2:1033–1042
Samuels HH, Forman BM, Horowitz ZD, Ye ZS (1988) Regulation of gene expression by thyroid hormone. J Clin Invest 81:957–967
Schachter BS, Johnson LK, Baxter JD, Roberts JL (1982) Differential regulation by glucocorticoids of proopiomelanocortin mRNA levels in the anterior and intermediate lobes of the rat pituitary. Endocrinology 110:1442–1444
Schoenenberg P, Kehrer P, Muller AF, Gaillard RC (1987) Angiotensin II potentiates corticotropin-releasing activity of CRF_{41} in rat anterior pituitary cells: mechanism of action. Neuroendocrinology 45:86–90
Schuster WA, Treacy MN, Martin F (1988) Tissue specific *trans*-acting factor interaction with proximal rat prolactin gene promoter sequences. EMBO J 7:1721–1733
Shupnik MA, Ardisson LJ, Meskell MJ, Bornstein J, Ridgway EC (1986a) Triiodothyronine (T_3) regulation of thyrotropin subunit gene transcription is proportional to T_3 nuclear receptor occupancy. Endocrinology 118:367–371
Shupnik MA, Greenspan SL, Ridgway EC (1986b) Transcriptional regulation of thyrotropin subunit genes by thyrotropin-releasing hormone and dopamine in pituitary cell culture. J Biol Chem 261:12675–12679
Shupnik MA, Weinmann CM, Notides AC, Chin WW (1989) An upstream region of the rat luteinizing hormone β gene binds estrogen receptor and confers estrogen responsiveness. J Biol Chem 264:80–86
Simard J, Labrie F, Gossard F (1986) Regulation of growth hormone mRNA and pro-opiomelanocortin mRNA levels by cyclic AMP in rat anterior pituitary cells in culture. DNA 5:263–270

Slater EP, Rabenau O, Karin M, Baxter JD, Beato M (1985) Glucocorticoid receptor binding and activation of a heterologous promoter by dexamethasone by the first intron of the human growth hormone gene. Mol Cell Biol 5:2984–2992
Sobel A, Tashjian Ah Jr (1983) Distinct patterns of cytoplasmic protein phosphorylation related to regulation of synthesis and release of prolactin by GH cells. J Biol Chem 258:10312–10324
Somasekhar MB, Gorski J (1988) Two elements of the rat prolactin 5'-flanking region are required for its regulation by estrogen and glucocorticoids. Gene 69:13–21
Stanley HF, Lyons V, Obonsawin MC, Bennie J, Carroll S, Roberts JL, Fink G (1988) Regulation of pituitary α subunit, β luteinizing hormone and prolactin messenger ribonucleic acid by gonadotropin-releasing hormone and estradiol in hypogonadal mice. Mol Endocrinol 2:1302–1310
Starzec A, Counis R, Jutisz M (1986) Gonadotropin-releasing hormone stimulates the synthesis of the polypeptide chains of luteinizing hormone. Endocrinology 119:561–565
Starzec A, Jutisz M, Counis R (1989) Cyclic adenosine monophosphate and phorbol ester, like gonadotropin-releasing hormone, stimulate the biosynthesis of luteinizing hormone polypeptide chains in a nonadditive manner. Mol Endocrinol 3:618–624
Steel JH, Hamid Q, Van Noorden S, Jones P, Denny P, Burrin J, Legon S, Bloom SR, Polak JM (1988) Combined use of in situ hybridisation and immunocytochemistry for the investigation of prolactin gene expression in immature, pubertal, pregnant, lactating and ovariectomised rats. Histochemistry 89:75–80
Supowit SC, Potter E, Evans RM, Rosenfeld MG (1984) Polypeptide hormone regulation of gene transcription: specific 5' genomic sequences are required for epidermal growth factor and phorbol ester regulation of prolactin gene expression. Proc Natl Acad Sci USA 81:2975–2979
Takahashi H, Hakamata Y, Watanabe Y, Kikuno R, Miyata T, Numa S (1983) Complete nucleotide sequence of the human corticotropin-β-lipotropin precursor gene. Nucleic Acids Res 11:6847–6858
Takahashi H, Nabeshima Y, Ogata K, Takeuchi S (1984) Molecular cloning and nucleotide sequence of DNA complementary to human decidual prolactin mRNA. J Biochem 95:1491–1499
Talmadge K, Vamvakopoulos NC, Fiddes JC (1984) Evolution of the genes for the β subunits of human chorionic gonadotropin and luteinizing hormone. Nature 307:37–40
Tan K-N, Tashjian Jr AH (1981) Receptor-mediated release of plasma membrane-associated calcium, and stimulation of calcium uptake by thyrotropin-releasing hormone in pituitary cells in culture. J Biol Chem 256:8994–9002
Tepper MA, Roberts JL (1984) Evidence for only one β-luteinizing hormone and no β-chorionic gonadotropin gene in the rat. Endocrinology 115:385–391
Tremblay Y, Tretjakoff I, Peterson A, Antakly T, Zhang CX, Drouin J (1988) Pituitary-specific expression and glucocorticoid regulation of a proopiomelanocortin fusion gene in transgenic mice. Proc Natl Acad Sci USA 85:8890–8894
Truong AT, Duez C, Belayew A, Renard A, Pictet R, Bell GI, Martial JA (1984) Isolation and characterization of the human prolactin gene. EMBO J 3:429–437
Tsukada T, Fink JS, Mandel G, Goodman RH (1987) Identification of a region in the human vasoactive intestinal polypeptide gene responsible for regulation by cyclic AMP. J Biol Chem 262:8743–8747
Uhler M, Herbert E, D'Eustachio P, Ruddle FD (1983) The mouse genome contains two nonallelic pro-opiomelanocortin genes. J Biol Chem 258:9444–9453
Usui T, Nakai Y, Tsukada T, Fukata J, Nakaishi S, Naitoh Y, Imura H (1989) Cyclic AMP-responsive region of the human proopiomelanocortin (POMC) gene. Mol Cell Endocrinol 62:141–146
Usui T, Nakai Y, Tsukada T, Takahashi H, Fukata J, Naito Y, Nakaishi S, Tominaga T, Murakami N, Imura H (1990) Expression of the human pro-opiomelanocortin gene introduced into a rat glial cell line. J Mol Endocrinol 4:169–175
Vale W, Spiess J, Rivier C, Rivier J (1981) Characterization of a 41-residue ovine hypothalamic peptide that stimulates secretion of corticotropin and β-endorphin. Science 213:1394–1397
Vale W, Vaughan J, Smith M, Yamamoto G, Rivier J, Rivier C (1983) Effects of synthetic ovine corticotropin-releasing factor, glucocorticoids, catecholamines, neurohypophysial peptides, and other substances on cultured corticotropic cells. Endocrinology 113:1121–1131

Virgin JB, Silver BJ, Thomason AR, Nilson JH (1985) The gene for the β subunit of bovine luteinizing hormone encodes a gonadotropin mRNA with an unusually short 5'-untranslated region. J Biol Chem 260:7072–7077

Vogel DL, Magner JA, Sherins RJ, Weintraub BD (1986) Biosynthesis, glycosylation, and secretion of rat luteinizing hormone α and β subunits: differential effects of orchiectomy and gonadotropin-releasing hormone. Endocrinology 119:202–213

Wark JD, Gurtler V (1986) Glucocorticoids antagonize induction of prolactin-gene expression by calcitriol in rat pituitary tumour cells. Biochem J 233:513–518

Watkins PC, Eddy R, Beck AK, Vellucci V, Leverone B, Tanzi RE, Gusella JF, Shows TB (1987) DNA sequence and regional assignment of the human follicle-stimulating hormone β subunit gene to the short arm of human chromosome 11. DNA 6:205–212

West BL, Catanzaro DF, Mellon SH, Cattini PA, Baxter JD, Reudelhuber TL (1987) Interaction of a tissue-specific factor with an essential rat growth hormone gene promoter element. Mol Cell Biol 7:1193–1197

White BA (1985) Evidence for a role of calmodulin in the regulation of prolactin gene expression. J Biol Chem 260:1213–1217

White BA, Bancroft FC (1983) Epidermal growth factor and thyrotropin-releasing hormone interact synergistically with calcium to regulate prolactin mRNA levels. J Biol Chem 258:4618–4622

White BA, Preston GM (1988) Evidence for a role of topoisomerase II in the Ca^{2+}-dependent basal level expression of the rat prolactin gene. Mol Endocrinol 2:40–46

Wierman ME, Rivier JE, Wang C (1989) Gonadotropin-releasing hormone-dependent regulation of gonadotropin subunit messenger ribonucleic acid levels in the rat. Endocrinology 124:272–278

Wight PA, Crew MD, Spindler SR (1988) Sequences essential for activity of the thyroid hormone responsive transcription stimulatory element of the rat growth hormone gene. Mol Endocrinol 2:536–542

Wondisford FE, Radovick S, Moates JM, Usala SJ, Weintraub BD (1988) Isolation and characterization of the human thyrotropin β subunit gene: differences in gene structure and promoter function from murine species. J Biol Chem 263:12538–12542

Wood DF, Docherty K, Ramsden DB, Sheppard MC (1987a) A comparison of the effects of bromocriptine and somatostatin on growth hormone gene expression in the rat anterior pituitary gland in vitro. Mol Cell Endocrinol 52:257–261

Wood WM, Gordon DF, Ridgway EC (1987b) Expression of the β subunit gene of murine thyrotropin results in multiple messenger ribonucleic acid species which are generated by alternative exon splicing. Mol Endocrinol 1:875–883

Yaffe BM, Samuels HH (1984) Hormonal regulation of the growth hormone gene: relationship of the rate of transcription to the level of nuclear thyroid hormone-receptor complexes. J Biol Chem 259:6284–6291

Yamashita S, Slanina S, Kado H, Melmed S (1986a) Autoregulation of pituitary growth hormone messenger ribonucleic acid levels in rats bearing transplantable mammosomatotrophic pituitary tumors. Endocrinology 118:915–918

Yamashita S, Weiss M, Melmed S (1986b) Insulin-like growth factor I regulates growth hormone secretion and messenger ribonucleic acid levels in human pituitary tumor cells. J Clin Endocrinol Metab 63:730–735

Ye ZS, Samuels HH (1987) Cell- and sequence-specific binding of nuclear proteins to 5'-flanking DNA of the rat growth hormone gene. J Biol Chem 262:6313–6317

Ye ZS, Forman BM, Aranda A, Pascual A, Park HY, Casanova J, Samuels HH (1988) Rat growth hormone gene expression: Both cell-specific and thyroid hormone response elements are required for thyroid hormone regulation. J Biol Chem 263:7821–7829

Use of Transgenic Animals in the Study of Neuropeptide Genes

A. FUKAMIZU[1] AND K. MURAKAMI[1]

Contents

1 Introduction

One of the important problems of neurobiology concerns the mechanisms underlying the ability of the nervous system to respond to external stimuli, such as hormones and growth factors. Since the nervous system maintains an extensive repertoire of reactions to stimuli, the biosynthesis of neurotransmitter molecules including neuropeptides has been widely investigated to date. The developing nervous system provides a particularly excellent model for studying cell differentiation, because neuronal and glial precursor cells in the mammalian central nervous system (CNS) are derived from neuroepithelial cells of the neural tube and their diversification begins in an early stage of embryogenesis. Change in neuronal electrical activity during development is one of the intermediate processes capable of translating external signals into changes in neuropeptide gene expression.

Control of behaviour in animals involves the transfer of information at chemical synapses and the conversion of neural signals into hormonal messages by the neuroendocrine systems, which represents one of the delivery pathways of neuropeptides to their target cells, similar to the endocrine systems; neuropeptides are released by neurons and act near or at a distance from the site of

[1] Institute of Applied Biochemistry, University of Tsukuba, Tsukuba City, Ibaraki 305, Japan

Current Topics in Neuroendocrinology, Vol. 11

production (Kow and Pfaff 1988). The other pathway is thought to be the neurocrine system, which is similar to the paracrine system (Fig. 1). The diverse cellular roles of neuropeptides are presumably all mediated by interactions between these peptides and the specific receptors located in the surface of their target cells. However, the nature of the intracellular signaling pathways that are likely to be triggered as a result of the interaction of neuropeptides with the receptor remains unclear. For understanding how these responses are regulated, it is therefore necessary to examine the function of the transmitter molecules in specific pathways in the brain and of the hormonal messenger molecules released by the neuroendocrine systems. In addition, it is very important to investigate how the gene expression and synthesis of neuropeptides are regulated.

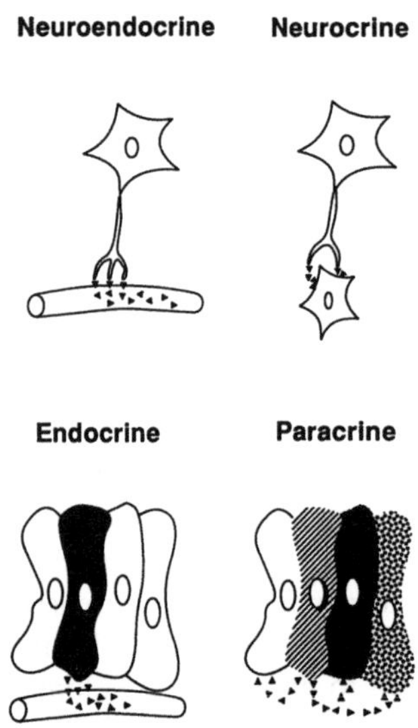

Fig. 1. Mechanisms of action of neuropeptides. The *closed triangles* represent neuropeptides

Over the past decades, a host of technical developments on protein purification have made it possible to isolate a large number of neuropeptides in mammalian brains. Additionally, marked advances in molecular biology have facilitated the structural determination of the genes and cDNAs for neuropeptides (Douglass et al. 1984; Sutcliffe 1988). At cellular levels, by advanced cell culture techniques, it is now possible to determine the level at which the regulation of neuropeptide gene expression occurrs. Although the mechanisms of neuropeptide action are still mostly unknown, an increasing body of recent evidence suggests that at least some neuropeptides have physiologically important actions on neural target cells. These neuropeptides have several interesting features; First, they have multiple functions and may serve as neurotransmitters or neuromodulators. Second, neuropeptides are not restricted to the central nervous system, being found also in the gastrointestinal tract and in the other organs in which the peptides were originally identified. Third, neuropeptides are generally synthesized as larger and biologically inactive precursors, and converted by a processing enzyme to the active form with subsequent cleavage at pairs of basic amino acids (e.g., Lys-Arg and Arg-Arg).

Introduction of cloned genes into animals by microinjection of fertilized eggs is a powerful approach to study gene regulation, because this permits functional studies of potential *cis*-acting DNA elements involved in gene regulation in vivo in a normal developmental environment. Transgenic mice expressing foreign genes of interest have thus been produced to test the regulatory DNA sequences containing promoter or enhancer, or sometimes a combination of both (Kelly and Darlington 1985). In addition, transgenic mice provide an attractive biological system for analyzing the roles of the introduced gene product. In neurobiology, a large number of transgenic mice have therefore been generated to investigate the function and regulation of neuropeptide genes. In the present short review, rather than attempt to survey all reports of the creation of transgenic mice carrying neuropeptide genes, we have focused on the functional analyses of growth hormone-releasing factor, somatostatin, gonadotropin-releasing hormone, calcitonin/calcitonin gene-related peptide, vasopressin, angiotensin II, and cAMP-responsive-element binding protein, the lattermost being a transcription factor controlling some of the neuropeptide genes, in transgenic mice.

2 Transgenic Animals

Although a detailed description of the methods employed in generating transgenic mice is beyond our scope in this review, the procedure is briefly illustrated in Fig. 2; pronuclear stage (one-cell) embryos are collected from female mice mated to fertile males after superovulation by sequential administration of follicle-stimulating hormone (FSH)/luteinizing hormone (LH) and human chorionic gonadotropin (hCG). The cloned DNA linearized by appropriate restriction endonucleases is injected into one of the male pronuclei with a microneedle. The manipulated embryo is then transferred into the oviduct of a foster mother (pseudopregnant recipient) previously mated with a sterile male. After birth, offspring obtained from these procedures are screened by DNA hybridization analysis to determine presence of the foreign gene (transgene). The more detailed technical applications are reviewed in excellent papers (Gordon et al. 1980; Gordon and Ruddle 1983; Hogan et al. 1986; Palmiter and Brinster 1986).

To date, most transgenic animals have been produced by using the above procedures. Analyses of the DNA in the adults developing from injected eggs reveal that up to 40% of such animals contain copies of the exogenous gene in somatic tissues, and these foreign genes are often transmitted through the germ-line to the next generation. To the best of our knowledge, the transgene is integrated into the chromosomes at random. Although the number of copies of the integrated DNA varies from one to a hundred, the integration usually occurs at only one chromosomal site in a given embryo. Once a transgenic line is established, these animals promise to be an extremely valuable tool for the analysis of transgene function during mammalian development. Therefore, transgenic mice have provided a number of model systems to study the actions of exogenous gene products such as hormones, growth factors, oncogenes, homeoboxes, and neuropeptides (Palmiter and Brinster 1986; Hanahan 1988, 1989; Rosenfeld et al. 1988; Connelly et al. 1989; Kessel and Gruss 1990).

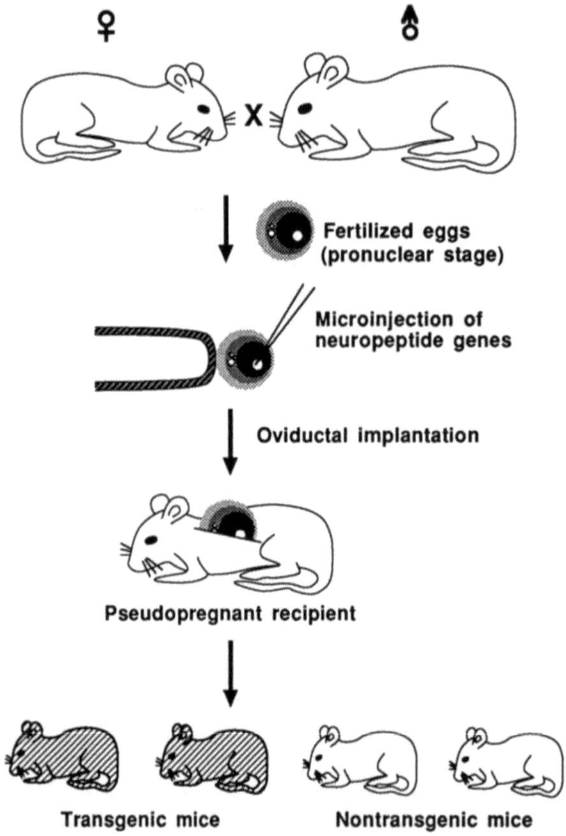

Fig. 2. Generation of transgenic mice by microinjection of genes into the fertilized eggs

2.1 Growth Hormone-Releasing Factor

The intracellular messenger system by which growth hormone (GH) exerts its biological activity has not been clearly established. It is generally accepted that GH must first bind to its specific receptors at the cell surface before it elicits its biological effects. It is also believed that insulin-like growth factor I is the major mediator for the somatogenic effects of GH. A great amount of interest has therefore been focused on the transfer of GH gene to accelerate growth of animals, since the effects of exogenous GH on the transgenic mice can be examined, at least in part, by phenotypic observation. Indeed, a number of transgenic animals that overexpress the product from the introduced rat, bovine, ovine, or human GH gene under control of the mouse metallothionein-I (MT-I) promoter/regulator have shown enhanced growth relative to that of control litter mates (Palmiter et al. 1982, 1983; Hammer et al. 1984, 1985a, b; Orian et al. 1989). The transgene expression was found in many organs, particularly liver and intestine. Based on these initial successes in controlling growth in mice by ectopic overproduction of foreign GH, several excellent investigations were carried out in neuropeptide biology (Ornits et al. 1985; Swanson et al. 1985; Behringer et al. 1988; Chandrashekar et al. 1988; Mathews et al. 1988; Borrelli et al. 1989; Steger et al. 1990,1991; Chen et al. 1991a, b; Orian et al. 1991).

In mammals, it is well known that somatic growth is regulated, in part, by controlling the secretion of GH from the anterior pituitary. This secretion is governed by a complex interaction of neural mechanisms and hormonal feedback regulation from the hypothalamus. Among them, hypothalamic growth hormone-releasing factor (GRF) and somatostatin are the most important factors. GRF is a 44-amino-acid neuropeptide which is believed to stimulate GH release from somatotrophs. For exploring in greater detail the in vivo role of GRF in controlling growth, it is important to generate the transgenic mice expressing GRF, because one would expect the overexposure of exogenous GRF to elicit an enhancement of mouse growth by stimulating the production of endogenous GH (Fig. 3).

To this end, Hammer et al. (1985c) established strains of transgenic mice carrying the human GRF gene under control of the mouse MT-I promoter and analyzed the effect of human GRF on somatic growth of mice. As expected, the GRF and elevated levels of GH were detected in serum and the transgenic mice grew to about 1.2 to 1.5 times the size of nontransgenic litter mates. Although MT-I promoter is known to function preferentially in liver, the human GRF gene expression was observed in the transgenic kidney, gut, and pancreas, as well as in

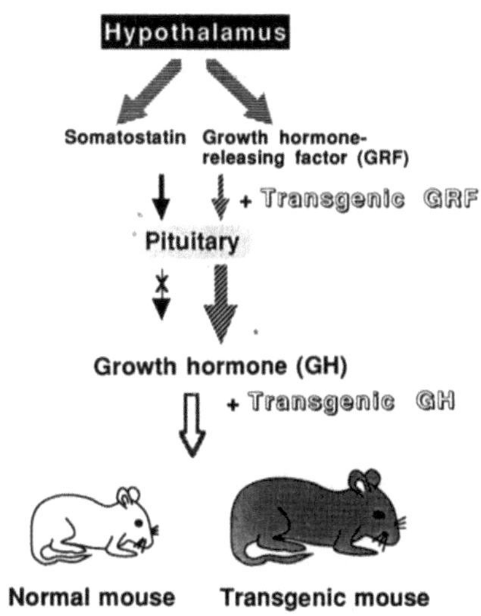

Fig. 3 Neuroendocrinological cascade of actions of growth hormone (*GH*) and growth hormone-releasing factor (*GRF*).

the liver. Additionally, the transgene was highly expressed in the pituitary and an unexpected proliferation of the somatotrophs was found, suggesting that GRF has an additional effect on somatotroph development.

2.2 Somatostatin

Almost all peptide hormones are synthesized in the cells as precursor proteins of higher molecular weight whose sequences have often been revealed using molecular biology techniques. The prohormones are usually biologically inactive and it has been recognized that they must undergo several posttranslational processing steps to yield the active peptides (Habener 1981; Docherty and Steiner 1982). Neuropeptides are also synthesized as a part of larger precursor, which may contain related or other active peptides (Lynch and Snyder 1986). The processing of neuropeptides proceeds through a limited and controlled proteolysis of precursor forms in a cascade fashion, catalyzed by specific sets of peptidases (prohormone-converting enzymes). Although there are common features to the generation of a wide variety of biologically active peptides, including neuropeptides, tissue-specific processing machinery clearly exists (Harris 1989). However, the precise mechanism of neuropeptide maturation, including the identification and purification of neuropeptide processing enzymes, has not yet been completely uncovered.

Somatostatin (SS) is one of the well-characterized neuropeptides and is expressed in hypothalamic and extrahypothalamic brains, pancreatic islets, gut mucosa, and parafollicular thyroid. It exists in two molecular forms in the body, as a tetradecapeptide (SS-14) and an octacosapeptide (SS-28), both of which have related physiological effects: inhibition of peptide hormone secretion, including insulin, growth hormone, and prolactin. The primary structure of SS has been determined by molecular cloning of SS cDNA and deduced to be a 92-amino-acid peptide (Shen et al. 1982; Goodman et al. 1983). Analysis of this precursor form revealed that SS-28 (the C-terminal 28-amino-acid residues of proSS) and SS-14 (the C-terminal fragment of SS-28) are processed from the same larger molecule. However, the difference of concentrations of processed fragments in the different tissues, where the SS gene is expressed, is remarkable. The pancreas contains almost entirely SS-14, whereas the predominant peptide of all small intestine is SS-28 (Trent and Weir 1981; Baskin and Ensinck 1984). Due to the fact that SS-28 is cleared more slowly in the circulation than SS-14 (Tannenbaum et al. 1982), it has been proposed that it takes part in the neurohumoral regulation, whereas SS-14 is involved as a neurotransmitter in the brain.

To examine a correct processing of SS produced from the exogenously introduced SS gene, Low et al. (1985) generated the transgenic mice carrying a MT-I promoter/rat SS fusion gene and analyzed the SS products in several tissues. Analysis of the SS peptide concentrations showed that the most active site of SS production was the transgenic anterior pituitary in which SS-14 was predominantly processed, although the pituitary does not normally synthesize SS. In contrast, this processing was observed neither in the liver nor kidney. From these findings, the authors concluded that in the case of SS, the processing is not limited to tissues that normally express SS.

2.3 Gonadotropin-Releasing Hormone

The gonadal polypeptide hormones LH and FSH influence mammalian reproductive function. The menstrual cycle is well known to be the result of complex interacting processes among the hypothalamus, pituitary gland, and ovaries. An abundance of experimental investigations has been made in order to understand the neuroendocrine events which control the menstrual cycle, particularly as to how the neural information signal is translated into endocrine secretion. This signal eventually results in the release of decaneuropeptide gonadotropin-releasing hormone (GnRH) by nerve terminals in the mediobasal hypothalamus, which is the final common pathway controlling the pulsatile secretion of LH and FSH from the anterior pituitary gland (Schally 1978; Fink 1979). Another important function of central GnRH may be to regulate the expression of sexual behaviour. In most rodent brains, immunoreactive GnRH neurons are diffusely distributed from the olfactory bulb to the caudal extent of the anterior hypothalamus, with the majority located in the medial preoptic area (Barry 1979; Merchenthaler et al. 1980; Jennes and Stumpf 1986).

It is of interest that there is a mouse which possesses a genetically mutated locus affecting the neuroendocrine control of reproduction, resulting in the failure of postnatal gonadal development. This hypogonadal (*hpg*) mouse was first described in 1977 by Cattanach et al. It has long been postulated that a lesion of GnRH gene is deeply involved in this autosomal recessive mutation, because the *hpg* mouse shows a failure to synthesize GnRH in hypothalamic neurons adequately. Thus, the genetic mutation in the *hpg* mouse provided a very interesting experimental model in the area of neurobiology.

For analyzing this defect in the *hpg* mice, Mason et al. (1986a) first isolated and determined the structure of the normal and *hpg* mouse GnRH genes. It has been shown that the normal GnRH gene consists of four exons interrupted by three introns. In contrast, the GnRH gene in the *hpg* mouse has a large deletion of at least 33.5 kb and lacks the third and fourth exons, resulting in a gene encoding only the signal peptide and GnRH decapeptide. Furthermore, the authors have demonstrated by in situ hybridization technique that GnRH messenger RNA (mRNA) with a lower level than that of the normal one is transcribed from the truncated GnRH gene in the *hpg* mouse, although the number of GnRH cell bodies was significantly reduced in the anterior hypothalamus. In contrast, immunoreactive GnRH was not found in the *hpg* mouse brains.

To rescue the genetic lesion in the *hpg* mice, Mason et al. (1986b) next tried to restore the reproductive function of the mutant mouse by generating the transgenic *hpg* mouse. First of all, they introduced into normal mice the 13.5-kb genomic DNA fragment containing the normal GnRH gene locus with its natural promoter and produced the transgenic mice carrying more than 20 copies of the foreign gene. By mating *hpg*/+ heterozygous mouse with these transgenic mice, they succeeded in generating a *hpg/hpg* homozygous transgenic mouse expressing the transgene in the hypothalamus. In this type of transgenic mouse, the females displayed normal estrous cycle and the male showed full mating behavior. This phenotypic observation emphasizes the physiological consequence of the restored levels of FSH and LH in the serum and pituitary of the *hpg* mice.

2.4 Calcitonin/Calcitonin Gene-Related Peptide

Alternative RNA splicing would appear to play an important role in the control of neuropeptide function as a post-transcriptional regulator because this processing is capable of generating two or more mRNA transcripts from a single gene, in a cell-type specific manner, resulting in the production of the related peptides with similar actions in the different tissues or distinct biological system (Rosenfeld et al. 1984; Leff et al. 1986; Padgett et al. 1986). Calcitonin (CT)/calcitonin gene-related peptide (CGRP) is one of the representative neuropeptides with respect to this RNA splicing choice. CT, synthesized predominantly in thyroid C cells (Jacobs et al. 1981), is a 32-amino-acid peptide hormone, which functions as a physiological regulator of calcium. In contrast, CGRP, a 37-amino-acid peptide produced mainly in the central and peripheral nervous systems by alternative processing of CT gene, is one of the first novel bioactive peptides to be discovered by recombinant DNA technique (Amara et al. 1982; Rosenfeld et al. 1983). Although CGRP is also expressed in the thyroid C cells, the ratio of CT and CGRP expression is thought to be 95:1 (Sabate et al. 1985). A number of biological activities of CGRP have been demonstrated, several of them in common with CT. The reported activities of CGRP correlate well with the presence of the peptide in selective areas of the CNS (Kawai et al. 1985; Shimada et al. 1985), supporting an idea that CGRP might be a neurotransmitter or neuromodulator.

To understand the regulatory mechanism of the alternative RNA splicing CT/CGRP gene, Crenshaw et al. (1987) have created the transgenic mice, bearing the rat CT/CGRP gene under the control of MT-I gene promoter. This promoter is advantageous, since it is active in a wide variety of tissues and can direct expression of many fusion genes in transgenic mice (Palmiter and Brinster 1986). Therefore, this strategy made it possible to ask whether CT/CGRP RNA is spliced in a tissue-specific or ubiquitous fashion. The authors have shown that the transgene is expressed in most visceral tissues (liver, kidney, lung, spleen, stomach, submandibular gland, and heart) and that CT mRNA is the primary transcript. In the brain, on the other hand, CGRP mRNA was the predominant neuronal transcript, and it also expressed in many neurons irrespective of an undetectable level of the endogenous CGRP expression. In addition to CGRP mRNA, CT transcript was contained in a small fraction of neurons. These results clearly demonstrated that although CT transcript is generated in a wide variety of tissues, CGRP splicing is restricted to neurons. Even though factors involved in the tissue-specific splicing remain to be identified, it is possible that selective RNA splicing machinery is required for CT/CGRP gene processing.

2.5 Vasopressin

The neurohypophyseal hormones vasopressin (VP) and oxytocin (OT) are encoded by highly homologous genes that are closely linked in the mammalian genomes, being separated by only about 10 kb (Richter 1988). The two genes are located in opposite orientations within about 18 kb, and each gene has three exons interrupted by two exons: Exon 1 encodes the signal peptide and the N-terminus of carrier molecules of these peptides, neurophysin II (NP-II) for VP or neurophysin I (NP-I) for OT. Exon 2 encodes most of the NP-I or NP-II. Exon 3 encodes the C-terminus of these carrier sequences and, in the case of VP, a glycopeptide whose function is unknown. Although nonapeptides VP and OT differ from each other by only two amino acids, these peptides hormones are well known for their distinct biological functions. VP plays a crucial role in the regulation of cardiovascular function. Its antidiuretic actions on kidney correct fluctuations in blood osmolality, and its pressor actions on vascular smooth muscle maintain peripheral resistance (Share 1988). On the other hand, OT has important functions in contraction of the smooth muscle of myoepithelial cells of the lactating mammary gland for the milk let-down reflex and in concentration of the parturient uterine muscle.

VP is synthesized in the magnocellular neurons of the supraoptic (SON) and paraventricular (PVN) nuclei of the hypothalamus. Many of the neurons in the SON and PVN project to the posterior pituitary and release VP into the peripheral circulation. In order to examine the effect of overexpressed VP on mouse physiology, Habener et al. (1989) produced the transgenic mice having MT-I promoter/rat VP gene in which the 35-bp 5'-flanking region of the VP gene immediately downstream of the MT-I sequences possesses a tentative neuronal cell-specific expression element, – GCCCAGCC – (Theill et al. 1987). In transgenic mice, levels of VP were markedly elevated in the liver, kidney, intestine, and pancreas, where MT-I promoter is functional. Also, levels of VP in the brain were elevated three to four times compared with those of the non-transgenic mice. Although the endogenous expression of MT-I gene has not been detected in PVN of hypothalamus, analysis of in situ hybridization histochemistry demonstrated that the transgene is expressed in the magnocellular neurons. This suggested that the above-cited octameric DNA sequence confers cell-specific expression of the transgene. A similar observation was reported that a metallothionein-growth hormone fusion gene was expressed in PVN of the transgenic mice (Swanson et al. 1985). Interestingly, serum levels of VP and serum osmolalities were elevated, consistent with a state of mild nephrogenic diabetes insipidus. However, the transgenic mice did not show any physiologic sign such as hypervasopressinemia.

2.6 Angiotensin II

The octapeptide angiotensin II (AII) is generated from a larger precursor, angiotensinogen, whose *M*r is 62K, by several processing steps: This precursor is mainly synthesized in the liver and secreted into the circulation, where it is cleaved by renin produced in the juxtaglomerular cells of the kidney to release a decapeptide angiotensin I (AI). This peptide, a physiologically inactive form, is subsequently processed to AII by angiotensin-converting enzyme in the lung (Fig. 4). The biologically active AII is a most potent vascoconstrictor and a factor that stimulates aldosterone secretion from the adrenal gland. Through these AII actions the circulating renin-angiotensin system plays an important role in the regulation of blood pressure and electrolyte homeostasis (Oparil and Haber 1974; Reid et al. 1978) and thus this system has been believed to be involved in the pathogenesis of hypertension (Ganten et al. 1991; Krieger and Dzau 1991).

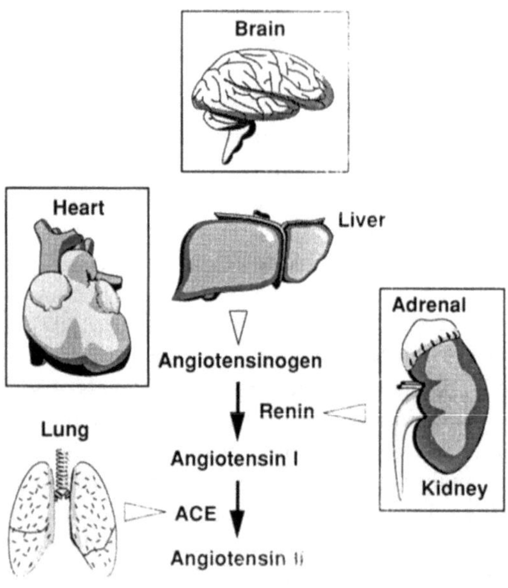

Fig. 4. The renin-angiotensin system. The organs that have been reported to possess all the components of the renin-angiotensin systems are indicated by *squares*. *ACE*, angiotensin-converting enzyme

In addition to the well-defined peripheral effects, AII has been known to exert a number of actions on the CNS. The central actions of AII includes effects upon thirst, fluid intake, blood pressure regulation, and release of pituitary hormones (Phillips 1987). Much evidence supports the existence of a functional renin-angiotensin system having all the components including renin, angiotensinogen, AI, AII, and angiotensin-converting enzyme in the brain (Ganten and Speck 1978; Ganten et al. 1983; Ganong 1984). Since plasma AII does not cross the blood-brain barrier, it follows that this neuropeptide must be generated in situ within the brain to work as neuroendocrine or neurocrine peptides or gain access to the brain in effective amounts via the circumventricular organs through the AII receptors. In fact, the capability of central AII synthesis has been shown in neuronal cultures from brains excluding the peripheral renin-angiotensin system (Raizada et al. 1983, 1984). Furthermore, specific AII-binding sites have also been found in the brain which were compatible with the immunohistochemical localization of AII (Sirett et al. 1977; Phillips et al. 1979; Mendelsohn et al. 1984; Lind et al. 1985): The demonstration of the existence of the brain renin-angiotensin system has raised the possibility that AII could have an important neuroendocrine role. However, the regulatory mechanisms of the renin-angiotensin system in the brain and of the gene expression of the components of this system are not understood completely.

Recently, a new idea has been proposed to augment the traditional concept of the endocrine renin-angiotensin system: the paracrine-autocrine effects of locally synthesized AII influence tissue functions (Campbell 1987; Jin et al. 1988). The existence of the local renin-angiotensin system is supported by an increasing body of evidence that renin, angiotensinogen, angiotensin-converting enzyme, and angiotensin receptor are widely distributed in tissues, such as the adrenal, brain, kidney, heart, ovary, and testis. Furthermore, AII has been reported to be a factor involved in new vessel formation (Fernandez et al. 1985), reproductive functions such as follicular development and ovulation (Pellicer et al. 1988), and cell growth (Dzau 1988; Dzau and Safar 1988). In particular, the effect of AII on blood vessel growth, e.g., hypertrophy and hyperplasia, should be noted; it has been shown that AII can influence blood vessel growth indirectly via norepinephrine (Loudon et al. 1983), which is reported to be capable of inducing the growth of cardiac monocytes and stimulating the expression of proto-oncogene c-*myc* in the monocytes (Starksen et al. 1986). It has also been demonstrated that AII can induce c-*myc* gene expression in rat vascular smooth muscle cells (Naftilan et al. 1989). Of particular interest is the fact that constitutive expression of c-*myc* mRNA in the heart of the transgenic mice carrying the c-*myc* gene results in enhanced hyperplastic growth during development and suggests a regulatory role for c-*myc* in cardiac myogenesis (Jackson et al. 1990). Although there is no direct evidence that AII can promote the growth of cardiac myocyte cells, it is interesting to speculate that AII has an important role in the heart cell development by controlling c-*myc*.

As initial contributions to evaluating the in vivo functions of the renin-angiotensin system, a number of transgenic animals carrying the renin or angiotensinogen genes have been created by several laboratories. Although expression of the foreign renin or angiotensinogen genes was regulated in tissue- or cell type-specific manners, most of the transgenic mice did not show any high blood pressure (Tronik et al. 1987; Mullins et al. 1988, 1989; Clouston et al. 1989; Fukamizu et al. 1989, 1991a, b; Miller et al. 1989; Seo et al. 1990; Takahashi et al. 1991) On the other hand, the two groups reported the generation of transgenic animals with elevated blood pressure:Ohkubo et al. (1990) produced transgenic mice carrying the rat renin or rat angiotensinogen gene, both of which were under control of mouse MT-I gene promoter. Only the transgenic mice possessing both of the introduced genes exhibited significantly elevated blood pressure, whereas those harboring either of the transgenes maintained the normal blood pressure. Mullins et al. (1990) also developed a transgenic rat bearing the mouse *Ren-2* gene with its natural promoter. In the rats, the transgene was highly expressed in the adrenal, but its expression was suppressed in the kidney, although the former is a minor source and the latter is a major source of renin production. Interestingly, these rats showed a fulminant hypertension, and this disorder was inheritable. More interestingly, in the case of both transgenic animal models, the specific angiotensin-converting enzyme inhibitor captopril was effective in reducing the elevated blood pressure, indicating that this elevation of blood pressure is mediated by the renin-angiotensin system. Therefore, these results may support the traditional and novel concepts, as described in the first and third paragraphs in this section, that the endocrine and paracrine/autocrine functions of the renin-angiotensin system play an important role in the maintenance of blood pressure and in the development of hypertension, although the mechanism of elevation of blood pressure in the transgenic animals has not yet been clarified.

Although we have previously suggested that the sequences 1.3-kb upstream from the putative transcription start site of the human angiotensinogen gene can function as a promoter in human hepatoma HepG2 cells (Fukamizu et al. 1990a), little is known about the molecular mechanism underlying expression of the gene. However, such information should be of major importance, not only for enhancing our understanding of this gene regulation but also as the basis for creating an animal model of human hypertension in the future. Thus, we have recently generated the two lines of transgenic mice, hAG2-5 and hAG3-2 – carrying the 14-kb genomic fragment containing the 1.3-kb ustream promoter region, exons 1 to 5, and 1.3-kb 3'-flanking sequences – and examined whether the genetic information of the isolated gene is sufficient to regulate this gene in vivo and in vitro (Takahashi et al. 1991). As summarized in Table 1, the level of the transgene mRNA in both lines was highest in the liver, consistent with the tissue specificity of the angiotensinogen gene expression. Surprisingly, a large quantity of mRNA for the transgene were found in the kidney at levels comparable to that in the liver, although the level of mouse angiotensinogen mRNA was very low in the mouse kidney as well as in human kidney (Fukamizu et al. 1990b).

One possible explanation could account for the lack of down regulation of the human angiotensinogen gene in the kidney of transgenic mice: suppression of the gene expression in the kidney is usually maintained by a negative regulatory

Table 1. Tissue distribution of angiotensinogen mRNA

Tissues	hAG2-5	hAG3-2	Human	NM
Liver	+++	+++	+++	+++
Kidney	+++	+++	+	+
Brain	+	+	NA	+
SMG	+	+	NA	+
Heart	++	++	NA	+
Lung	+	+	NA	–
Spleen	–	–	NA	–

+ + +, strong expression; + +, moderate expresssion; +, weak expression; –, not detected; NA, data not available; hAG, human angiotensinogen; SMG, submandibular gland; NM, normal mouse.

sequence that is missing in the microinjected DNA fragment. The other is that a mouse negative regulatory factor (or factors) can not recognize the human gene sequences. To test these hypotheses, we constructed a fusion reporter gene containing the 1.3-kb 5'-flanking region of the human angiotensinogen gene linked to a bacterial chloramphenicol acetyltransferase (CAT) gene: we transfected it into two human cell lines, HepG2 cells that express angiotensinogen and embryonic kidney 293 cells that do not express angiotensinogen; and we examined the promoter activity of the fragment. We detected high levels of CAT activity in both cell lines, suggesting that the upstream sequences are also functional in renal cells. If there is a negative regulatory sequence (or sequences) within the human angiotensinogen gene used for the microinjection, one might expect that a cloned DNA fragment can repress the CAT transcription from the reporter gene in 293 cells because the endogenous transcription of the human angiotensinogen gene is not observed in the cells. To this end, we devided the 12.7-kb transgenic construct, exept for the 1.3-kb promoter region, into three subfragments, ligated them into the 3' end of the above-mentioned CAT gene, and examined their ability to repress CAT transcription from the fusion genes. The transient transfection analysis showed that neither DNA fragment represses the CAT transcription. From these observations, we concluded that the 1.3-kb 5'-flanking sequences are essential for efficient expression of the angiotensinogen gene in hepatic and renal cells and that the microinjected DNA construct lacks key control elements that normally function to repress the gene expression in renal cells. Our concepts are summarized in Fig. 5

Angiotensinogen has an additional property: it is an acute-phase protein. This has been supported by several lines of evidence: The deduced amino acid sequences from the cDNA (Ohkubo et al. 1983) of the large C-terminal portion of angiotensinogen are found to be significantly related to those of α_1-antitrypsin and α_1-antichymotrypsin (Doolittle 1983), which are classified as acute-phase proteins. It has also been shown that (a) the structure of the angiotensinogen gene shares an identical structural organization with the α_1-antitrypsin and α_1-antichymotrypsin genes with respect to the number and positions of introns (Tanaka et al. 1984; Gaillard et al. 1989; Fukamizu et al. 1990a), (b) rat angiotensinogen mRNA in the liver increases rapidly following the administration of *Escherichia coli* lipopolysaccharide (LPS) (Kageyama et al. 1985), and (c) the induction of angiotensinogen mRNA by acute-phase response is regulated at the transcriptional level and mediated by a *cis*-acting DNA element located upstream of the transcription initiation site of the rat angiotensinogen gene (Ron et al. 1990). Furthermore, Clouston et al. (1989) demonstrated that a mouse angiotensinogen minigene including the 0.75-kb 5'-flanking region, exon 1, parts of exon 2 and 4, and exon 5 contains sufficient genetic information to mediate induction by glucocorticoid and LPS. Since human angiotensinogen has been also reported to be an acute-phase protein (Nielsen and Knudsen 1987), it is a logical extension of this to explore the effect of LPS on expression of the human angiotensinogen gene in transgenic mice. After the administration of LPS and glucocorticoid into the hAG3-2 transgenic mouse, we analyzed the level of hepatic angiotensinogen mRNA by Northern blot hybridization (Takahashi et al. 1992). Unexpectedly, we

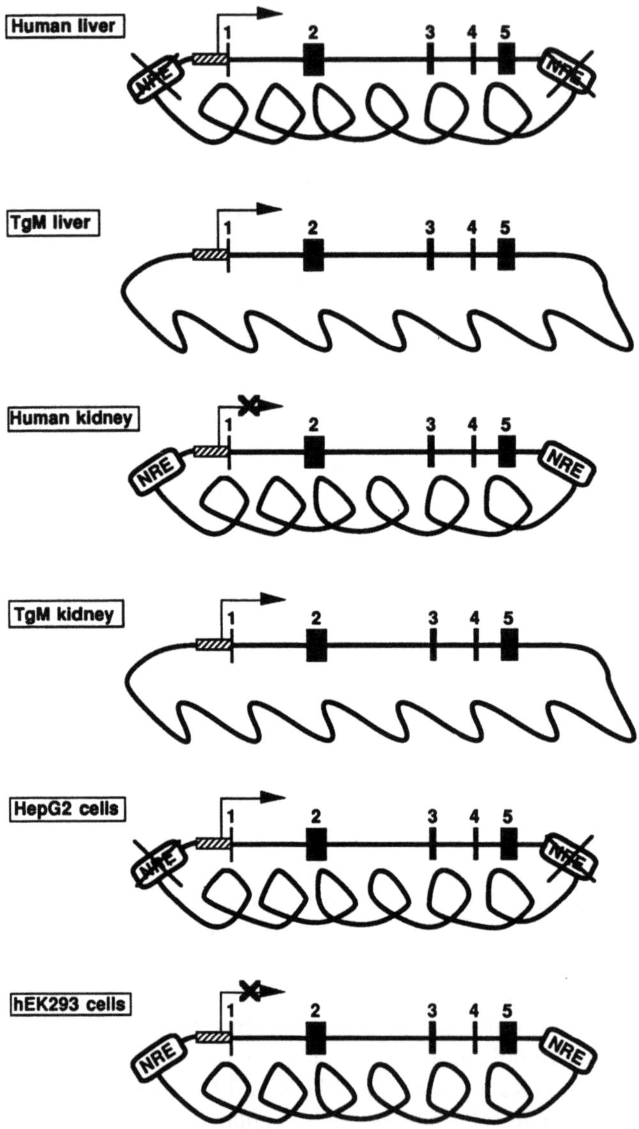

Fig. 5. Schematic representation of possible regulatory mechanisms of the human angiotensinogen gene. The *hatched boxes* and the *black boxes* with sequential numbers represent the 1.3-kb promoter sequences and exons of the human angiotensinogen gene, respectively. The *looped* and *wavy* lines represent the human and mouse chromosomes, respectively. The direction of transcription was indicated by *arrows*. *NRE*, negative regulatory element; *TgM*, transgenic mouse; *hEK*, human embryonic kidney

failed to induce human angiotensinogen mRNA in the transgenic mice by this treatment, although the possible acute phase-responsive DNA element was identified in the 5'-flanking sequences of the gene (Fukamizu et al. 1990a). In contrast, the level of the endogenous mouse angiotensinogen mRNA was significantly increased in the same transgenic mice. It has been reported that the expression pattern of the acute-phase proteins varies from one species to another (Heinrich et al. 1990): serum amyloid P, for example, shows dramatic elevation during the acute phase in mouse, but neither in rat or nor human; α_2-macroglobulin is the acute-phase protein with the most spectacular changes in rat, but not in human, suggesting that synthesis of acute-phase proteins are controlled in a species-specific fashion during the acute-phase response. Thus, we provided evidence that expression of the mouse and human angiotensinogen genes is differentially regulated during the acute-phase response in transgenic mice.

3 Neuropeptide Gene Expression and cAMP-Responsive-Element Binding Protein

The molecular events underlying cellular response to external stimuli, triggered by the interaction of peptide hormones or growth factors with cell surface receptors, represent a major focus of the investigation of cell differentation and proliferation. In many systems, the flow of information from the cell surface to the nucleus is mediated by two major biochemical pathways: one is related to the turnover of phosphatidylinositol with subsequent protein kinase C (PKC) activation (Nishizuka 1984) and the other is via the cyclic AMP (cAMP)-dependent protein kinase A (PKA) (Roesler et al. 1988). These kinases are activated by second messengers, such as diacylglycerol (DG) and cAMP, leading to the phosphorylation of physiologically important cyctoplasmic or nuclear proteins. Ultimately, the external signals cause profound changes in cellular functions by altering the basic pattern of gene expression through such second messengers. Indeed, transcription of a number of cellular genes is known to be activated by phorbol esters such as 12-*O*-tetradecanoylphorbol-13-acetate (TPA) and cAMP analogues, which can mimic the receptor-mediated signal transduction. It has been generally accepted that gene expression is regulated through sequence-specific binding of proteins to *cis*-acting DNA elements (Maniatis et al. 1987; Jones et al. 1988; Mitchell and Tjian 1989), and a large number of regulatory DNA sequences have been studied to date. The palindromic DNA sequence TGA(C/G) TCA mediates transcriptional activation in response to TPA, which stimulates the PKC-dependent signaling pathway instead of DG (Angel et al. 1987). The closely related sequence TGACGTCA mediates transcriptional activation in response to cAMP, which stimulates the PKA-dependent signal transduction pathway (Habener 1990). Thus, because of their functional properties, the former and the latter sequences are designated as the TPA-responsive element (TRE) and the cAMP-responsive element (CRE), respectively.

It is well known that neuronal cells respond to environmental signals by a change in neuronal electrical activity, which is one of the intermediate processes capable of translating external signals into changes in neuropeptide gene expression (Black et al. 1987; Comb et al. 1987). Since neuronal cell types are defined in part by the individual neuropeptides that they produce, examination of the mechanism underlying the ability of the nervous system to regulate neuropeptide gene expression in response to extracellular stimuli could enhance our knowledge of how neuronal cell phenotype in this system is controlled. One approach for studying the regulation of neuropeptide genes is to employ neuroendocrine tumor cell lines. These are particular valuable for the analysis of *cis*-acting DNA elements in the genes as well as relevant transcription factors. One of these cell lines, PC12, which was originally derived from a rat pheochromocytoma and synthesizes a variety of neuropeptides, including somatostatin and other neuroendocrine markers, has been widely used as a good model for studies of neuropeptide gene regulation. Using this cell line, a functional CRE was first identified in the 5'-flanking region of the gene for somatostatin (Montminy et al. 1986), and it has now been recognized that many neuropeptide genes possess CRE or CRE-like sequences within 150 bp of the transcription start site in their promoter regions (Goodman 1990; Montminy et al. 1990).

In order to further explore the regulatory mechanism of eukaryotic gene expression by cAMP, Montminy and Bilezikjian (1987) purified a CRE-binding protein (CREB) with *M*r 43K from PC 12 cells using DNA-affinity chromatography and showed that the purified CREB binds to CRE and is a phosphorylatable protein. Like some other transcription factors, CREB exists in monomeric and dimeric forms (Fig. 6), with the dimer having a ten fold higher affinity for the CRE than the monomer (Yamamoto et al. 1988). Dephosphorylation of CREB dramatically reduced the ability of the dimer to bind to CRE, but had no effect on the binding activity of the monomer. Of particular interest is the fact that CREB is efficiently phosphorylated by PKA and PKC. Although CREB binding activity was unchanged by PKA, phosphorylation of CREB by PKC induced formation of the dimer capable of binding to CRE. Analyses of amino acid sequence by molecular cloning and phosphorylation site by peptide mapping revealed that CREB consists of 341 amino acid residues, in which Ser-133 is critical to the *trans*-activation by PKA (Gonzalez et al. 1989a,b). Intertstingly, further mutation analysis of this Ser to Ala demonstrated that the mutant CREBM1 abolished the ability to activate transcription through CRE (Gonzalez et al. 1989b), although this mutant appears to be able to maintain the basal activity of transcription (Yamamoto et al. 1990). Therefore, these findings provide evidence that the extracellular stimuli translate into the intracellular information through PKA and PKC signaling pathways, resulting in the gene activation by, at least in part, CREB phosphorylation.

cAMP has been shown to have a unique mitogenic activity on somatotrophic cells of the anterior pituitary (Billestrup et al. 1986). If CREB plays an important role in proliferation of the somatotrophic cell by cAMP, the overexpressed CREBM1 would antagonize CREB activity and prevent the normal development of these cells. To demonstrate an involvement of CREB in the somatotroph development, Struthers et al. (1991) have generated transgenic mice harboring the CREBM1 gene under control of the rat growth hormone promoter, which directs somatotroph-dominant expression of transgenes in the pituitary (Behringer et al. 1988). As they expected, the transgenic mice exhibited a dwarf phenotype with the deficient pituitary showing normal posterior and intermediate lobes but atrophied anterior lobes. Although the mechanisms by which this dwarfism is induced remains unclear, it is evident that the overexpressed CREBM1 without the phosphoacceptor site of PKA blocks normal CREB activity. This result provides the novel concept that the constitutive phosphorylation and activation of CREB are required for somatotroph proliferation.

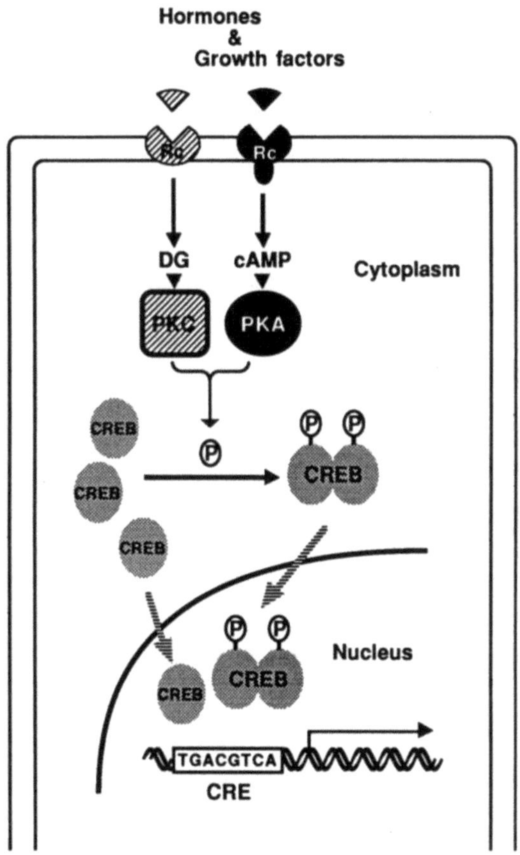

Fig. 6. Signal transduction pathways via the phosphorylation of cAMP-responsive element binding protein (*CREB*) by activation of protein kinase C (*PKC*) and protein kinase A (*PKA*). *Rc*, receptor; *DG*, diacylglycerol; *P*,phosphorylation; *CRE*, cAMP-responsive element

4 Concluding Remarks

Over 10 years have passed since the first successful generation of transgenic mice by pronuclear microinjection was reported by Gordon et al. (1980). Until now, a huge number of transgenic animals have been produced for the functional studies of not only neuropeptides, but also of hormones, growth factors, enzymes, oncogenes, homeoboxes, transcription factors, and virus proteins. One of the most important aspects of transgenic technology by introduction of genes to be examined involves the fact, that the transgene can be overexpressed under tissue-specific or nonspecific conditions by using its natural or chimeric promoters in animals and that it is thereby possible that physiologic consequences of the oversynthesized transgene products can be observed phenotypically. In fact, these animals have contributed to the development of therapeutic agents of diseases. For such purposes, the potential for creating animal models of human illness is likely to continue to increase.

Although the recombinational reactions of foreign gene transfer into animals are usually nonhomologous, cells also possess an enzymatic machinery required for homologous recombination. This type of recombination between chromosomal DNA and exogenous gene is referred to as gene targeting, and it offers an additional dimension to transgenic technology (Capecchi 1989). Gene targeting permits the transfer of genetic alterations created in vitro to precise sites within the cellular genome. Particularly, mouse blastocyst-derived embryonic stem (ES) cells that are pluripotent provide us with an excellent opportunity to investigate gene function in the context of developing and adult animals through germline transmission. This strategy of creating animals with specific genomic changes has immense potential in medicine. In the near future, homologous recombination technology should promote our understanding of the functional roles of neuropeptide genes.

Acknowledgement. We thank Mr. Keiji Tanimoto for his help with the preparation of Figs. 4 and 5.

References

Angel P, Imagawa M, Chiu R, Stein B, Imbra RJ, Rahmsdorf HJ, Jonat C, Herrlich P, Karin M (1987) Phorbol ester-inducible genes contain a common cis element recognized by a TPA-modulated trans-acting factor. Cell 49:729–739

Amara SG, Jonas V, Rosenfeld MG, Ong ES, Evans RM (1982) Alternative RNA processing in calcitonin gene expression generates mRNAs encoding different polypeptide products. Nature 298:240–244

Barry J (1979) Immunohistochemistry of luteinizing hormone-releasing hormone-producing-neurons of the vertebrates. Int Rev Cytol 60: 179–221

Baskin DG, Ensinck JW (1984) Somatostatin in epithelial cells of intestinal mucosa is present primarily as somatostatin 28. Peptides 5:615–621

Behringer RR, Mathews LS, Palmiter RD, Brinster RL (1988) Dwarf mice produced by genetic ablation of growth hormone-expressing cells. Genes Dev 2:453–461

Billestrup N, Swanson LW, Vale WW (1986) Growth hormone-releasing factor stimulates proliferation of somatotrophs in vitro. Proc Natl Acad Sci USA 83:6854–6857

Black JB, Adler JE, Dreyfus CF, Friedman WF, LaGamma EF, Roach AH (1987) Biochemistry of information storage in the nervous system. Science 236:1263–1268

Borrelli E, Heyman RA, Arias C, Sawchenko PE, Evans R (1989) Transgenic mice with inducible dwarfism. Nature 339:538–541

Campbell DJ (1987) Circulating and tissue angiotensinogen systems. J Clin Invest 79:1–6

Capecchi MR (1989) Altering the genome by homologous recombination. Science 244:1288–1292

Cattanach BM, Iddon CA, Charlton HM, Chiappa SA, Fink G (1977) Gonadotropin-releasing hormone deficiency in a mutant mouse with hypogonadism. Nature 269:338–340

Chandrashekar V, Bartke A, Wagner TE (1988) Endogenous human growth hormone (GH) modulates the effect of gonadotropin-releasing hormone on pituitary function and the gonadotropin response to the negative feedback effect of testosterone in adult male transgenic mice bearing human GH gene. Endocrinology 123:2717–2722

Chen WY, Wight DC, Chen NY, Colman TC, Wagner TE, Kopchick JJ (1991a) Mutations in the third alpha-helix of bovine growth hormone dramatically affect growth hormone secretion in vitro and growth enhancement in transgenic mice. J Biol Chem 266:2252–2258
Chen WY, White ME, Wagner TE, Kopchick JJ (1991b) Functional antagonism between endogenous mouse growth hormone (GH) and a GH analog results in dwarf transgenic mice. Endocrinology 129:1402–1408
Clouston WM, Lyons IG, Richards RI (1989) Tissue-specific and hormonal regulation of angiotensinogen minigenes in transgenic mice. EMBO J 8:3337–3343
Comb M, Hyman SE, Goodman HM (1987) Mechanisms of trans-synaptic regulation of gene expression. Trends Neurosci 10:473–478
Connelly CS, Fahl WE, Iannaccone PM (1989) The role of transgenic animals in the analysis of various biological aspects of normal and pathologic states. Exp Cell Res 183:257–276
Crenshaw EB, Russo AF, Swanson LW, Rosenfeld MG (1987) Neuron-specific alternative RNA processing in transgenic mice expressing a metallothionein-calcitonin fusion gene. Cell 49:389–398
Docherty K, Steiner DF (1982) Posttranslational proteolysis in polypeptide hormone biosynthesis. Annu Rev Physiol 44:625–638
Doolittle RF (1983) Angiotensinogen is related to the antitrypsin-antithrombin ovalbumin gene. Science 222:417–419
Douglass J, Olivier C, Herbert E (1984) Polyprotein gene expression: generation of diversity of neuroendocrine peptides. Annu Rev Biochem 53:665–715
Dzau VJ (1988) Circulating versus local renin-angiotensin system in cardiovascular homeostasis. Circulation 77 Suppl 1:I–4I–13
Dzau VJ, Safar ME (1988) Large conduit arteries in hypertension: role of the vascular renin-angiotensin system. Circulation 77:947–954
Fernandez LA, Twickler J, Mead A (1985) Neovascularization produced by angiotensin II. J Lab Clin Med 105:141–145
Fink G (1979) Neuroendocrine control of gonadotropin secretion. Br Med Bull 35:155–160
Fukamizu A, Seo MS, Hatae T, Yokoyama M, Nomura T, Katsuki M, Murakami K (1989) Tissue-specific expression of the human renin gene in transgenic mice. Biochem Biophys Res Commun 165:826–832
Fukamizu A Takahashi S, Seo MS, Tada M, Tanimoto K, Uehara S, Murakami K (1990a) Structure and expression of the human angiotensinogen gene: identification of a unique and highly active promoter. J Biol Chem 265:7576–7582
Fukamizu A, Takahashi S, Murakami K (1990b) Expression of the human angiotensinogen gene in human cell lines. J Cardiovasc Pharmacol 16 Suppl 4:S11–S13
Fukamizu A, Hatae T, Kon Y, Sugimura M, Hasegawa T, Yokoyama M, Nomura T, Katsuki M, Murakami K (1991a) Human renin in transgenic mouse kidney is localized to juxtaglomerular cells. Biochem J 278:601–603
Fukamizu A, Uehara S, Sugimura K, Kon Y, Sugimura M, Hasegawa T, Yokoyama M, Nomura T, Katsuki M, Murakami K (1991b) Cell type-specific expression of the human renin gene. J Biol Regul Homeost Agents 5:112–116
Gaillard I, Clauser E, Corvol P (1989) Structure of human angiotensinogen gene. DNA 8:87–99
Ganong WF (1984) The brain renin-angiotensin system. Annu Rev Physiol 46:17–31
Ganten D, Speck G (1978) The barin renin-angiotensin system: a model for the synthesis of peptides in the brain. Biochem Pharmacol 27:237G–238G
Ganten D, Hermann K, Bayer C, Unger T, Lang RE (1983) Angiotensin synthesis in the brain and increased turnover in hypertensive rats. Science 221:869–871
Ganten D, Lindpaintner K, Ganten U, Peters J, Zimmerman F, Bader M, Mullins J (1991) Transgenic rats: new animal models in hypertension research. Hypertension 17:843–855
Gonzalez GA, Yamamoto KK, Fischer WH, Karr D, Menzel P, Biggs W, Vale WW, Montminy MR (1989a) A cluster of posphorylation sites on the cAMP-regulated nuclear factor CREB predicted by its sequence. Nature 337:749–752
Gonzalez GA, Montminy MR (1989b) Cyclic AMP stimulates somatostatin gene transcription by phosphorylation of CREB at serine 133. Cell 59:675–680
Goodman RH (1990) Regulation of neuropeptide gene expression. Annu Rev Neurosci 13:111–127

Gooodman RH, Aron DC, Roos BA (1983) Rat pre-prosomatostatin: structure and processing by microscopical membranes. J Biol Chem 258:5570–5573
Gordon JW, Ruddle FH (1983) Gene transfer into mouse embryos: production of transgenic mice by pronuclear injection. Methods Enzymol 101:411–433
Gordon JW, Scangos GA, Plotkin DJ, Barbosa JA, Ruddle FH (1980) Genetic transformation of mouse embryos by microinjection of purified DNA. Proc Natl Acad Sci USA 77:7380–7384
Habener JF (1981) Regulation of parathyroid hormone secretion and biosynthesis. Annu Rev Physiol 43:211–223
Habener JF (1990) Cyclic AMP response element binding proteins: a cornucopia of transcription factors. Mol Endocrinol 4:1087–1094
Habener JF, Cwikel BJ, Hermann H, Hammer RE, Palmiter RD, Brinster RL (1989) Metallothionein-vasopressin fusion gene expression in transgenic mice: nephrogenic diabetes insipidus and brain transcripts localized to magnocellular neurons. J Biol Chem 264:18844–18852
Hammer RE, Palmiter RD, Brinster RL (1984) Partial correction of murine hereditary disorder by germ-line incorporation of a new gene. Nature 311:65–67
Hammer RE, Brinster RL, Palmiter RD (1985a) Use of gene transfer to increase animal growth. Cold Spring Harbor Symp Quant Biol 50:379–387
Hammer RE, Pursel VG, Rexroad CE, Wall RJ, Bolt DJ, Ebert KM, Palmiter RD, Brinster RL (1985b) Production of transgenic rabbits, sheep and pigs by microinjection. Nature 315:680–683
Hammer RE, Brinster RL, Rosenfeld MG, Evans RE, Mayo KE (1985c) Expression of human growth hormone-releasing factor in transgenic mice results in increased somatic growth. Nature 315:413–416
Hanahan D (1988) Dissecting multistep tumorigenesis in transgenic mice. Annu Rev Genet 22: 479–519
Hanahan D (1989) transgenic mice as probes into complex systems. Science 246: 1265–1275
Harris RB (1989) Processing of pro-hormone precursor proteins. Arch Biochem Biophys 275:315–333
Heinrich PC, Castell JV, Andus T (1990) Interleukin-6 and the acute phase response. Biochem J 265:621–636
Hogan BLM, Constantini F, Lacy E (1986) Manipulation of the mouse embryo: a laboratory manual. Cold Spring Harbor Laboratory, Cold Spring Harbor
Jackson T, Allard MF, Sreenan CM, Doss LK, Bishop SP, Swain JL (1990) The c-myc proto-oncogene regulates cardiac development in transgenic mice. Mol Cell Biol 10:3709–3716
Jacobs JW, Goodman RH, Chin WW, Dee PC, Habener JF, Bell NH, Potts JT (1981) Calcitonin messenger RNA encodes multiple polypeptides in a single precursor. Science 213:457–458
Jennes L, Stumpf WE (1986) Gonadotropin-releasing hormone immunoreactive neurons with access to fenestrated capillaries in mouse brain. Neuroscience 18:403–416
Jin M, Wilhelm MJ, Lang RE, Unger T, Lindpaintner K, Ganten D (1988) Endogenous tissue renin-angiotensin systems: from molecular biology to therapy. Am J Med 84 Suppl 3A:28–36
Jones NC, Rigby DWJ, Ziff EB (1988) Trans-acting protein factors and the regulation of eukaryotic transcription: lessons from studies on DNA tumor viruses. Genes Dev 2:267–281
Kageyama R, Ohkubo H, Nakanishi S (1985) Induction of rat liver angiotensinogen mRNA following acute inflammation. Biochem Biophys Res Commun 129:826–832
Kawai Y, Takami K, Shisaka S, Emson PC, Hylliard CJ, Girgis S, McIntyre I, Tohyama M (1985) Topographic localization of calcitonin gene-related peptide in the rat brain: an immunohisto-chemical analysis. Neuroscience 15:747–763
Kelly JH, Darlington GJ (1985) Hybrid genes: molecular approaches to tissue-specific gene regulation. Annu Rev Genet 19:273–296
Kessel M, Gruss P (1990) Murine developmental control genes. Science 249:374–379
Kow LM, Pfaff DW (1988) Neuromodulatory actions of peptides. Annu Rev Pharmacol Toxicol 28:163–188
Krieger JE, Dzau VJ (1991) Molecular biology of hypertension. Hypertension 18 Suppl1:I3–I17
Leff SE, Rosenfeld MG, Evans RM (1986) Complex transcription unit: diversity in gene expression by alternative RNA processing. Annu Rev Biochem 55:1091–1117
Lind RW, Swanson LW, Ganten D (1985) Organization of angiotensin II immunoreactive cells and fibers in the rat nervous system. Neuroendocrinology 40:2–24

Loudon M, Bing RF, Thurston H, Swales JD (1983) Arterial wall uptake of renal renin and blood pressure control. Hypertension 5:629–634
Low MJ, Hammer RE, Goodman RH, Habener JF, Palmiter RD, Brinster RL (1985) Tissue-specific posttranslational processing of the pre-prosomatostatin encoded by metallothionein-somatostatin fusion gene in transgenic mice. Cell 41:211–219
Lynch DR, Snyder SH (1986) Neuropeptides: multiple molecular forms, metabolic pathways, and receptors. Annu Rev Biochem 55:773–799
Maniatis T, Goodburn S, Fisher JA (1987) Regulation of inducible and tissue-specific gene expression. Science 236:1237–1245
Mason AJ, Hayflick JS, Zoeller RT, Young WS, Phillips HS, Nikolics K, Seeburg PH (1986a) A deletion truncating the gonadotropin-releasing hormone gene is responsible for hypogonadism in the hpg mouse. Sciene 234:1366–1371
Mason AJ, Pitts SL, Nikolics K, Szonyi E, Wilcox JN, Seeburg PH, Stewart TA (1986b) The hypogonadal mouse: reproductive functions restored by gene therapy. Science 234:1372–1378
Mathews LS, Hammer RE, Brinster PL, Palmiter RD (1988) Expression of insulin-like growth factor I in transgenic mice with elevated levels of growth hormone is corrected with growth. Endocrinology 123:433–437
Mendelsohn FAO, Quirion R, Saavedra JM, Aquilera G, Catt KJ (1984) Autoradiographic localization of angiotensin II receptors in rat brain. Proc Natl Acad Sci USA 81:1575–1597
Merchenthaler I, Kovacs G, Lovasz G, Setalo G (1980) The preopticoinfundibular LH-RH tract of the rat. Brain Res 198:63–74
Miller CCJ, Carter AT, Brooks JI, Lovell-Badge RH, Brammar WJ (1989) Differential extra-renal expression of the mouse renin genes. Nucleic Acids Res 17:3117–3128
Mitchell PJ, Tjian R (1989) Transcriptional regulation in mammalian cells by sequence specific DNA binding proteins. Science 245:371–378
Montminy MR, Bilezikjian LM (1987) Binding of a nuclear protein to the cyclic-AMP response element of the somatostatin gene. Nature 328:175–178
Montminy MR, Sevarino KA, Wagner JA, Mandel G, Goodman RH (1986) Identification of a cyclic-AMP-responsive element within the rat somatostatin gene. Proc Natl Acad Sci USA 83:6682–6686
Montminy MR, Gonzalez GA, Yamamoto KK(1990) Regulation of cAMP-inducible genes by CREB. Trends Neurosci 13:184–188
Mullins JJ, Sigmund CD, Kane-Haas C, Wu C, Pacholec F, Zeng Q, Gross KW (1988) Studies on the regulation of renin genes using transgenic mice. Clin Exp Hypertens [A]10:1157–1176
Mullins JJ, Sigmund CD, Kane-Haas C, Gross KW (1989) Expression of the DBA/2 Ren-2 gene in the adrenal gland of transgenic mice. EMBO J 8:4065–4072
Mullins JJ, Peters J, Ganten D (1990) Fulminant hypertension in transgenic rats harbouring the mouse Ren-2 gene. Nature 344:541–544
Naftilan AJ, Pratt RE, Dzau VJ (1989) Induction of platelet-derived growth factor A-chain and c-myc gene expression by angiotensin II in cultured rat vascular smooth muscle cells. J Clin Invest 83:1419–1424
Nielsen AH, Knudsen F (1987) Angiotensinogen is an acute phase protein in man. Scand J Clin Lab Invest 47:175–178
Nishizuka Y (1984) The role of protein kinase C in cell surface signal transduction and tumor promotion. Nature 308:693–698
Ohkubo H, Kageyama R, Ujihara M, Inayama S, Nakanishi S (1983) Cloning and sequence analysis of cDNA of rat angiotensinogen. Proc Natl Acad Sci USA 80:2196–2200
Ohkubo H, Kawakami H, Kakehi Y, Takumi T, Arai H, Iwai M, Tanabe Y, Masu H, Hata J, Iwao H, Okamoto H, Yokoyama M, Nomura T, Katsuki M, Nakanishi S (1990) Generation of transgenic mice with elevated blood pressure by introduction of the rat renin and angiotensinogen genes. Proc Natl Acad Sci USA 87:5153–5157
Oparil S, Haber E (1974) The renin-angiotensin system. I. N Engl J Med 291:389–401
Orian JM, Lee CS, Weiss LM, Brandon MR (1989) The expression of a metallothionein-ovine growth hormone fusion gene in transgenic mice does not impair fertility but results in pathological lesions in the liver. Endocrinology 124:455–463
Orian JM, Snibson K, Stevenson JL, Brandon MR, Herington AC (1991) Elevation of growth

hormone (GH) and prolactin receptors in transgenic mice expressing ovine GH. Endocrinology 128:1238–1246
Ornits DM, Palmiter RD, Hammer RE, Brinster RL, Swift GH, MacDonald R (1985) Specific expression of an elastase-human growth hormone fusion gene in pancreatic acinar cells of transgenic mice. Nature 313:600–602
Padgett RA, Grabowski PJ, Konarska MM, Seiler S, Sharp PA (1986) Splicing of messenger RNA precursors. Annu Rev Biochem 55:1119–1150
Palmiter RD, Brinster RL (1986) Germ-line transformation of mice. Annu Rev Genet 20:465–499
Palmiter RD, Brinster RL, Hammer RE, Trumbauer ME, Rosenfeld RG, Birnberg NC, Evans RM (1982) Dramatic growth of mice that develop from eggs microinjected with metallothionein growth hormone fusion genes. Nature 300:611–615
Palmiter RD, Norstedt G, Gelinas RE, Hammer RE, Brinster RL (1983) Metallothionein-human GH fusion genes stimulate growth of mice. Science 222:809–814
Pellicer A, Palumbo A, DeCherney AH, Naftolin F (1988) Blockage of ovulation by angiotensin antagonist. Science 240:1660–1661
Phillips MI (1987) Functions of angiotensin in the central nervous system. Annu Rev Physiol 49:413–435
Phllips MI, Weyhenmeyer J, Felix D, Ganten D (1979) Evidence for an endogenous brain renin angiotensin system. Fed Proc 38:2260–2266
Raizada MK, Phillips MI, Gerndt JS (1983) Primary cultures from fetal rat brain incorporate ^{3}H-isoleucine and ^{3}H-valine into immuprecipitable angiotensin II. Neuroendocrinology 36:64–67
Raizada MK, Stenstrom B, Phillips MI, Summers C (1984) Angiotensin II in neuronal cultures from brains of normotensive and hypertensive rats. Am J Physiol 247:C115–C119
Reid IA, Morris BJ, Ganong WF (1978) The renin-angiotensin system. Annu Rev Physiol 40:377–410
Richter D (1988) Molecular events in expression of vasopressin and oxytocin and their cognate receptors. Am J Physiol 255:F207–F219
Roesler WJ, Vanderbark GR, Hanson RW (1988) Cyclic AMP and the induction of eukaryotic gene expression. J Biol Chem 263:9063–9066
Ron D, Brasier AR, Wright KA, Tate JE, Habener JF (1990) An inducible 50 kilodalton NFκB-like protein and a constitutive protein both bind the acute-phase response element of the angiotensinogen gene. Mol Cell Biol 10:1023–1032
Rosenfeld MG, Merod JJ, Amara SG, Swanson LW, Sawchenko PE, Rivier J, Vale WW, Evans RM (1983) Production of a novel neuropeptide encoded by the calcitonin gene via tissue-specific RNA processing. Nature 304:129–135
Rosenfeld MG, Amara SG, Evans RM (1984) Alternative RNA processing: determining neuronal phenotype. Science 225:1315–1320
Rosenfeld MG, Crenshaw EB, Lira SA, Swanson L, Borelli E, Heyman R, Evans RM (1988) Transgenic mice: applications to the study of the nervous system. Annu Rev Neurosci 11:353–372
Sabate MI, Stolarsky LS, Polak JM, Bloom SR, Vardell IM, Ghatei MA, Evans RM, Rosenfeld MG (1985) Regulation of neuroendocrine gene expression by alternative RNA processing: co-localization of calcitonin and calcitonin gene-related peptide (CGRP) in thyroid C-cells. J Biol Chem 260:2589–2592
Schally AV (1978) Aspects of hypothalamic regulation in the pituitary gland. Science 200:18–28
Seo MS, Fukamizu A, Nomura T, Yokoyama M, Katsuki M, Murakami K (1990) The human renin gene in transgenic mice. J Cardiovasc Pharmacol 16 Suppl 4:S8–S10
Seo MS, Fukamizu A, Saito T, Murakami K (1991) Identification of a previously unrecognized production site of human renin. Biochim Biophys Acta 1129:87–89
Share L (1988) Role of vasopressin in cardiovascular regulation. Physiol Rev 68: 1248–1284
Shen LP, Pichet RL, Rutter WJ (1982) Human somatostatin I: sequence of the cDNA. Proc Natl Acad Sci USA 79:4575–4579
Shimada S, Shiosaka S, Emson PC, Hylliard CJ, Girgis S, McIntyre I, Tohyama M (1985) Calcitonin gene-related peptidergic projection from parabranchial area to the forebrain and diencephalon in the rat: an immunohistochemical analysis. Neuroscience 16:607–616
Sirett NE. McLeanAS, Bray JJ, Hubbard JI (1977) Distribution of angiotensin II receptors in rat brain. Brain Res 122:299–312

Starksen NF, Simpson PC, Bishopric N, Coughlin SR, Lee WMF, Escobedo JA, Williams LT (1986) Cardiac myocyte hypertrophy is associated with c-myc protooncogene expression. Proc Natl Acad Sci USA 83:8348–8350

Steger RW, Bartke A, Parkening TA, Collins T, Yun JS, Wagner TE (1990) Neuroendocrine function in transgenic male mice with human growth hormone expression. Neuroendocrinology 52:106–111

Steger RW, Bartke A, Parkening TA, Collins T, Buonomo FC, Tang K, Wagner TE, Yun JS (1991) Effects of heterologous growth hormones on hypothalamic and pituitary function in transgenic mice. Neuroendocrinology 53:365–372

Struthers RS, Vale WW, Arias C, Sawchenko PE, Montminy MR (1991) Somatotroph hypoplasia and dwarfism in transgenic mice expressing a non-phosphorylatable CREB mutant. Nature 350: 622–624

Sutcliffe JG (1988) mRNA in the mammalian central nervous system. Annu Rev Neurosci 11:157–198

Swanson LW, Simmons DM, Arizza J, Hammer R, Brinster R, Rosenfeld MG, Evans RM (1985) Novel developmental specificity in the nervous system of transgenic animals expressing growth hormone fusion genes. Nature 317:363–366

Takahashi S, Fukamizu A, Hasegawa T, Yokoyama M, Nomura T, Katsuki M, Murakami K (1991) Expression of the human angiotensinogen gene in transgenic mice and cultured cells. Biochem Biophys Res Commun 180:1103–1109

Takahashi S, Fukamizu A, Sugiyama F, Kajiwara N, Yagami K, Murakami K (1992) Species-specific induction of angiotensinogen mRNA in transgenic mouse liver during acute phase reaction. J Vet Med Sci 54:367–369

Tanaka T, Ohkubo H, Nakanishi S (1984) Common structural organization of the angiotensinogen and the α_1-antitrypsin genes. J Biol Chem 259:8063–8065

Tannenbaum GS, Ling N, Brazeau P (1982) Somatostatin-28 is longer acting and more selective than somatostatin-14 on pituitary and pancreatic hormone release. Endocrinology 111:101–107

Theill LE, Wiborg O, Vuust J (1987) Cell-specific expression of the human gastrin gene: evidence for a control element located downstream of the TATA box. Mol Cell Biol 7:4329–4336

Trent DF, Weir GC (1981) Heterogeneity of somatostatin-like peptides in rat brain, pancreas, and gastrointestinal tract. Endocrinology 108:2033–2037

Tronik D, Dreyfus M, Babinet C, Rougeon F (1987) Regulated expression of Ren-2 gene in transgenic mice derived from parental strains carrying only the Ren-1 gene. EMBO J 6:983–987

Yamamoto KK, Gonzalez GA, Biggs WH, Montminy MR (1988) Phosphorylation-induced binding and transcriptional efficacy of nuclear factor CREB. Nature 334:494–498

Yamamoto KK, Gonzalez GA, Menzel P, Rivier J, Montminy MR (1990) Characterization of a bipartite activator domain in transcriptional factor CREB. Cell 60:611–617

Molecular Alterations in Nerve Cells: Direct Manipulation and Physiological Mediation

M. G. Kaplitt[1], S. D. Rabkin[2], and D. W. Pfaff[1]

Contents

1 Introduction

There are at least three classes of technical approaches to molecular neurobiology: first, the well-recognized approach of cloning and sequencing novel cDNAs; second, the use of molecular hybridization techniques to study expression and regulation of specific messenger RNAs (mRNAs); and third, the emerging field of direct manipulation of neuronal expression in vivo. The first two approaches are, in a manner of speaking, "passive" with respect to molecular alterations in nerve cells, in that, following physiologically mediated changes, transcripts known to be present are quantitatively observed, while the third approach actively alters neuronal cell physiology.

As an example of the second approach, consider in situ hybridization. For studies in the central nervous system, procedures which maintain high spatial resolution are desirable, because it is clear that different nerve cells, even adjacent neurons, can express different genes and can be regulated selectively. Methods for applying in situ hybridization to brain tissue have been discussed in detail (Uhl

[1] Laboratory of Neurobiology and Behavior, Rockefeller University, 1230 York Avenue, New York, N.Y. 10021 USA
[2] Program in Molecular Biology, Memorial Sloan Kettering Cancer Center, New York, N.Y. 10021 USA

Current Topics in Neuroendocrinology, Vol. 11

1986), including those useful for cDNA probes (McCabe et al. 1986; McCabe and Pfaff 1989), those useful for ribonucleotide probes (Gibbs et al. 1989), those for combining in situ hybridization with immunocytochemistry (Shivers et al. 1986), and methods for detailed quantitative analysis of in situ hybridization results (McCabe et al. 1989). Currently, for example, combinations of retrograde neuroanatomical markers with in situ hybridization are being used to see if oxytocin-expressing neurons which send their axons to the lower brain stem and spinal cord are affected selectively by hormone treatment (Chung et al. 1991a). With such protocols, gene expression for a transcription factor, the progesterone receptor, is shown to be driven by estradiol treatment (Romano et al. 1989a) in a tissue-specific and sex-specific (Lauber et al. 1991) fashion. mRNA levels for an abundant neuropeptide such as enkephalin are regulated precisely by estradiol (Romano et al. 1988), in genetic females but not in males (Romano et al. 1990), rapidly, within 1 hour (Romano et al. 1989b). These changes in proenkephalin mRNA can be related quantitatively to female reproductive behavior (Lauber et al. 1990). The effects of estrogen and progestin on oxytocin mRNA (Chung et al. 1991b; Kawata et al. 1991) are especially interesting, since various stages of hormone actions on oxytocinergic systems may multiply each other (Pfaff 1988). Finally, some hormone effects may not act directly in the cell nucleus of the neuropeptide-expressing neuron, but instead may be indirect or even subsequent to neuropeptide release (Rothfeld et al. 1989; Rosie et al. 1990; Park et al. 1990). These and similar studies serve to effectively document physiologically mediated changes.

In contrast, the third approach to molecular neuroscientific questions involves direct, active manipulation of the genetic readout from individual nerve cells. An increasingly popular procedure, the production of transgenic mice, addresses developmental questions effectively (Oberdick et al. 1990). Investigations are currently limited, however, in the choice of species amenable to transgenic manipulations. Additionally, for analyses of molecular mechanisms in the adult brain the transgenic approach has had the disadvantages of a stochastic process with respect to the circumstances under which the transfected gene would be expressed and the locations and functions of the nerve cells in which it is expressed. Tissue-specific promoters have been used quite effectively in order to limit this problem and generate a degree of specificity (Forss-Petter et al. 1990; Oberdick et al. 1990). Even with this degree of selectivity, however, expression in all cell types in which a given promoter is active is still unavoidable.

For work in the adult brain, selective local microapplication of the genes under study may offer the advantages of a more deterministic, precise methodology. Local applications of naked DNA might work (Wolff et al. 1990; Moffat 1990), but we would expect that lipofection (see below) would deliver the same sequences of interest more efficiently. Incorporation of the gene under study in a viral construct has many advantages (see below) and probably offers greater flexibility than simply transplanting cells into the central nervous system which hopefully will express the gene of interest. For example, fetal transplantation can be quite effective at compensating for certain defects such as Parkinson's disease (Lindvall et al. 1990). Yet, for studying most basic neurobiological functions at the molecular level as well as for correcting defects in regions which are resistant to transplantation, direct manipulation of endogenous adult neurons in vivo would be desirable. Thus, throughout the rest of this chapter, we emphasize the literature and/or our own work on lipofection, and on a defective herpesvirus-based vector.

2 Lipofection

"Liposome" is the term commonly applied to phospholipid vesicles of various types (Straubinger and Papahadjaopoulos 1983). The first liposomes to be created in the laboratory were called "smectic mesophases" due to their properties as liquid crystals (Bangham 1963). While the usefulness of these particles as vehicles for delivering substances was contemplated during the following several years, liposomes were primarily employed to study the properties of biological membranes. In 1972, however, G. Gregoriadis and B. E. Ryman injected liposomes into animals in order to determine their ability to deliver molecules into cells *in vivo* (Gregoriadis and Ryman 1972). This study simply utilized albumin as the marker, yet it opened the possibilities for the eventual use of liposomes as carriers of biologically active agents.

For the past decade, there has been increasing use of liposomes for encapsulation and transfection of nucleic acids in vitro. Many early procedures utilized common commercially supplied lipids and required labor-intensive procedures in order to create vesicles which were both unilamellar and large enough to adequately entrap and protect nucleic acids (reviewed by Straubinger and Papahadjopoulos 1983). Multilamellar vesicles were created easily, but they could not adequately protect nucleic acids and synthesis often required agitation which could shear an intact molecule (Straubinger and Papahadjopoulos 1983). Similarly, smaller unilamellar vesicles, which are now used for delivering small molecules (Presant et al. 1988), required sonication during synthesis which could harm nucleic acids. Certain lipids also resulted in liposomes which were potentially toxic to mammalian cells (Hannun and Bell 1989).

Synthetic cationic lipids have recently been designed in an effort to limit the difficulties and dangers encountered with liposome-mediated transfection (Felgner et al. 1987). These are based upon theories of lipid self-assembly advanced by Israelachvilli (Israelachvilli et al. 1977). Basically, molecules with large polar head groups and nonpolar tails tend to form micelles, while molecules with more evenly sized head and tail groups tend to form bilayers, which can then entrap molecules such as nucleic acids into vesicles. It has been further postulated that the use of cationic lipids promotes encapsulation of negatively charged nucleic acids, and also promote fusion with negatively charged cell membranes (Felgner and Ringold 1989). Synthetic cationic liposomes have successfully delivered functional DNA (Felgner et al. 1987; Mackey et al. 1988) and RNA (Malone et al. 1989) into mammalian cells in culture. Liposomes based upon these lipids are reported to be more effective at delivering functional nucleic acids into cells with less toxicity than conventional transfection methods (Felgner et al. 1987; Felgner and Ringold 1989).

Advances in liposome technology present the possibility of transfecting nucleic acids into animal cells in vivo. Other in vitro transfection procedures such as electroporation or calcium phosphate precipitation would be quite dangerous, as they are often toxic to cells in culture and could cause similar damage to animal tissue. This would be particularly problematic in the nervous system, as damaged neurons might not be replaced. Even if neuronal death did not occur, it would most likely be necessary to optimize protocols for different cell types. Lipofection with a mild reagent, therefore, appears to be the most likely transfection method which could provide a standardized, effective protocol adapted for in vivo use.

Lipofection into the nervous system in vivo has recently been achieved (Holt et al. 1990). A plasmid containing the cDNA for luciferase was transfected into embryonic neurons in the brain of *Xenopus*. Immunocytochemistry with a guinea pig antiluciferase antibody demonstrated that successful introduction and expression of the luciferase gene had occurred. The authors further observed that co-transfection of this vector with a vector containing the chloramphenicol acetyltransferase (CAT) gene resulted in coexpression of the two genes in more than 85% of cells expressing luciferase. While most of this work was performed on isolated embryonic heads which can only survive 4–5 days, lipofection into the eyes or via injection into the lumen of embryonic brains in whole animals yielded significant luciferase expression through 28 days following surgery. Finally, the authors found that skin covering the embryonic eyes and brain was fairly resistant to transfection utilizing liposomes. This observation was exploited for the purpose of targeting plasmid uptake, as removal of skin covering the eye bud resulted in selective transfection and expression of the luciferase gene in the retina.

The *Xenopus* study suggests the possible utility of liposomes for the transfer of genetic information into the nervous system of an adult mammal in vivo. Liposomes have already been used to deliver substances into other cells of higher mammals. In humans, for instance, liposomes have been used for several years in attempts to target radionuclides to cancer cells so as to improve imaging of tumors in radiologic procedures (Presant et al. 1988, 1990). Others have employed liposomes in order to improve delivery of chemotherapeutic agents (Rahman et al. 1980; Gabizon et al. 1982). Furthermore, expression of insulin (Nicolau et al. 1983) and preproinsulin (Soriano et al. 1983) cDNAs have been demonstrated following liposome-mediated transfection into rats in vivo. These studies reveal that in vivo transfection of DNA with liposomes is more than simply a theoretical possibility.

The use of lipofection to introduce exogenous genes into the nervous system in vivo would have certain practical advantages. Molecular targeting of liposomes to specific cells represents one area of benefit. It is clear that directing transfection to limited neuronal types would be particularly advantageous in a tissue with such cellular heterogeneity as the brain. Lipids may be chemically coupled to other molecules for the purpose of targeting. For example, this has been demonstrated in vitro when protein A, which selectively recognizes mouse IgG2A antibodies, was coupled to liposomes (Mackey et al. 1988). In vitro, these liposomes specifically targeted their genetic contents to B cells which expressed mouse IgG2A antibodies. Were this extended to targeting in vivo, ligands might be used which would be otherwise difficult. Proteins such as antibodies, which are large and complex, or nonpeptide molecules could be coupled to liposomes chemically. Other systems such as viral vectors would require a gene encoding a protein which must be properly expressed and targeted to nuclear or cell membranes to achieve the same end, and even if this were achieved, nonpeptide molecules would still be excluded. Liposome technology would thereby expand the possible cell types and ligands available to permit molecular targeting of exogenous genetic information. Finally, it should be noted that the development of liposome-mediated in vivo gene transfer might be more easily utilized by neurobiologists with limited tissue culture and molecular biological experience.

3 Herpesvirus Vectors

Many animal viruses have been used as shuttle vectors for in vitro and in vivo expression (Gluzman 1988). Expression in the central nervous system could theoretically be achieved using any neurotropic virus (rabies, pseudorabies, etc.). Practically, however, the genetic composition of the virus (nucleic acid type, size), host range, and toxicity serve to limit the number of attractive candidates. Retroviruses are generally used only when infecting cultured cells prior to transplantation (Shimohama et al. 1989) or for infecting fetal brain directly (Jaenisch 1980), since the necessity of a cell division for viral integration precludes successful infection of quiescent adult brain cells. Additionally, were a retrovirus to infect an adult neuron, the requisite integration might destroy or transform the cell. Due to these considerations, therefore, vectors based upon the neurotropic herpes simplex virus type 1 (HSV1) are the most promising candidates in the near future for a broad host range system for the safe delivery of functional genetic material into the adult mammalian brain.

3.1 Herpes Simplex Virus

There is a wide host range for HSV1, including humans, rats, and mice (Roizman and Sears 1987). HSV1 is a DNA virus consisting of a 150-kb linear duplex genome (Hayward 1986) which has been completely sequenced (McGeoch et al. 1988). This consists of the long (L) and short (S) genetic elements, which are covalently attached to each other (Fig. 1). Each element contains a unique region, termed U_L and U_S. U_L is flanked by inverted repeats (TR_L and IR_L) which contain two subunits designated ab and b'a'. The short element has a similar pattern; however the repeats flanking U_S (termed IR_S and TR_S) are subdivided into a'c' and ac units respectively. Three origins of replication are found in the HSV1 genome (Spaete and Frenkel 1985). Ori_L is the replication origin which resides in the unique region of the L element, thus present only once in the HSV genome. Ori_S is present in the inverted repeat of S, within the c region, and is therefore present twice in the HSV genome. Replication of HSV1 DNA is believed to proceed through a circular intermediate, formed through the association of the terminals repeats of the L and S elements (Vlazny and Frenkel 1981; Jacob et al. 1979; Varmuza and Smiley 1985). Concatamers of head-to-tail HSV1 genomes which are accumulated during replication are processed into genome-length units and then packaged into virions. A cleavage/packaging signal, residing in the a sequence of the inverted repeats, is required for this process (Spaete and Frenkel 1985; Stow et al. 1983). Identities and exact roles for many of the necessary proteins are unknown.

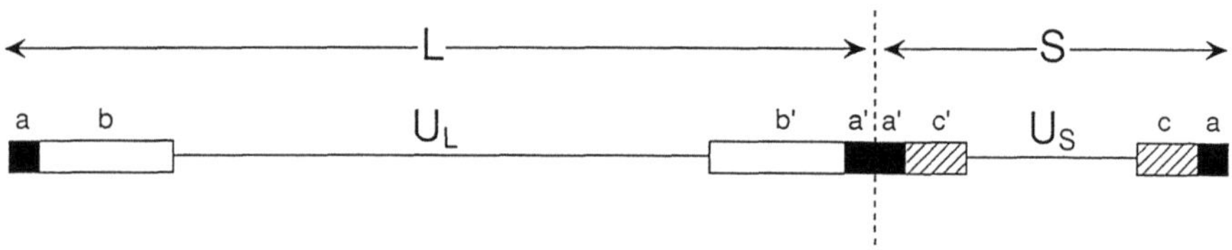

Fig. 1 Organization of the HSV1 genome. Ori_L is located near the middle of the U_L region, while Ori_S, present twice, resides within the c and c' regions. The cleavage/packaging signal maps within the a region of the inverted repeats (Adapted from Roizman 1979)

Three classes of mRNA are transcribed from the HSV1 genome (reviewed by Wagner 1985). These classes, labeled alpha, beta, and gamma, reflect the temporal expression of the respective messages, with their corresponding translation products mostly following the same pattern. The alpha genes are considered immediate-early, as they are expressed soon after infection and well before viral DNA replication. This group contains five genes, encoding polypeptides which primarily regulate expression of later HSV transcripts. The beta transcripts are early, as they follow alpha gene expression but preceed HSV DNA replication. This group encodes a more complex set of proteins which includes those proteins involved in viral DNA replication. Seven viral gene products, including an HSV1-encoded DNA polymerase, are known to be required in *trans* for DNA replication to occur (Challberg and Kelly 1989; Hay and Russell 1989). Finally, the gamma or late transcripts appear simultaneously with DNA replication and largely encode structural proteins necessary for virion assembly.

The HSV life cycle may follow two paths after inoculation into a host. Natural infections often occur through mucous membranes or skin (Roizman and Batterson 1984). At this stage the virus will pursue a replicative course, which may result in lesions or dissemination. When the clinical manifestations of infection subside, however, the virus may persist in a latent state (Wagner et al. 1988). Latency is most frequently established in peripheral ganglia of the nervous system, although nonneuronal latency has been observed (Roizman and Sears 1987). In the absence of disseminated infection, replication usually occurs only in ganglia, while the virus can persist in a latent state in other parts of the nervous system. Reactivation of infection from the latent state may occur following certain stimuli (Hill et al. 1978).

3.2 Functional Components of HSV1 Neurotropism

HSV1 neurotropism can be divided conceptually into at least two stages: adsorption and neuroinvasion. Neuroinvasion has been studied but is still incompletely characterized. The initial HSV1 association with most cell types is believed to occur through interactions with cell surface heparan sulfate (WuDunn and Spear 1989). While this may partially account for the broad host and tissue ranges attributed to HSV1, more specific interactions between viral surface glycoproteins and cellular components seem to be necessary. At least eight glycoproteins are encoded by HSV1 (Spear 1985), and glycoproteins B, C, and D have been implicated in virus adsorption to host cells (Fuller and Spear 1985; Johnson et al. 1984).

An early study utilizing gC− mutants suggested a role for glycoprotein C in the process of neuroinvasion (Kümel et al. 1985). Further analysis by the same laboratory, however, yielded the conclusion that gC is not a virulence determinant (Sunstrum et al. 1988). Marker rescue experiments, which entail recombining a mutant strain with cloned fragments from a wild-type virus in order to "rescue" the wild-type phenotype, demonstrated that gC+ strains which had been rescued had not developed wild-type levels of neurovirulence. The only virulent strains isolated after recombination retained the gC− phenotype. Some other change must have occurred, therefore, which rescued the neuroinvasive phenotype. It is possible that the process of transfecting gC− DNA into cells induced mutations, as has been observed (Thompson et al. 1986). While the nature of this neurovirulence mutation remains unknown, the study also highlights difficulties which may be encountered when interpreting the results of virulence studies using recombinant virus prepared via transfection.

Recent evidence suggests that glycoprotein D may be an important mediator of HSV1 neuroinvasion. Mutant viruses lacking gD can adsorb to nonneuronal host cells in culture but appear to be unable to invade such cells (Ligas and Johnson 1988; Johnson and Ligas 1988). This supports the concept of a functional separation of adsorption and invasion in any herpesvirus infection, which may involve other molecules in addition to heparan sulfate and gD. Other studies utilizing an HSV1 strain with reduced neuroinvasiveness, termed ANG, have furthered the understanding of the nature of HSV1 invasion in neurons (Goodman et al. 1989; Izumi and Stevens 1990). ANG appears to be nonlethal for mice when injected peripherally, although it is completely neurovirulent when injected intracranially. The mutation has been mapped to a single base change at amino acid position 84 of the ANG gD gene (Izumi and Stevens 1990). This implies that gD is important for invasion of peripheral neurons but is less crucial for invasion of CNS neurons. The authors have speculated that gD may be specific for nerve terminal invasion, since peripheral invasion must initiate at nerve terminals while infection after direct intracranial inoculation may occur at neuron somas. As will be discussed, however, disseminated brain infection most likely occurs through transneuronal spread, largely requiring nerve terminal invasion. Thus, some variation in HSV1 glycoprotein interactions with peripheral versus central neuron components must be considered.

3.3 HSV1 Neurovirulence

"Neurovirulence" is a term commonly employed to characterize all aspects of productive viral infections of the nervous system. For the purposes of this discussion, however, neurovirulence refers only to those functions which influence HSV1 infection of neurons after adsorption and invasion have occurred. One gene clearly implicated in this process has been the thymidine kinase (TK) gene (Field and Darby 1980; Sibrack et al. 1982; Chrisp et al. 1989). TK is not essential for viral replication in dividing cells (Dubbs and Kit 1964); however quiescent neurons are unable to substantially support viral DNA replication without viral TK activity (Field and Darby 1980). Thus decreased neurovirulence in TK− mutants may reflect decreased viral replication after invasion, resulting in limited viral dissemination. The ribonucleotide reductase gene has also been shown to be necessary for viral replication in resting cells (Preston et al. 1988). The effects of ribonucleotide reductase mutations on neurovirulence have not been studied; however the association with replication in quiescent cells suggests such a possibility. A role for HSV1 DNA polymerase in viral spread from peripheral to central sites has been postulated, although this has not been conclusively demonstrated (Day et al. 1988). Other regions of the viral genome have been implicated in neurovirulence through marker rescue experiments (Rosen et al. 1985; Thompson et al. 1985). Corresponding proteins are as yet unknown, and as stated earlier, such experiments must be viewed carefully, particularly when a functional correlation has not been assigned.

Finally, the nature of HSV1 dissemination in the CNS should be considered. The major route of spread for HSV1 appears to be transneuronal following intra-axonal transport, although local cell-to-cell spread has been observed (Bak et al. 1977; McLean et al. 1989; Ugolini et al. 1989; Kuypers and Ugolini 1990). After invasion, HSV1 is transported to the neuron soma where replication occurs. This is followed by transneuronal transfer to a neuron terminal impinging upon the infected cell (Kuypers and Ugolini 1990). The amplification and retrograde transport which has been observed has led to the exploitation of HSV1 as a transneuronal tracing marker (Ugolini et al. 1989; McLean et al. 1989). While it has been suggested that retrograde transport predominates over anterograde transport for most HSV1 strains, a recent study has demonstrated that the nature of HSV1 spread may be strain specific (Zemanick and Strick 1990). This indicates that the strain of HSV1 to be used may be an important factor to consider when designing vectors for specific tasks in the adult mammalian brain.

3.4 HSV1 Vectors

HSV1 has long been proposed as a potential vector for expression of exogenous genes in mammalian cells (Arsenakis et al. 1987). The large genome of this virus contains many genes which are dispensable, thus permitting the insertion of substantial amounts of DNA into a recombinant herpesvirus. The virulence of wild-type HSV strains presents difficulties, however. Despite replacement of certain sequences with foreign genes, the remaining wild-type genes may still subvert the genetic machinery of the host cell. While establishment of latency by a recombinant virus is desirable, the potential for reactivation provides an additional dilemma. Strategies to avoid such situations have been developed. Several attenuated recombinant virus strains have been created which do not reactivate nor do they destroy their host cell (Longnecker et al. 1988). They can also confer upon the host resistance to subsequent HSV infections. While attenuation and manipulation of HSV strains may prove to be an effective means for creating vectors, another strategy has been developed to create cloning-amplifying vectors based upon HSV which appears to be simpler and provides a greater degree of flexibility in designing useful vectors.

3.5 Defective Interfering Viruses

The HSV amplicon is a eukaryotic cloning and expression vector system based upon the concept of HSV defective interfering viruses (Spaete and Frenkel 1982). Naturally occurring defective interfering viruses derive from a wild-type virus but are altered such that they contain several copies of a small subset of the parental genome (Huang 1973; Huang and Baltimore 1977). They are defective due to their inability to replicate in the absence of the parental virus as helper, and they then interfere with the replication of that parental helper virus. When undiluted HSV is serially passaged, defective viruses appear and begin to accumulate (Frenkel 1981; Frenkel et al. 1981). The titer of the parental stock will decrease with subsequent passages while the percentage represented by defective virus will increase, as measured by Southern analysis or restriction analysis of radiolabeled total cellular DNA. The percentage of defective particles in the viral stock will eventually peak, and the titer of the parental strain will reach a nadir. It should be noted that while the percentage of virus represented by defective particles increases, there is still an absolute decrease in defective virus along with the more rapid decrease in helper virus. After the highest ratio of defective/helper virus has been achieved, which can reach more than 90%, continued propagation will yield decreasing amounts of defective virus and increasing parental titers. This cyclic pattern will continue through subsequent passages. At high helper virus titers, the parental strain is presumed to provide proteins in *trans* which are necessary for replication of the defective virus, while the defective particle will then more efficiently replicate. It is likely that when the titers of both the helper and defective viruses decrease, some cells infected with defective virus are not coinfected with helper, preventing defective virus replication. Other cells may only be infected with helper, which can independently replicate. The result is the cyclic increase and decrease in titers and ratios of defective/helper virus which is commonly observed.

The identification of defective HSV particles provided new opportunities for investigating and manipulating the HSV genome. Studies of defective viral DNA revealed heterogeneities in the gene subsets retained in different populations of defective viruses, yet certain genes were found to be consistently present within these subsets (Graham et al. 1978; Kaerner et al. 1979; Locker and Frenkel 1979; Frenkel 1981; Frenkel et al. 1981; Locker et al. 1982). It was further shown that defective viruses replicate in a rolling circle manner, similar to the postulated mechanism for intact HSVs (Becker et al. 1978). These observations, coupled with the assumption that at least some of the required sequences within a defective genome were *cis*-acting, led to the identification of replication origins (Vlazny and Frenkel 1981) and cleavage/packaging signals (Vlazny et al. 1982) within defective DNA. This not only served to increase understanding of defective viral DNA, but also resulted in identification of the sites and functions of these sequences in wild-type HSV DNA. With the identification of the primary *cis*-acting signals necessary for replication of defective viruses, exploitation of these sequences in a novel manner became feasible.

3.6 HSV1 Amplicons

The HSV amplicon, as originally conceived by N. Frenkel, is a eukaryotic cloning and expression vector capable of being packaged into HSV virions (Frenkel 1981; Spaete and Frenkel 1982). The potential application of recombinant viral based vectors had been envisioned earlier (Jackson et al. 1972), followed by the successful transfer and expression of a rat β-globin gene in monkey kidney cells via an SV40-derived vector which packaged into SV40 virions (Mulligan et al. 1979). Due to the approximately 5-kb limitation for packaging DNA into an SV40 virion, Frenkel and colleagues developed a vector to package into HSV virions, which can accept approximately 150 kb of DNA (Spaete and Frenkel 1982). For this discussion, it is also significant to remember that vectors packaged into an HSV virion are of greater value as agents for in vivo gene transfer due to the wide host and tissue range of the HSV. The initial vector was derived from the bacterial pKC7 plasmid. Retention of a bacterial replication origin and β-lactamase gene (which confers resistance to ampicillin) permitted propagation and manipulation of the vector in a bacterial system, as with any plasmid. The vector also contained those HSV sequences found to be essential for maintenance of naturally occurring defective viruses. When the vector was then cotransfected with wild-type HSV DNA, defective genomes composed of repeats of the HSV-pKC7 plasmid were observed after several passages, with the most abundant passages containing more than 90% defective virus. The cyclic increases and decreases in defective and helper virus were also observed. This demonstrated that two *cis*-acting regions of the HSV genome (replication origin and cleavage/packaging site) could be recombined into a plasmid and still function properly in the presence of helper factors in *trans* (Spaete and Frenkel 1982).

Further development of transient HSV plasmid-based systems resulted in the precise elucidation of the *cis*-acting replication origins and cleavage/packaging signals. Deletion mapping of Ori_S coupled with functional analysis of the sequence in a recombinant plasmid localized the origin function to a 100-bp sequence (Stow and McMonagle 1983). Similar procedures resulted in the precise localization of both the Ori_L sequence (Spaete and Frenkel 1985) and the cleavage/packaging signal (Stow et al. 1983; Varmuza and Smiley 1985). Additionally, it has been determined that an amplicon of up to 15 kb can be stably propagated as a defective virus (Kwong and Frenkel 1984). Larger amplicons appear to undergo recombinations which result in heterogeneous populations including defective virions containing smaller repeated units derived from the original amplicon.

The potential for employing amplicons as eukaryotic expression vectors in cultured mammalian cells was achieved by Kwong and Frenkel (1985). The cDNA for a chicken ovalbumin gene was placed under the control of an HSV immediate-early gene promoter, and both were cloned into an amplicon containing an HSV DNA replication origin, a cleavage/packaging signal, and plasmid sequences for bacterial propagation. In these studies, however, the helper virus DNA employed was from a mutant virus termed *vhs 1*. This mutant was deficient in certain virion-associated functions mediating shutoff of host mRNA translation (as opposed to host shutoff functions encoded during infection), but overproduced the immediate early gene products. As a helper, therefore, the HSV *vhs 1* served to drive expression of the chimeric gene in the amplicon under the control of the immediate early promoter. Immunoprecipitation of labeled proteins from cells infected with defective virus stocks demonstrated successful expression of the ovalbumin gene in the host cells in tissue culture. This experiment highlighted the utility of HSV-based amplicons for transfer and expression of exogenous genes into mammalian cells.

3.7 Use of HSV1 Vectors for Neuronal Expression

The development of HSV-based amplicons for expression of cloned genes is of particular use to the neurobiologist due to the inherent neurotropism of the herpes simplex virus. Defective virions retain the ability to infect neurons in tissue culture (Geller and Breakefield 1988; Geller and Freese 1990). These authors developed a vector similar to that of Kwong and Frenkel, with the major exception being the replacement of ovalbumin with β-galactosidase as a reporter gene. In addition, they utilized an HSV strain, designated tsK, with a temper-ature-sensitive mutation in the ICP4 gene (Marsden et al. 1976; Davison et al. 1984). This virus supplied the helper functions at the permissive temperature of 32°, but was unable to replicate at nonpermissive temperatures above 37°. This study demonstrated that HSV-based amplicons can infect neurons in tissue culture and will result in the stable expression of a functional product from an exogenous gene.

An amplicon has been developed in our laboratory which successfully expresses a foreign gene in the mammalian brain in vivo (Kaplitt et al. 1991). This vector, labeled pHCL, contains the two HSV *cis*-acting sequences necessary for DNA replication and packaging, as well as bacterial sequences from the pT7-3 plasmid (Fig. 2). The human cytomegalovirus (CMV) immediate-early promoter was utilized because it is a strong promoter for neuronal expression (Werner et al. 1990). β-galactosidase was chosen as a reporter gene for initial studies with the vector. Finally, we have used the tsK mutant as a helper virus and have thus far observed it to be nonvirulent in rat brain. This is quite advantageous when compared with other virulent HSV based systems which have been attempted (Palella et al. 1989).

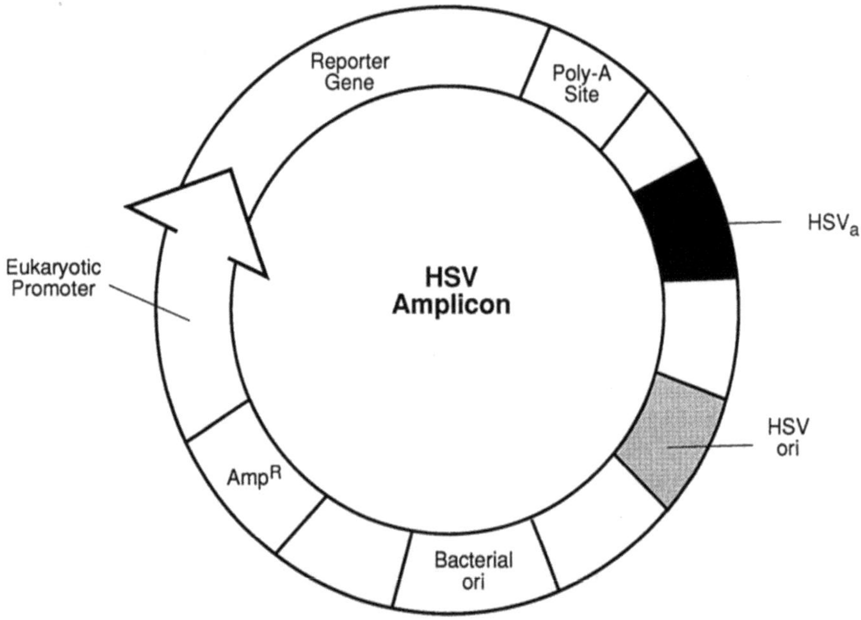

Fig. 2. Prototype HSV amplicon. The two HSV *cis*-acting sequences, HSV cleavage/packaging (*HSV*a) and HSV replication origin (*HSV ori*) are required for replication and packaging as a herpes-virus. There are no position or orientation constraints upon where these sequences may be placed within the amplicon genome. The bacterial replication origin (*Bacterial ori*) and β-lactamase gene (*Amp*^{R}) permit manipulation of the amplicon as a plasmid in bacterial systems. A splicing signal may be placed between the reporter gene and the promoter or polyadenylation site (*Poly-A site*) if so desired (not shown)

Following transfection of the vector into Vero cells and superinfection with the temperature-sensitive tsK strain, the resulting stock was propagated for several passages (Figs. 3 and 4). β-galactosidase activity in individual cells at the non-permissive temperature, as determined by cleavage of the substrate X-Gal into a blue precipitate, served as an assay for defective virus. Helper virus was titered by plaque assay at the permissive temperature.

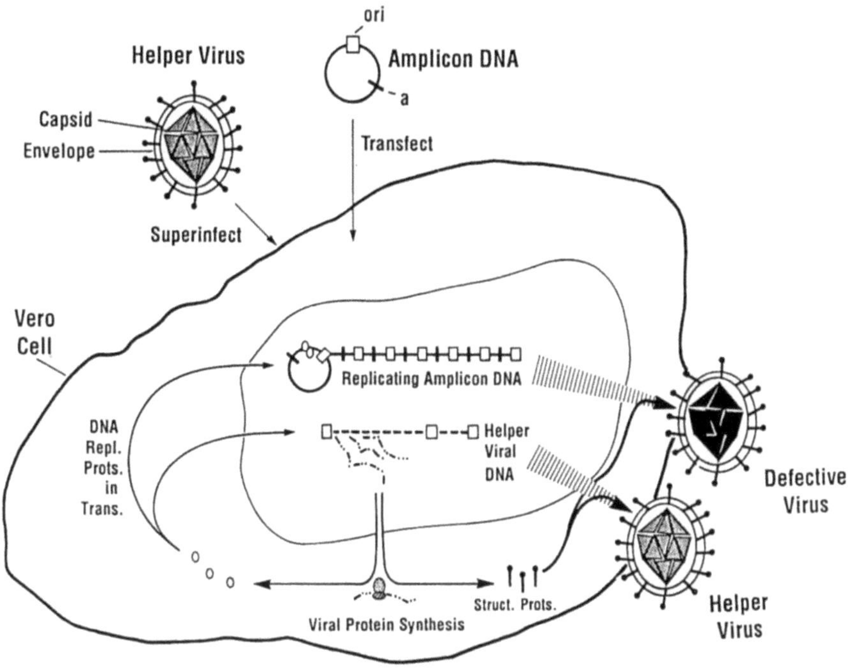

Fig. 3. Synthesis of amplicon-containing defective virus particles. As the amplicon (Fig. 2) does not encode any herpes-virus proteins, only the helper virus DNA will provide the necessary *trans*-acting replication factors. Both the amplicon and the helper virus genomes are replicated in the nucleus and approximately 150 kb units are cleaved and packaged into virions. The helper virus provides structural and other virion-associated proteins; the defective and helper virus progeny will differ only in the genomes contained within their virions. Upon subsequent passaging of the resulting viral stock, only those cells infected with both the defective and helper virus will support further defective virus replication. Viral stocks are propogated on HSV1-susceptible tissue culture cells (e.g., Vero kidney cells). *Rep.Prots.*, replicating proteins; *Struct.Prots.*, structural proteins

Surgical introduction of our defective virus into the rat brain in vivo resulted in successful expression of the β-galactosidase gene (Fig. 5). Our initial studies have repeatedly demonstrated site-specific activity of the exogenous gene product in cells at or near the site of stereotaxic microinjection. These experiments have used the ventromedial hypothalamus and hippocampus as targets, but reproducible expression is expected to be possible in other brain regions as well. Control animals, microinjected with a similar titer of tsK alone, did not reveal any blue cells upon staining for β-galactosidase histochemistry under appropriate conditions (see below). Our observations demonstrate the utility of an HSV1-based defective viral vector system for delivery and functional expression of exogenous genes in the adult mammalian brain in vivo.

As β-galactosidase is commonly used as a reporter gene in mammalian expression systems, it is necessary to note certain precautions which must be taken when assaying for this enzyme in the brain. β-galactosidase is a bacterial enzyme which cleaves lactose, and the X-Gal substrate is a modified sugar which yields a blue color upon cleavage. There are endogenous mammalian lysosomal enzymes, however, which are capable of cleaving X-Gal at acidic pH. This requires that the assay be performed slightly above neutral pH, which is optimal for the bacterial enzyme. Even at optimal pH, however, nonspecific blue staining will be observed if brain sections are overstained. Our experience and that of others demonstrates that 3–4 h is sufficient to allow the β-galactosidase reaction to occur while limiting nonspecific staining.

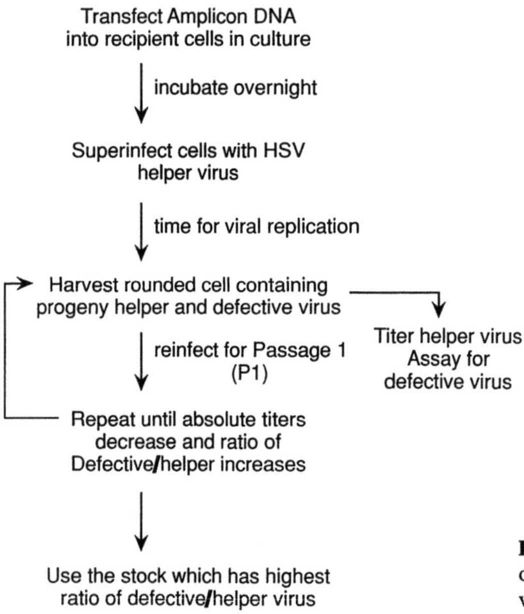

Fig. 4. General protocol for creation of an amplicon-containing defective virus stock

Despite all of these efforts, we observed a small degree of background staining in control brain sections which appeared to be confined to vascular endothelial cells and occasional neurons. This phenomenon had also been noted by Emson et al. (1990). They described a similar pattern of background staining, which was then eliminated by inclusion of the chelator ethylene glycol tetra-acetic acid (EGTA) in the fixative. EGTA preferentially chelates calcium ions, and presumably this sufficiently decreases the activity of endogenous enzymes so as to eliminate background staining. Upon implementation of this fixation protocol, we could eliminate background staining (Fig. 5). While others have surmised that background staining can be an insurmountable problem (Shimohama et al. 1989), we feel that the protocol of Emson et al., when followed precisely, renders the X-Gal staining technique a reliable and accurate assessment of the activity of exogenous, bacterial β-galactosidase in the adult mammalian brain.

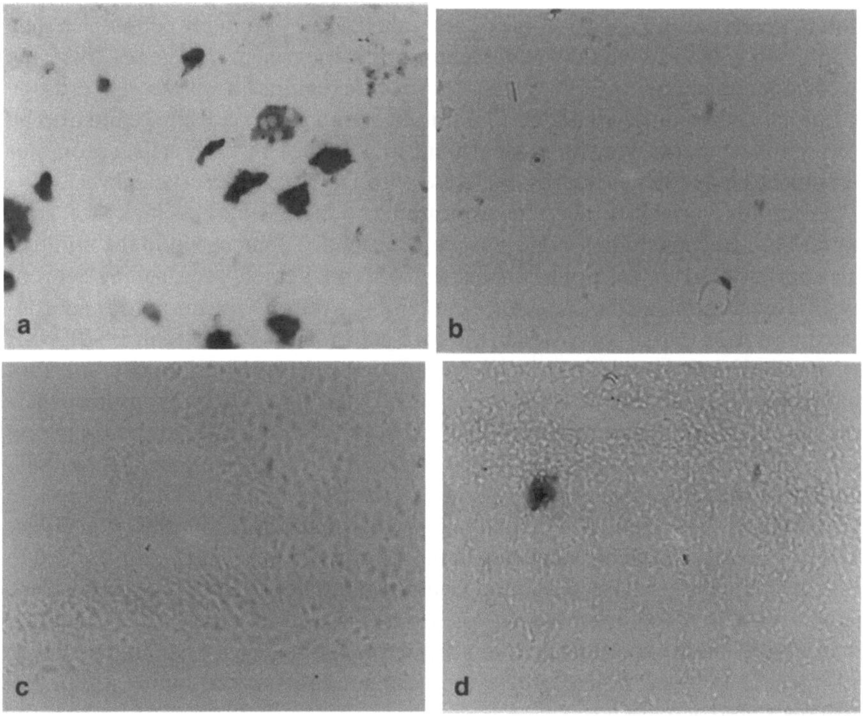

Fig. 5. Expression of β-galactosidase-containing amplicon as detected by X-Gal histochemical assay. **a** Monolayer of Vero cells in culture, transfected with the amplicon. Every blue cell visible is expressing β-galactosidase, and differences in staining intensity are evident. **b** Section through rat hypothalamus infected with tsK helper virus only, as a control, illustrating false-positive staining when suboptimal tissue preparation was used (see text). Note endothelial cell stained in *lower right* and other potentially confounding stained cells. **c** Section through mediobasal rat hypothalamus infected with tsK helper virus only, as a control. Here, optimal tissue preparation was used, including EGTA in the fixative (see text). Note absence of blue cells. Third ventricle (*upper left*) and brain bottom are visible. **d** Section through rat hippocampus, infected with the pHCL-based defective virus (see text). A positive hippocampal pyramidal cell strongly expressing β-galactosidase is seen. In this experiment, control tissue was totally unstained

3.8 Designing Improved Amplicons for Neurobiology

There are currently two aspects of the defective herpesvirus system which may be manipulated by the neurobiologist in order to improve in vivo expression of a transferred gene. First, there is the promoter used to drive expression of the desired gene. As the brain is a heterogeneous organ, some promoters may be ineffectual in certain regions or cells of the brain. For example, a promoter which is highly specific for neurons may not function properly in glial cells. Alternatively, it would be unlikely that a gene under the control of an estrogen response element (ERE) would be expressed in neurons which did not themselves express estrogen receptors. Transient analysis has demonstrated a wide variability in natural promoter activity in cultured neurons among several eukaryotic promoters (Werner et al. 1990). It is possible, therefore, that many cells which may be infected with virus would be unable to express the transferred gene. This must be considered when designing vectors for specific regions or cell types. The use of in situ hybridization as an added tool for analyzing gene transfer should also be considered, since assaying for protein alone may at least partially reflect promoter activity and may not necessarily correlate with DNA transfer efficiency.

Promoter variability may be exploited for increasing specificity of gene expression. Recent studies with transgenic mice have demonstrated the utility of this approach. In an attempt to further understand elements controlling neuron-specific expression, a DNA sequence containing a putative promoter for neuron-specific enolase (NSE) was fused to the *lacZ* gene (Forss-Petter et al. 1990). NSE is expressed in most postmitotic neurons and some neuroendocrine cells, but is not readily detected in other cells (Schmechel and Morangos 1983). Transgenic mice carrying the fusion gene revealed β-galactosidase activity only in the brain and testes, and expression within the brain was limited to neurons. A similar approach employed the promoter for the gene *L7*, expression of which is limited to cerebellar Purkinje cells and retinal rod bipolar neurons (Oberdick et al. 1988, 1990). A fusion transgene consisting of most of the *L7* gene and *lacZ* was used to transform mouse germ lines, and the resulting animals expressed β-galactosidase activity only in those brain regions previously shown to express *L7* (Oberdick et al. 1990). These experiments demonstrate the feasibility of obtaining tissue- or cell-specific expression of a foreign gene using amplicons containing promoters with restricted activity.

The choice of HSV1 strain-supplying helper functions is the second area of concern when generating amplicon-based defective viruses for transfer of genes into the brain in vivo. One of the benefits of this system is the potential of generating a viral stock with more than 90% defective particles. In any stock, however, some small number of infectious helper virions will be retained and inoculated into the host along with the defective virions. Additionally, any potentially disruptive proteins which are included in the virus particle will be part of both the defective and helper virions even though they are only encoded by the helper. The choice of a helper virus must then address: (1) the ability of even a smaller number of virulent particles to cause a productive, lethal infection, and (2) the ability of certain virion-associated proteins to disrupt both host and viral gene expression.

An inability to replicate under in vivo conditions is a very desirable trait for a helper virus. The helper virus must supply those replication functions necessary for creation of a defective virus, but they must only be supplied in the culture setting and must become inactive in vivo. There are two means by which this may be achieved. First, there are temperature-sensitive mutants, such as tsK, which will replicate at 31°C but not at physiological temperatures. In the case of tsK, there is a mutation in the viral ICP4 gene, which encodes one of the immediate early gene products required for transactivation of viral genes encoding replication functions (Davison et al. 1984). As a result, this strain will not proceed past the early stages of infection at the higher, non-permissive temperature. One caveat when considering temperature-sensitive strains is the fact that often the mutation is a point mutation, which carries the risk of reversion to wild type. While infrequent, the possibility of such an event does exist and stocks should be monitored carefully for the appearance of revertants. Another replication-deficient strain which may be used is a host-range restricted mutant. This is a strain with a complete deletion of a necessary gene, thus allowing replication only in cell lines which complement for the missing function. Virions infecting non-complementing cells, such as cells in vivo, will not replicate. While these strains cannot revert to wild type from a simple point mutation, they carry the risk of reversion from recombination with the cell line containing the complementing gene.

Shutoff of host protein synthesis by helper virus-encoded functions may theoretically be problematic for in vivo expression studies. Two virally encoded functions serve to shut down host protein synthesis. One is a late activity expressed after the onset of viral transcription (Wagner 1985). Presumably this should not be of concern when dealing with strains containing immediate-early gene mutations, as they will not transactivate the appropriate gene. There is, however, an earlier host shutoff function (Schek and Bachenheimer 1985). This is known as vhs because it is associated with the virion, and mutant HSV strains defective in the vhs function have been isolated (Read and Frenkel 1983). The defect in the vhs mutant genome has been mapped (Kwong et al. 1988), although a corresponding protein has not been isolated.

The vhs function appears to involve destabilization and degradation of mRNA. HSV infection has long been associated with disaggregation of host polysomes (Sydiskis and Roizman 1967). After the vhs function was identified, it was determined that virion associated host shutoff involved degradation of cellular mRNA (Schek and Bachenheimer 1985). Recently, however, virion-associated mRNA degradation has been shown to include both host and viral mRNAs (Kwong and Frenkel 1987; Oroskar and Read 1989). This may at least in part serve to temporally regulate the cascade of viral mRNA synthesis. As the protein associated with vhs has not been isolated, it is not known whether this represents a virally encoded ribonuclease or viral potentiation of cellular degradation factors.

As the vhs function is virion associated, use of a helper virus which contains a wild-type vhs gene will result in both progeny helper and defective virions capable of mRNA degradation. Any function encoded by the helper virus and packaged into the virion will be contained in the defective virus. Thus, in addition to the surface glycoproteins, vhs is among a group of additional virion-associated functions which will become part of both virions. This could serve to limit expression of the desired message, particularly at early time points. In this instance, amplicon-type vectors may have a selective advantage over other HSV1 vectors. If normal immediate-early gene expression is not shut off to the same degree as the host by virtue of greater levels of expression, then a virion with multiple copies of the same gene would be desirable. Vectors utilizing most of the HSV1 genome will only contain one copy of the desired insert for each virion, and expression may become limited particularly in cells which do not strongly induce the chosen promoter. As defective virions will contain 10–20 copies of the amplicon, the same promoter-reporter combination will be expressed at much greater levels, potentially allowing for expression, particularly at early times, which might not be as easily detected with other vectors.

A helper virus combining several mutations would seem to be most advantageous for maximizing expression and limiting potential damage in vivo. Ideally, creation of a cell line which could package amplicon DNA into an HSV1 virion without virion-associated factors, such as host shutoff functions, would eliminate the helper virus and associated problems. While cell lines have been created in order to package retroviral vectors (Mann et al. 1983; Cone and Mulligan 1984), the far greater size and complexity of the HSV1 genome would make this difficult. As stated earlier, there is also an imcomplete understanding of all of the factors involved in HSV1 DNA replication and virion assembly. At present, therefore, it is more feasible to design improved helper virus strains. Although strains exist which contain the individual mutations discussed above, multiple mutant strains have not yet been created. There would be obvious benefits, for example, in obtaining a virus with both the TsK and vhs mutations, thereby limiting both replication and virion associated host shutoff. As the biology of the herpes simplex virus becomes clearer, there will be greater opportunities to design safer helper virus strains as well as to contemplate construction of an HSV1-packaging cell line.

References

Arsenakis M, Poffenberger KL, Roizman B (1987) Novel herpes simplex virus genomes: construction and application. UCLA Symp Mol Cell Biol New Ser 43:427

Bangham AD (1963) Physical structure and behavior of lipids and lipid enzymes. Adv Lipid Res 1:65–104

Bak IJ, Markham CH, Cook ML, Stevens JG (1977) Intraaxonal transport of herpes simplex virus in the rat central nervous system. Brain Res 136:415–429

Becker Y, Asher Y, Weinberg-Zahlering E, Rabkin S (1978) Defective herpes simplex virus DNA: circular and circular-linear molecules resembling rolling circles. J Gen Virol 40:319–335

Behr JP, Demeneix B, Loeffler JP, Perez-Mutul J (1989) Efficient gene transfer into mammalian primary endocrine cells of lipopolzamine-coated DNA. Proc Natl Acad Sci USA 86:6982–6986

Challberg MD, Kelly TJ (1989) Animal virus DNA replication. Annu Rev Biochem 58:671–717

Chrisp CE, Sunstrum JC, Averill DR Jr, Levine M, Glorioso JC (1989) Characterization of encephalitis in adult mice induced by intracerebral inoculation of herpes simplex virus type 1 (KOS) and comparison with mutants showing decreased virulence. Lab Invest 60(6):822–830

Chung SK, McCabe JT, Pfaff DW (1991a) Estrogen influences on oxytocin mRNA expression in preoptic and anterior hypothalamic regions studied by in situ hybridization. J Comp Neurol 307:281–295

Chung SK, Haldar J, Pfaff DW (1991b) Combination of in situ hybridization with retrograde marker to study hormone effect on oxytocin mRNA-expressing neurons which project to spinal cord. Society for Neuroscience Abstracts 17:547

Cone RD, Mulligan RC (1984) High-efficiency gene transfer into mammalian cells: generation of helper-free recombinant retrovirus with broad mammalian host range. Proc Natl Acad Sci USA 81:6349–6353

Davison MJ, Preston VG, McGeoch DJ (1984) Determination of the sequence alteration in the DNA of the herpes simplex virus type 1 temperature-sensitive mutant tsK. J Gen Virol 65:859–863

Day SP, Lausch RN, Oakes JE (1988) Evidence that the gene for herpes simplex virus type 1 DNA polymerase accounts for the capacity of an intertypic recombinant to spread from eye to central nervous system. Virology 163:166–173

Dubbs DR, Kit S (1964) Mutant strains of herpes simplex deficient in thymidine kinase-induced activity. Virology 22:493–502
Emson PC, Shoham S, Feler C, Buss T, Price J, Wilson CJ (1990) The use of a retroviral vector to identify foetal striatal neurones transplanted into adult striatum. Exp Brain Res 79:427–430
Felgner PL, Ringold GM (1989) Cationic liposome mediated transfection. Science 337:387–388
Felgner PL, Gadek TR, Holm M, Roman R, Chan HW, Wenz M, Northrop JP, Ringold GM, Danielson M (1987) Lipofection: a highly efficient, lipid mediated DNA-transfection procedure. Proc Natl Acad Sci USA 84:7413–7417
Field HJ, Darby G (1980) Pathogenicity in mice of strains of herpes simplex virus which are resistant to acyclovir in vitro and in vivo. Antimicrob Agent Chemother 17:209
Forss-Petter S, Danielson PE, Catsicas S, Battenberg E, Price J, Nerenberg M, Sutcliffe JG (1990) Transgenic mice expressing beta-galactosidase in mature neurons under neuron-specific enolase promoter control. Neuron 5:187–197
Frenkel N (1981) Defective interfering herpesviruses. In: Nahmias AJ, Dowdle WR, Schinazy RS (eds) Human herpesviruses. An interdisciplinary perspective. Elsevier/North-Holland, New York, pp 91–120
Frenkel N, Locker H, Vlazney, D (1981) Structure and expression of class I and class II defective interfering HSV genomes. In: Becker Y (ed) Herpes virus DNA. Nijhoff, Boston, pp 149–184
Fuller AO, PG Spear (1985) Specificities of monoclonal and polyclonal antibodies that inhibit adsorption of herpes simplex virus to cells and lack of inhibition by potent neutralizing antibodies. J Virol 55:475–482
Gabizon A, Dagan A, Goren D, Barenholz Y, Fuks Z (1982) Liposomes as in vivo carriers of adriamycin: reduced cardiac uptake and reserved antitumor activity in mice. Cancer Res 42:4734–4739
Geller AI, Breakefield XO (1988) A defective HSV-1 vector expresses Escherichia coli β-galactosidase in cultured peripheral neurons. Science 241:1667–1669
Geller AI, Freese A (1990) Infection of cultured central nervous system neurons with a defective herpes simplex virus 1 vector results in stable expression of Escherichia coli β-galactosidase. Proc Natl Acad Sci USA 87:1149–1153
Gibbs RB, McCabe JT, Buck CR, Chao MV, Pfaff DW (1989) Expression of NGF receptor in the rat forebrain detected with in situ hybridization and immunohistochemistry. Mol Brain Res 6:275–287
Gluzman Y (ed) (1988) Viral vectors. Cold Spring Harbor Laboratory, Cold Spring Harbor
Goodman JL, Cook ML, Sederati F, Izumi K, Stevens JG (1989) Identification, transfer, and characterization of cloned herpes simplex virus invasiveness regions. J Virol 63:1153–1161
Graham BJ, Bengali Z, Vande Woude GF (1978) Physical map of the origin of defective DNA in herpes simplex virus type 1 DNA. J Virol 25:878–887
Gregoriadis G, Ryman BE (1972) Fate of protein-containing liposomes injected into rats. Eur J Biochem 24:485–571
Hannun YA, Bell RM (1989) Functions of sphingolipids and sphingolipid breakdown products in cellular regulation. Science 243:500
Hay RT, Russell WC (1989) Recognition mechanisms in the synthesis of animal virus DNA. Biochem. J 258:3–16
Hayward GS (1986) Herpes virus. I. Genome structure and regulation. Cancer Cells 4:59–76
Hill TJ, Blyth WA, Harbour DA (1978) Trauma to the skin causes recurrence of herpes simplex in the mouse. J Gen Virol 39:21–28
Holt CE, Garlick N, Cornel E (1990) Lipofection of cDNAs in the embryonic vertebrate central nervous system. Neuron 4:203–214
Huang AS (1973) Defective interfering viruses. Annu Rev Microbiol 27:101–117
Huang AS, Baltimore D (1977) Defective interfering animal viruses. In: Fraenkel-Conrat H, Wagner R (eds) Regulation and genetics: viral gene expression and integration. Plenum, New York, pp 73–116 (Comprehensive virology treatise, vol 10)
Israelachvilli JN, Mitchell JD, Ninham BW (1977) Theory of self-assembly of lipid bilayers and vesicles. Biochim Biophys Acta 470:185–201
Izumi KM, Stevens JG (1990) Molecular and biological characterization of a herpes simplex virus type 1 (HSV-1) neuroinvasiveness gene. J Exp Med 172:487–496

Jackson PA, Symons RH, Berg P (1972) Biochemical method for inserting new genetic information into DNA of simian virus 40: circular SV40 DNA molecules containing lambda phage genes and the galactose operon of Escherechia coli. Proc Natl Acad Sci USA 69:2904–2909

Jacob RJ, Morse LS, Roizman B (1979) Anatomy of herpes simplex virus DNA XII. Accumulation of head-to-tail concateners in nuclei of infected cells and their role in the generation of four isomeric arrangements of viral DNA. J Virol 29:448–457

Jaenisch R (1980) Retroviruses and embryogenesis: microinjection of Moloney leukemia virus into midgestation mouse embryos. Cell 19:181–188

Johnson DC, Ligas MW (1988) Herpes simplex viruses lacking glycoprotein D are unable to inhibit virus penetration: quantitative evidence for virus-specific cell surface receptors. J Virol 62(12):4605–4612

Johnson DC, Wittels M, Spear PG (1984) Binding to cells of virosomes containing herpes simplex virus type 1 glycoproteins and evidence for fusion. J Virol 52:238–247

Kaerner HC, Maickle IB, Ott A, Schroder CH (1979) Origin of two different classes of defective HSV-1 Angelotti DNA. Nucleic Acid Res 6:1467–1478

Kaplitt MG, Pfaus JG, Kleopoulos SP, Hanlon BA, Rabkin SD, Pfaff DW (1991) Expression of a functional foreign gene in adult mammalian brain following in vivo transfer via a herpes simplex virus type 1 defective viral vector. Mol Cell Neurosci 2:320–330

Kawata M, McCabe JT, Chung SK, Dutt A, Yuri K, Hirakawa M, Kumamoto K, Hirayma Y, Pfaff DW (1991) The effect of progesterone on oxytocin messenger RNA in hypothalamic neurons of estrogen-treated female rats studied with quantitative in situ hybridization histochemistry. Biomed Res 12:405–415

Kieff ED, Bachenheimer SC, Roizman B (1971) Size, composition and structure of deoxyribonucleic acid of herpes simplex virus subtypes 1 and 2. J Virol 8:125

Kumel G, Kaerner HC, Levine M, Schroder CH, Glorioso JC (1985) Passive immune protection by herpes simplex virus-specific monoclonal antibodies and monoclonal antibody-resistant mutants altered in pathogenicity. J Virol 56:930–937

Kuypers HGJM, Ugolini G (1990) Viruses as transneuronal tracers. Trends Neurosci 13(2):71–75

Kwong AD, Frenkel N (1984) Herpes simplex virus amplicon: effect of size on replication of constructed defective genomes containing eukaryotic DNA sequences. J Virol 51:595–603

Kwong AD, Frenkel N (1985) The herpes simplex virus amplicon IV. Efficient expression of a chimeric chicken ovalbumin gene amplified within defective virus genomes. Virology 142:421–425

Kwong AD, Frenkel N (1987) Herpes simplex virus-inflected cells contain a function that destabilizes both host and viral mRNAs. Proc Natl Acad Sci USA 84:1926–1930

Kwong AD, Kruper JA, Frenkel N (1988) Herpes simplex virus virion host shutoff function. J Virol 62:912–921

Lauber AH, Romano GJ, Mobbs CV, Howells RD, Pfaff DW (1990) Estradiol induction of proenkephalin messenger RNA in hypothalamus: dose-response and relation to reproductive behavior in the female rat. Mol Brain Res 8:47–54

Lauber AH, Romano GJ, Pfaff DW (1991) Sex difference in estradiol regulation of progestin receptor mRNA in rat mediobasal hypothalamus as demonstrated by in situ hybridization. Neuroendocrinol 53:608–613

Ligas MW, Johnson DC (1988) A herpes simplex virus mutant in which glycoprotein D sequences are replaced by beta-galactosidase sequences binds to but is unable to penetrate into cells. J Virol 62:1486–1494

Lindvall O, Brundin P, Widner H, Rehncrona S, Gustavii B, Frackowlak R, Leenders KL, Sawle G, Rothwell JC, Marsden CD, Bjorklund A (1990) Grafts of fetal dopamine neurons survive and improve motor function in Parkinson's Disease. Science 247:574–577

Locker H, Frenkel N (1979) Structure and origin of defective genomes contained in serially passaged herpes simplex virus type 1 (Justin). J Virol 29:1065–1077

Locker H, Frenkel N, Halliburton I (1982) Structure and expression of class II defective herpes simplex virus genomes encoding the infected cell polypeptide number 8. J Virol 43:574–593

Longnecker R, Roizman B, Meignier B (1988) Herpes simplex viruses as vectors: properties of a prototype vaccine strain suitable for use as a vector. In: Gluzman Y, Hughes SH (eds) Viral vectors. Cold Spring Harbor Laboratory, Cold Spring Harbor, pp. 68–72

Mackey P, Lewis F, McMillan L, Jonah ZL (1988) Gene transfer from targeted liposomes to specific lymphoid cells by electroporation. Proc Natl Acad Sci USA 85:8027–8031
Malone RW, Felgner PC, Verma IM (1989) Cationic liposome-mediated RNA transfection. Proc Natl Acad Sci USA 86:6077–6081
Mann R, Mulligan RC, Baltimore D (1983) Construction of a retrovirus packaging mutant and its use to produce helper-free defective retrovirus. Cell 33:153–159
Marsden HS, Crombie IK, Subak-Sharpe JH (1976) Control of protein synthesis in herpesvirus-infected cells; analysis of the polypeptides induced by wild type and sixteen temperature-sensitive mutants of HSV strain 17. J Gen Virol 31:347–372
McCabe J, Pfaff DW (1989) In situ hybridization: a methodological guide. Methods Neurosci 1:98–126
McCabe JT, Morell JI, Ivell R, Schmale H, Richter D, Pfaff DW (1986) In situ hypridization technique to localize rNRA and mRNA in mammalian neurons. J Histochem Cytochem 34:45–50
McCabe JT, Desharnais RA, Pfaff DW (1989) Graphical and statistical approaches to data analysis for in situ hybridization. Methods Enzymol 168:822–848
McGeoch DJ, Dalrymple MA, Davison AJ, Dolan A, Frame MC, McNab D, Perry LJ, Scott JE, Taylor P (1988) The complete DNA sequence of the long unique region in the genome of herpes simplex virus type 1. J Gen Virol 69:1531–1574
McLean JH, Shipley MT, Bernstein DI (1989) Golgi-like, transneuronal retrograde labelling with CNS injections of herpes simplex virus type 1. Brain Res Bull 22:867–881
Moffat AS (1990) Animal cells transformed in vivo. Science 248:1493
Mulligan RC, Howard BH, Berg P (1979) Synthesis of rat β-globin in cultured monkey kidney cells following infection with an SV40 β-globin recombinant genome. Nature 277:108–114
Nicolau C, Le Pape A, Soriano P, Fargette F, Juhel M-F (1983) In vivo expression of rat insulin after intravenous administration of the liposome-entrapped gene for rat insulin I. Proc Natl Acad Sci USA 80:1068–1072
Oberdick J, Levinthal F, Levinthal C (1988) A Purkinje cell differentiation marker shows a partial DNA sequence homology to the cellular sis/PDGF2 gene. Neuron 1:367
Oberdick J, Smeyne RJ, Mann JR, Zackson S, Morgan JI (1990) A promoter that drives transgene expression in cerebellar purkinje and retinal bipolar neurons. Science 248:223–226
Oroskar AA, Read GS (1989) Control of mRNA stability by the virion host shutoff function of herpes simplex virus. J Virol 63(5):1897–1906
Pallela TD, Hidaka Y, Silverman LJ, Levine M, Glorioso J, Kelley WN (1989) Expression of human HPRT mRNA in brains of mice infected with a recombinant herpes simplex virus-1 vector. Gene 80:137–144
Park O-K, Gugneja S, Mayo KE (1990) Gonadotropin-releasing hormone gene expression during the rat estrous cycle: effects of pentobarbital and ovarian steroids. Endocrinology 127(1):365–372
Pfaff DW (1988) Multiplicative responses to hormones by hypothalamic neurons. In: Yoshida S, Share L (eds) Recent progress in posterior pituitary hormones. Elsevier/Excerpta Medica, Amsterdam, pp 257–267 (International congress series, no 797)
Presant CA, Proffitt RT, Turner AF, Williams LE, Winsor D, Werner JL, Kennedy P, Wiseman C, Gala K, McKenna RJ, Smith JD, Bouzaglou SA, Callahan RA, Baldeschwieler J, Crossley RJ (1988) Successful imaging of human cancer with radiolabelled phospholipid vesicles. Cancer 62:905–911
Presant CA, Blayney D, Proffitt RT, Franklin Turner A, Williams E, Nadel HI, Kennedy P, Wiseman C, Gala K, Crossley RJ, Preiss SJ, Ksionski GE, Presant SL (1990) Preliminary report: imaging of Kaposi sarcoma and lymphoma in AIDS with indium 111-labelled liposomes. Lancet 335:1307–1309
Preston VG, Darling AJ, McDougall IM (1988) The herpes simplex virus type 1 temperature-sensitive mutant ts1222 has a single base pair deletion in the small subunit of ribonucleotide reductase. Virology 167:458–467
Price J, Turner D, Cepko C (1987) Lineage analysis in the vertebrate nervous system by retrovirus-mediated gene transfer. Proc Natl Acad Sci USA 84:156–160
Rahman A, Kessler A, More N, Sikie B, Rowden G, Woolley P, Schein PS (1980) Liposomal protection of adriamycin-induced cardiotoxicity in mice. Cancer Res 40:1532–1537

Read GS, Frenkel N (1983) Herpes simplex virus mutants defective in the virion-associated shutoff of host polypeptide synthesis and exhibiting abnormal synthesis of alpha (immediate-early) polypeptides. J Virol 46:498–512

Roizman B (1979) The structure and isomerization of herpes simplex virus genomes. Cell 16:481–494

Roizman B, Batterson W (1984) The replication of herpes viruses. In: Fields B (ed) General virology. Raven, New York, pp 497–526

Roizman B, Sears AE (1987) An inquiry into the mechanisms of herpes simplex virus latency. Annu Rev Microbiol 41:543–571

Romano GJ, Harlan RE, Shivers BD, Howells RD, Pfaff DW (1988) Estrogen increases proenkephalin messenger ribonucleic acid levels in the ventromedial hypothalamus of the rat. Mol Endocrinol 2:1320–1328

Romano GJ, Krust A, Pfaff DW (1989a) Expression and estrogen regulation of progesterone receptor mRNA in neurons of the mediobasal hypothalamus: an in situ hybridization study. Mol Endocrinol 3:1295–1300

Romano GJ, Mobbs CV, Howells RD, Pfaff DW (1989b) Estrogen regulation of proenkephalin gene expression in the ventromedial hypothalamus of the rat: temporal qualitites and synergism with progesterone. Mol Brain Res 5:51–58

Romano GJ, Mobbs CV, Lauber A, Howells R, Pfaff DW (1990) Sex difference in enkephalin gene expression in rat hypothalamus. Brain Res 536:63–68

Rosen A, Gelderablom H, Darai G (1985) Transduction of virulence in herpes simplex virus type 1 from a pathogenic to an apathogenic strain by a cloned viral DNA fragment. Med Microbiol Immunol 173:257

Rosie R, Thomson E, Fink G (1990) Estrogen positive feedback stimulates the synthesis of LHRH mRNA in neurones of the rostral diencephalon of the rat. J Endocrinol 124:258–289

Rothfeld J, Hejtmancik JF, Conn PM, Pfaff DW (1989) In situ hybridization for LHRH mRNA following estrogen treatment. Mol Brain Res 6:121–125

Schek N, Bachenheimer SL (1985) Degradation of cellular mRNAs induced by a virion-associated factor during herpes simplex virus infection of Vero cells. J Virol 55:601–610

Schmechel DE, Marangos PJ (1983) Neuron specific enolase as a marker for differentiation in neurons and neuroendocrine cells. Current Methods Cell Neurobiol 1:1–62

Shimohama S, Rosenberg MB, Fagan AM, Wolff JA, Short MP, Breakefield XO, Friedmann T, Gage FH (1989) Grafting genetically modified cells into the rat brain: characteristics of E. coli β-galactosidase as a reporter gene. Mol Brain Res 5:271–278

Shivers BD, Schachter BS, Pfaff DW (1986) In situ hybridization for the study of gene expression in the brain. Methods Enzymol 124:497–510

Sibrack CD, Gutman LT, Wilfert CM, McLaren C, St Clair MH, Keller PM, Barry DW (1982) Pathogenicity of Acyclovir-resistant herpes simplex virus type 1 from an immunodeficient child. J Infect Dis 146(5)673–682

Soriano P, Dijkstra J, Legrand A, Spanjer H, Londos-Gangliardi D, Roerdink F, Scherphof G, Nicolau C (1983) Targeted and nontargeted liposomes for in vivo transfer to rat liver cells of a plasmid containing the preproinsulin I gene. Proc Natl Acad Sci USA 80:7128–7131

Spaete RR, Frenkel N (1982) The herpes simplex virus amplicon: a new eucaryotic defective-virus cloning-amplifying vector. Cell 30:295–304

Spaete RR, Frenkel N (1985) The herpes simplex virus amplicon: analysis of cis-acting replication functions. Proc Natl Acad Sci USA 82:694–698

Spear PG (1985) Glycoproteins specified by herpes simplex virus. In: Roizman B (ed) The herpesviruses, vol 3. Plenum, New York, pp 45–104

Stow ND, McMonagle EC (1983) Characterization of the TRs/IRs origin of DNA replication of herpes simplex virus type 1. Virology 130:427–438

Stow ND, McMonagle EC, Davidson AJ (1983) Fragments from both termini of herpes simplex virus type 1 genome contain signals required for the encapsidation of viral DNA. Nucleic Acids Res 11(23):8205–8220

Straubinger RM, Papahadjopoulos B (1983) Liposomes as carrier for intracellular delivery of nucleic acids. Methods Enzymol 101:512–527

Sunstrum JC, Chrisp CE, Levine M, Glorioso JC (1988) Pathogenicity of glycoprotein C negative mutants of herpes simplex virus type 1 for the mouse central nervous system. Virus Res 11:17–32

Sydiskis RJ, Roizman B (1967) The disaggregation of host polysomes in productive and abortive infection with herpes simplex virus. Virology 32:678–686
Thompson RL, Devi-Rao GB, Stevens JG, Wagner EK (1985) Rescue of a herpes simplex virus type 1 neurovirulence function with a cloned DNA fragment. J Virol 55:504
Thompson RL, Cook ML, Devi-Rao, GB, Wagner EK, Stevens JG (1986) Functional and molecular analyses of the avirulent wild-type herpes simplex virus type 1 strain KOS. J Virol 58:203–211
Ugolini G, Kuypers HGJM, Strick PL (1989) Transneuronal transfer of herpes virus from peripheral nerves to cortex and brainstem. Science 243:89–91
Uhl GR (ed) (1986) In situ hybridization in brain. Plenum, New York
Varmuza SL, Simley JR (1985) Signals for site specific cleavage of HSV DNA: maturation involving two separate cleavage events at sites distal to the recognition sequences. Cell 41:793–802
Vlazny DA, Frenkel N (1981) Replication of herpes simplex virus DNA: localization of replication recognition signals within defective virus genomes. Proc Natl Acad Sci USA 78:742–746
Vlazny DA, Kwong A, Frenkel N (1982) Site specific cleavage/packaging of herpes simplex virus DNA and the selective maturation of nucleocapsids containing full length viral DNA. Proc Natl Acad Sci USA 79:1423–1427
Wagner EK (1985) Individual HSV transcripts: characterization of specific genes. In: Roizman B (ed) The herpesviruses, vol 3. Plenum, New York, pp 45–104
Wagner EK, Devi-Rao G, Feldman LT, Dobson AT, Zhang Y, Flanagan WM, Stevens JG (1988) Physical characterization of the herpes simplex virus latency-associated transcript in neurons. J Virol 62(4):1194–1202
Werner M, Madreperla S, Lieberman P, Adler R (1990) Expression of transfected genes by differentiated, postmitotic neurons and photoreceptors in primary cell cultures. J Neurosci Res 25:50–57
Wolff JA, Malone RW, Williams P, Chong W, Acsadi G, Jani A, Felgner PL (1990) Direct gene transfer into mouse muscle in vivo. Science 247:1465–1468
WuDunn D, Spear PG (1989) Initial interaction of herpes simplex virus with cells is binding to heparan sulfate. J Virol 63(1):52–58
Zemanick MC, Strick PL (1990) Retrograde and anterograde transneuronal transport of HSV1 in the primate motor system. Soc Neurosci Abstr 16:425

Electrophysiological Methods for Studying Endocrine Cells and Neuronal Activity from the Hypothalamo-hypophyseal System

P.-M. LLEDO, L.-A. KUKSTAS, and J.-D. VINCENT

Contents

Institut Alfred Fessard, CNRS, Avenue de la terrasse, 91 198 Gif-sur-Yvette Cédex, France

Current Topics in Neuroendocrinology, Vol. 11

1 Introduction

The hypothalamo-hypophyseal system brings nerve and endocrine cells together in one anatomical entity such that the nervous system and the glandular cells of the anterior hypophysis communicate. The two structures share common properties. They both secrete peptidergic hormones (releasing and inhibiting hypothalamic factors, and the hypophyseal stimulins), and both show electrical properties such as excitability, with production of action potentials. Thus, electrophysiological techniques, which were previously reserved for studies of nerve and muscle cells, can be applied to the hypothalamo-hypophyseal system in both its nervous and endocrine structures (for reviews, see Vincent et al. 1980; Vincent and Dufy 1982; Vincent 1983; Poulain and Vincent 1987; Israel and Vincent 1990).

The electrophysiological properties of these cells reveal:

(1) the stimulus-secretion coupling, in particular at the neurosecretory terminals in the posterior hypophysis, and the median eminence, as well as in the endocrine cells of the anterior hypophysis;

(2) the modifications in membrane electrical properties brought about by the liaison of different regulatory factors with their receptors. These observations are used to explain the modulatory mechanism of membrane properties brought into play by each factor in order to enhance or inhibit hormone secretion.

Thus the electrical properties play a central role, which has until recently been ignored by endocrinologists, in the regulation of endocrine secretion in the anterior hypophysis. These electrical properties, common to nerve and endocrine cells, are linked to changes in membrane permeability to different ionic species; Ca^{2+}, Na^{+}, K^{+}, and Cl^{-} (see Ionic Channels of Excitable Membranes edited by Hille 1992). Electrophysiological techniques allow the study of the mechanisms underlying the changes in membrane ionic permeability brought about by different regulatory factors.

With the patch-clamp technique, we can directly investigate a single ion channel, or indeed a population of channels in a cell membrane. Different experimental approaches can then be used to determine the coupling mechanisms between receptors and ionic conductances, which take place via membrane coupling proteins, and different intracellular second messenger systems.

Recently, the role of genomic expression has been shown in the modulation of conductances and membrane receptors. Finally, changes in membrane electrical properties result from the sum of mechanisms on the hypothalamo-hypophyseal complex. It would be impossible to try to report all of the data accumulated concerning this subject, and we shall use a few examples of work carried out in our laboratory to describe the experimental electrophysiological paradigms that will be widespread in the future.

2 Preparations

2.1 In Vivo Systems

Most of the insight into behavior of excitable membranes has been gained from studies of a few nonmammalian preparations that are highly suitable for the two conventional electrophysiological techniques: extracellular recordings, where an electrode is placed outside the cell membrane, and intracellular recordings using a microelectrode to penetrate the cell. Extracellular recordings supply information about whether a cell is firing action potentials or is quiescent in its rate of firing. On the other hand, intracellular recordings are used to obtain information about the processes of excitation and inhibition and the mechanisms that initiate nerve impulses (see *From Neuron to Brain* edited by Kuffler et al. 1984). The in vivo system which has been the most widely used is the group of neurons located in the abdominal ganglion of aplysia. These cells have fascinated electrophysiologists for many years because of their endogenous rhythmic pattern of electrical activity (Strumwasser 1965), which is modulated by a variety of hormones, neurotransmitters, and synaptic stimuli (for an exhaustive review, see Adams and Benson 1985). Most extracellular recordings of neurosecretory neurons, such as magnocellular neurons, were obtained from unanesthetized monkeys in vivo (Hayward and Vincent 1970; Vincent and Hayward 1970; Vincent et al. 1972; Arnauld et al. 1974, 1975) or from anesthetized rats (Novin et al. 1970; Yamashita et al. 1970, Koizumi and Yamashita 1971, 1972; Mason and Leng 1984). However, serious difficulty is encountered when recording from mammalian tissues with external or intracellular microelectrodes, namely, ascertaining the exact position of the tip, which cannot be seen and controlling the environmental conditions of the tested cell. Because of these disadvantages, the majority of electrophysiological experiments are now performed on more suitable preparations such as the organ en bloc, brain slices, cell cultures, or oocytes.

2.2 Organ En Bloc and Brain Slices

The preparation described by Llinàs et al. (1981) consisted of an isolated guinea pig brain stem-cerebellum en bloc maintained in vitro by perfusion with a Krebs-Ringer solution. This approach allows stable intracellular recordings to be made and the regulation of physiological parameters such as temperature, pH, and perfusate chemistry. Pharmacological analyses could thus be carried out on single cells, and studies could be made of neural network properties in the mammalian brain. This technique was later extended to the whole brain in vitro (Walton and Llinàs 1982), but serious limitations with long-term survival were encountered.

In addition, well-organized areas, such as cerebellar cortex, cerebral cortex, or hippocampus can be sectioned in one plane to make a slice between 0.2 and 0.5 mm thick (for a review, see MacVicar and O'Beirne 1988); this preparation can then be maintained in a dish for several hours or days. The isolated brain slice method was first developed for biochemical studies by McIlwain in the early 1960s (McIlwain 1961). Later, Yamamoto (Yamamoto and McIlwain 1966) utilized this technique to obtain the first electrophysiological recordings from in vitro brain slices. By the mid to late 1970s, the use of isolated brain slices had become very widespread and analysis of the elctrophysiological properties of neurosecretory systems in the mammal has been greatly facilitated by the use of this technique (for reviews, see Jahnsen and Laursen 1983; Dingledine 1984). The advantages of brain slice techniques are manyfold, and similar to those for dissociated cell cultures. Firstly, the extracellular ionic environment around neurons can be rigorously controlled and easily altered. This is extremely important for electrophysiological analysis because it allows for determination of the ionic basis of currents and the rapid introduction of pharmacological agents. This makes it a useful technique for extra- and intracellular recordings from neurons in brain slices (Dingledine et al. 1980). Another factor is the ease of obtaining intracellular impalements; unlike in vivo preparations this method makes obtaining intracellular impalements easier because of the absence of vibrations or pulsations caused by vascular and respiratory systems, and allows the introduction of several microelectrodes into a brain slice. Furthermore, a single slice may contain many elements of the initial circuit, and with precise localization of the electrode tip, dendrites as well as cell bodies can be impaled. A recent advance comes from the partial enzymatic digestion of brain slices by collagenase or papain to expose cell bodies that are still viable (Gray and Johnston 1985). This allows for the use of patch-clamp technology to obtain high-resistance seals from the exposed cells for whole-cell voltage clamping and for patch-clamping analysis of ionic currents (Edwards et al. 1989; Sakmann et al. 1989). Since in these preparations most of the synaptic contacts are intact, the method is particularly interesting for the investigation of the molecular mechanisms of synaptic transmission at identified central neurons (for a review, see Konnerth 1990).

Viable brain slices have been prepared from many areas of the brain, including the hippocampus, hypothalamus, cortex, thalamus, striatum, substantia nigra, cerebellum, the brain stem, and inferior olive. It is usually straightforward to determine anatomical structures within a slice, because, when transilluminated, cell body areas are more translucent than dendritic areas, and nerve fiber tracts are the most opaque. Therefore it is possible to visually place microelectrodes in the desired anatomically defined location. Even in the brain stem (Llinàs and Yarom 1981) or thalamus (Jahnsen and Llinàs 1984), different nuclei can be reliably located. It is difficult to see individual cells using normal optical techniques, and three techniques have been developed for their visualization in intact brain slices. Each of these has certain advantages and disadvantages, the most important limitation arising from the thickness of the slice. The use of Normarski optics was first applied to 80-μm cerebellar slices by Yamamoto and Chujo (1978). This technique allows excellent visualization of individual neurons and the microelectrode. It cannot be used, however, in slices that are over 100-μm thick. Llinàs and Sugimori (1980) have developed a Hoffman modulation contrast system to visualize cerebellar neurons in slices 100–200 μm thick. This is particularly useful when working with cerebellar slices, because the Purkinje cell dentritic field is restricted to a two-dimensional plane. Therefore Purkinje neurons and dendrites will be intact in a 200-μm slice. Most neurons do not have such convenient morphological features and would have most of their dendrites shorn off in such a thin slice. MacVicar (1984) has developed a system using an inverted microscope and infrared video microscopy to visualize neurons in slices up to 400 μm thick. Although infrared optical systems do not have the resolution that shorter wavelength systems have, they can be used to visualize neurons in thick organ slices because tissue is more transparent in the infrared. Cell bodies as well as microelectrodes can be seen in slices 400 μm thick.

2.3 Cell Culture Systems

Tissue culture started with Harrison's experiments (1907) showing that small pieces of frog neural tube explanted into clots of frog lymph could be kept in vitro for several weeks. Culture techniques have been considerably refined, and it is now possible to isolate organs, tissues, and cells to initiate cell cultures that can be maintained sometimes indefinitely. There are two categories of tissue culture: primary cultures and cell lines; each can be further divided into several subgroups (Fedoroff 1977; Fig.1). Because of certain limitations, several studies have also been performed on acutely dissociated cells.

2.3.1 Acutely Dissociated Cells

It is possible to acutely isolate cells from tissue by mechanical and enzymatic treatment. Recently, neurons from brain slices have been isolated by enzymatic treatment (Gray and Johnston 1987; Mody et al. 1989; Legendre and Westbrook 1990). This preparation has the advantage of making available differentiated adult neurons after this treatment for patch-clamp recording techniques (Akaike et al. 1988). However, the neurons lack much of their dendritic tree, and it is unknown what effect the enzymatic treatment has on cell physiology. In our laboratory, we have used acutely dissociated pituitary cells in order to study their electrical activity (Kato et al. 1991). Anterior pituitaries were dissected from male Wistar rats under ether anesthesia. Minced pituitaries were incubated in divalent cation-free phosphate buffered saline (PBS) containing 0.2% trypsin. Pituitaries were then transferred to PBS containing 0.1% bovine serum albumin and trypsin inhibitor, and mechanically dispersed by triturating with a pipette. The dissociated cells were plated on poly-L-lysine-coated petri dishes and incubated in Dulbecco's Eagle's medium with 10% fetal calf serum. Cultured cells were used for tight-seal whole-cell recording (Hamill et al. 1981) within 24 h.

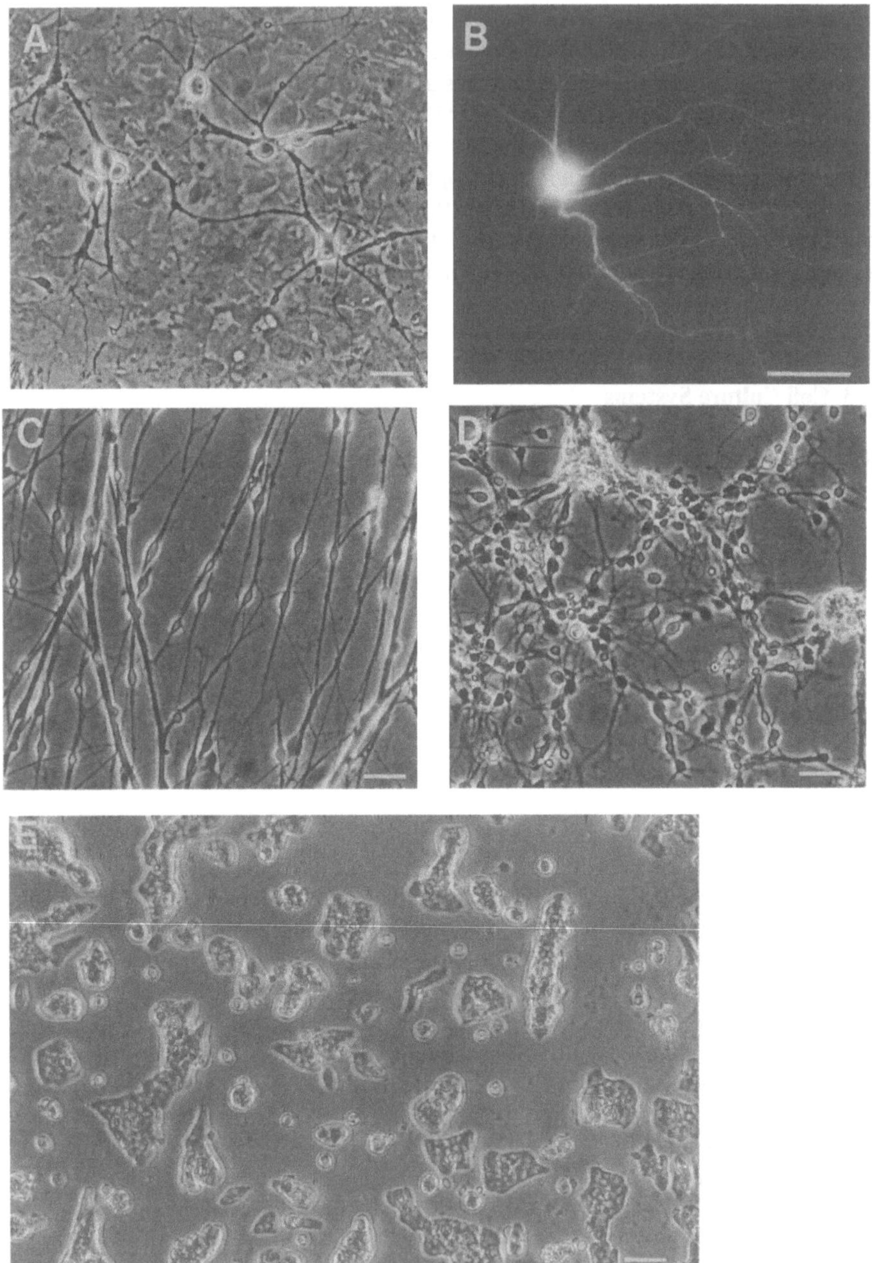

Fig. 1A-E. Phase-contrast micrographs of **A** spinal cord neurons maintained 3 weeks in culture; **B** a spinal cord neuron injected with Lucifer yellow; **C** rat Schwann cells from dorsal root ganglia after 31 days in culture (from Amédée et al. 1991, with permission); **D** acutely dissociated neurons from the rat striatum (24 h in culture); and **E** rat pituitary cells (15 days in culture). Observed under Normasky optics *Bar* = 50μm

2.3.2 Primary Cultures

Primary cultures are started with samples taken directly from the organism. There are three types which differ in their degree of organization, i.e., organ culture, tissue culture, and primary cell culture.

In organ culture, the goal is to achieve a high degree of organization, either by using cells that are immobile and do not divide or by preventing cell migration and division. The tissue culture method also originates from an explanted piece of tissue, but cell proliferation and migration from the margins of the explant are not halted, and disorganization occurs. Primary cell cultures lack organization, since the tissue is intentionally dissociated into individual cells by mechanical and enzymatic disruption. This method has been sucessful both with unspecialized cells such as fibroblasts and specialized cells such as neurons, endocrine cells, and various muscle tissues. The main advantage of primary cultures of specialized cells is that they provided suitable tools for morphological, electrophysiological and pharmacological studies and are extensively exploited in such areas (for a review, see Nelson and Lieberman 1981). With a view to learning the extent to which endogenous mechanisms or synaptic driving determines the bursting activity associated with application of peptides, in particular vasopressin, we have recorded intracellularly the electrical activity from differentiated neurons in primary cultures of dispersed, 14-day-old fetal mouse hypothalamus (Legendre et al. 1982; Theodosis et al. 1983). However, we have recently shown that ionic channel activity, and therefore membrane excitability, can change with the age of such cultures. Using the whole-cell configuration of the patch-clamp technique, we have studied primary cultures of rat pituitary lactotrophs (prolactin-secreting pituitary cells) and recorded two types of voltage dependent Ca^{2+} currents and two types of voltage-dependent K^+ currents. The results indicate that the amplitude of K^+ currents remains constant with the age of culture, whereas the amplitude of both types of Ca^{2+} currents increases markedly between days 1 and 12 (a 50% increase). These findings were obtained from normal prolactin-secreting cells which are inefficient in dividing when maintained in culture, suggesting that this change in excitability occurs because of culture conditions rather than because of a putative cell cycle (Lledo et al. 1991b).

Because the anterior pituitary is composed of several types of endocrine cells, apparently distributed in a disorderly way, particularly in rodents, it was quite impossible to record from a given cell type within the gland and know what type it was. Furthermore, because any given cellular type hardly exceeds 10%–20% of the total cell population, it was necessary to enrich cultures of the cell type to be studied. Throughout all of our experiments, anterior pituitary cells were obtained from Wistar rats, and the animals used were 14-day-old females, 14-day-old males, adult males, or lactating females, according to the desired cell type. Among the various techniques available, including Percoll gradient (Hall et al. 1982), cytofluorimetric (Denechaud et al. 1987; Hatfield and Hymer 1985), density gradient centrifugation, and separation by affinity methods (for a review, see Hymer and Hatfield 1984), we decided to separate the cells using a continuous gradient of bovine serum albumin (BSA) at unit gravity (Israel et al. 1983, 1987).

Briefly, this technique involved enzymatic and chemically defined media treatments followed by a mechanical dissociation with a flame-polished Pasteur pipette (Denef et al. 1978). Yields varied according to the donor and were about 1.1×10^6 cells per pituitary gland of 14-day-old females, up to 2.5×10^6–3×10^6 cells per gland of adult males, and lactating females yielded 4×10^6–7×10^6 viable cells per pituitary. The cell separation method, which has been described previously (Denef et al. 1978; Israel et al. 1983, 1987), consisted of layering cells onto a continuous density gradient of BSA at unit gravity and allowing the cells to sediment. The degree of enrichment in the different fractions of the gradient was assessed by immunofluorescent staining of original cultures. In general, a single population of a given cell type is located in one layer of the gradient but, when pituitaries where obtained from rat lactating females, lactotrophs were located in both the light fractions of the gradient (up to 95% lactotrophs), with a second population identified in the heavy fractions (up to 75%).

In order to identify single lactotroph cells in primary culture, prior to electrophysiological studies, the reverse hemolytic plaque assay (RHPA) has been shown to be very useful (Fig.2; Neill and Frawley 1983; Lledo et al. 1990b). We have also used the RHPA technique to quantify hormone secretion from individual cells. Indeed, this technique allows the determination of two parameters: (1) the percentage of plaque-forming cells, which is the percentage of secretors of a given hormone within a population; and (2) the plaque area, which has been shown to be linearly related to the release of radioimmunoassayable hormone (Luque et al. 1986). From lactating rats, analysis of the sizes of plaques produced by lactotrophs revealed a bimodal frequency distribution consisting of small and large modes. The large plaques were 2.7 times greater in area than the small plaques (the two average modes were 1500 and 4000 μm^2). Light fraction cells produced mainly large plaques (65% of plaque-forming cells) and heavy fraction

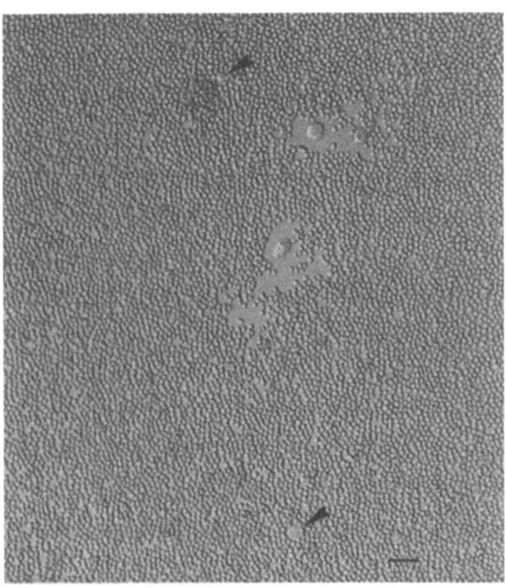

Fig. 2. The reverse hemolytic plaque assay of prolactin-producing cells. Sheep red blood cells surrounding prolactin-producing cells are lysed. Note cells that are not plaque-forming (*arrows*). *Bar* = 50μm

cells mostly small plaques (70% of plaque-forming cells), suggestimg a functional heterogeneity of lactotrophs (Kukstas et al. 1990; Lledo et al. 1991a).

2.3.3 Cell Lines

There are two types of cell lines which differ according to their origin: cell lines derived from primary cultures and those originating from neoplasic tissues.

The types of cells capable of cell division in primary cultures are at present rather limited and usually consist of undifferentiated cells (e.g., fibroblast-like cells). It is therefore not surprising that only a few cell lines of neurobiological interest (e.g., neuroblastoma cells) have been established so far, from normal animal tissue. Clonal cell lines and cell lines derived from tumors have been extensively used in neurobiological research partly because of the need for large quantities of homogenous cells, and for their capacity for adaptation to culture conditions (see for a review, Lendahl and McKay 1990). However, care must be taken in interpreting results obtained with these systems, since the cells are genetically unstable and may show degeneration of the genome during in vitro life. Furthermore, the properties of tumoral cells often differ from those of their normal counterparts. It is worth noting in this regard that GH_3 cells, a clonal line of rat pituitary tumor cells which secrete both prolactin and growth hormone (Bancroft 1981), have a modal chromosome number 69, compared with 42 for normal rat pituitary cells (Bancroft 1981).

Kidokoro (1975) first reported eletrophysiological results obtained from rat pituitary tumor cells, which were later confirmed by several groups (Dufy et al. 1979, 1980; see also Ozawa 1985 for a review). Most electrophysiological studies have been carried out on animal tissues, with a limited number on human tumors. Human material is obtained after surgery in which pituitary fragments are removed from patients. We have shown that different cell types displayed distinct electrical properties (Dufy et al. 1982; Israel et al. 1985). In particular, prolactinoma cells surprisingly did not respond either to dopamine or to thyrotropin-releasing hormone (TRH; Dufy et al. 1982), whereas cells from a prolactin microadenoma were responsive to dopamine (Israel et al. 1985), and nonfunctioning tumor cells showed responses to all of the substances tested.

2.4 Eggs and Oocytes

One of the most popular approaches to the study of the structure and function of membrane proteins that regulate neuronal membrane activity consisted of microinjection of foreign messenger RNAs (mRNA) into living cells such as the oocyte, eggs (the term “oocyte” refers to the cell before meiotic maturation, “egg”, after maturation), and somatic cells (see for reviews, Lane 1983; Soreq 1985; Snutch 1988). Because of its large size, which makes microinjection a relatively simply process, the first and still most widely used type of living cell that serves this purpose is the oocyte of the frog *Xenopus laevis* (Gurdon et al. 1971). In this respect, oocytes provide an excellent assay system to perform in vivo translation for a wide variety of exogenous mRNA in studies of mammalian neurotransmitters, hormone receptor and ion channels.

X. laevis ovarian follicles (Fig.3), manually extracted from the ovary, consist of an oocyte surrounded by a vitelline envelope, follicular cells, and thecal tissues, all of which are enveloped within the inner ovarian epithelium. It is a cell specialized for the production and storage of proteins (for later use during embryogenesis) exibiting extensive subcellular systems involved in the production of proteins. The oocyte is, therefore, a suitable preparation through which to understand the mechanisms involved in neuronal activity and modulation because:(1) they faithfully express many functional neurotransmitter receptors and voltage-gated ion channels after injection of mRNA from an appropriate tissue source; and (2) they can be easily recorded from electrophysiological experiments by techniques including the two-microelectrodes for the whole-cell recording and the patch-clamp (i.e., cell-attached configuration) for single channel recording. Nanogram quantities of poly(A) + mRNA are injected into individual *X. laevis* oocytes, and, after an 8- to 36-h incubation period, electrophysiological recordings can be performed in order to detect the synthesis of neurotransmitter receptors and/or ion channels. Table 1 summarizes some of the synthesized receptors and ion channels that have been induced in *X. laevis* oocytes by the injection of mRNA isolated from whole rat brain. There

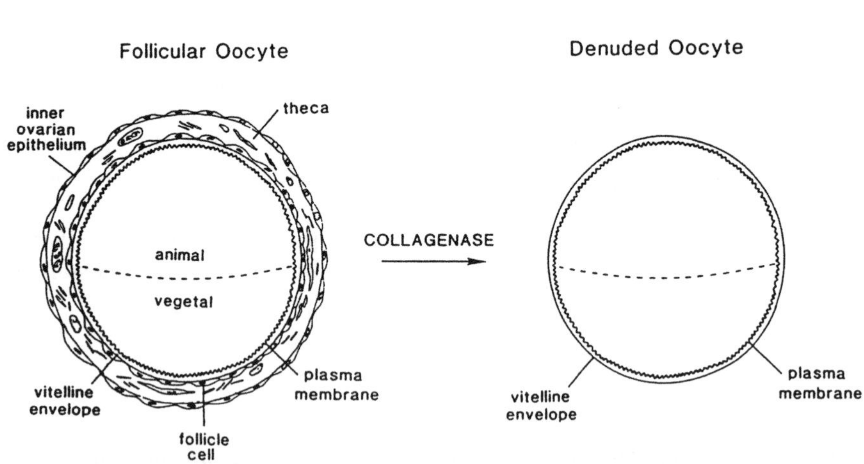

Fig. 3. Schematic diagram of *X. laevis* oocyte and surrounding tissues. The oocyte is surrounded by a number of cellular and noncellular layers and is termed a follicular oocyte. The oocyte plasma membrane is covered by a noncellular, fibrous mat called the vitelline membrane. In the follicular oocyte, several additional cellular layers flank the vitelline membrane, including: a monolayer of follicle cells; the theca, which is a fibrous layer containing blood vessels, nerve, and fibroblasts; and the inner ovarian epithelium, a continuum of the ovary wall. All endogenous receptors and the GTP-binding proteins and adenylate cyclase are localized to the follicle cell layer: Prior to patch-clamp analysis, the vitelline membrane is removed. (from Snutch 1988, with permission)

are four types of voltage-activated ion channels: one Na^+, one Ca^{2+}, and two K^+ channels (the "A" and delayed rectifier K^+ channels). Both inhibitory [e.g., γ-aminobutyric acid (GABA) and glycine] and excitatory amino acid receptors [e.g., *N*-methyl-*D*-aspartate (NMDA), quisqualate, and kainate] have also been expressed in oocytes. Finally, several metabotrope receptors for the neuropeptides (neurotensin and substance P), amines (the serotonin 5-HT_{1c}), cholinergic (M_1) and glutamate (quisqualate) subtypes have been characterized in mRNA-injected oocytes (for a review, see Snutch 1988). Nowadays this table is very incomplete, for example, it has also been shown that oocytes microinjected with mRNA extracted from GH_3 acquired responsiveness to TRH (Oron et al. 1987). The principal steps of the signal transmission pathway which occur in the oocyte after activation of those metabotropic receptors, have been elucidated (Kaneko et al. 1987) and are summarized in Fig. 4. All members of this class of receptor involve an ionositol-1,4,5-triphosphate ($InsP_3$)/Ca^{2+} second messenger pathway that acts in the oocyte to open a Ca^{2+}-dependent Cl^- channel. Nevertheless, with this technique, one should be aware of a number of potential problems. Why have so few voltage-gated ionic channels been detected in oocytes injected with rat brain mRNA while more than ten voltage-gated ionic channels have been described in neurons? Why have only specific receptor subtypes been detected in oocytes (5-HT_{1c} or M_1 subtypes) whereas there are at least five other serotonergic and other muscarinic receptor subtypes? These questions evoke the selectiveness of mRNA expression in oocytes which might be due to unknown translational and/or posttranslational processes, characteristic of the *X. laevis* oocyte. The second point that should be emphasized is related to the heterogeneity of mRNAs originated from whole organs (brains, heart, glands) which are injected together into oocytes. This statement involves that different neosynthesized polypeptides which never interact in vivo (e.g., a Na/K ATPase from astrocyte, and Ca^{2+} channels from hypothalamic neurons) may produce an unpredictable combination. In this way, MacDonald et al. (1989) have detected a K^+ current through expressed channels which resembles both transient and the delayed rectifier K^+ currents reported in mammalian neurons.

However, the large number of studies performed on oocytes show that many advances in basic neuroscience have been achieved by mRNA microinjected into this cell. In addition, the diversity of questions that can be profitably posed is now extended by the discovery that these cells can use, in addition to mRNA, injected DNA. This approach has been sucessful for the cDNA cloning of several neuronal ion channels and receptors, including the voltage-dependent Na^+ channel (Noda et al. 1986), the muscarine acetylcholine receptor (Kubo et al. 1986), and the GABA receptor/channel (Schofield et al. 1987). This opens a new field of investigation which will undeniably lead to better understanding in molecular neurobiology.

Table 1. Receptors and ion channels induced in *X. laevis* oocytes by injection of rat brain mRNA (From Snutch 1988, with permission)

Agonist	Permeation	Kinetics	Notes
Neurotensin	Cl^-	Long latency, transient spikes, oscillations	As per native receptor Densensitizes
	?	Smooth	Cross-desensitization with muscarinic ACh response Mimicked by $InsP_3$ injection Blocked by intracellular EGTA
Substance P	Cl^-	Long latency, transient spikes, oscillation	Pharmacology not reported Desensitizes
	?	Smooth	
Acetylcholine	Cl^-	Long latency, transient spikes, oscillations	Muscarinic receptor pharmacology Mimicked by $InsP_3$ injection Blocked by intracellular EGTA Desensitizes Cross-desensitization with 5-HT_{1c}responce
5-HT	Cl^-	Long latency, transient spikes, oscillations	Only 5-HT_{1c} receptor subtype detected Mimicked by $InsP_3$ injection Blocked by intracellular EGTA Desensitizes Cross desensitization with muscarinic ACh response
Glutamate Quisqualate	Cl^-	Long latency, transient spikes, oscillations	No block with Joro spider toxin Mimicked by $InsP_3$ injection Cross desensitization with muscarinic ACh response Attenuated by pertussis toxin EGTA
Kainate	Na^+/K^+	Short latency, smooth	No effect of pertussis toxin, cholera toxin, intracellular EGTA Blocked by Joro spider toxin Nondesensitizing
NMDA	Na^+/K^+	Short latency, smooth	Nondesensitizing Potentiated by glycine Voltage-dependent Mg^{2+} block
GABA	Cl^-	Short latency, smooth	Desensitizes $GABA_A$ subtype pharamacology No effect of pertussis toxin, cholera toxin
Glycine	Cl^-	Short latency, smooth	Blocked by strychnine Desensitizes

Table 1. *Continued*

Agonist	Permeation	Kinetics	Notes
Voltage-gated	Na^+	transiet, Peak current ~-10 to 0 mV	K_d~ 10n*M* for TTX Scorpion toxin removes inactivation Different inactivation properties with unfractionated and fractionated mRNA
Voltage-gated	Ca^{2+}	Partial inactivation, peak current~+15 mV	Recorded as Ba^{2+} current No effect of dihydropyridines Enhanced by phorbol esters
Voltage-gated	K^+	Transient	"A"-type current Blocked by 4-aminopyridine Insensitive to external TEA
		Noninactivating	"Delayed rectifier" type Blocked by external TEA May be more than one component

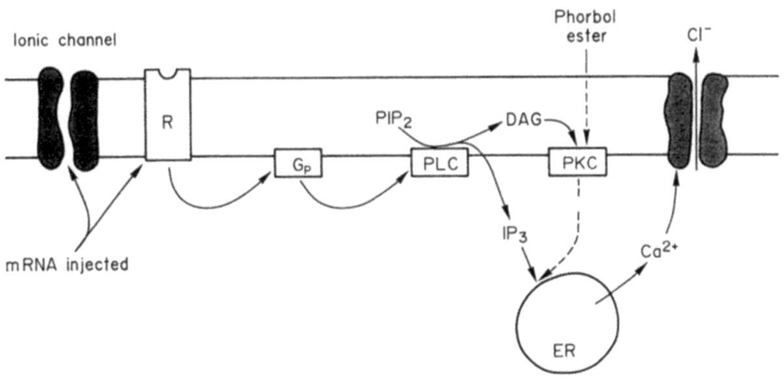

Fig. 4. Schematic diagram of receptors coupled to endogenous inositol triphosphate second messenger-mediated pathway. Injection of mRNA isolated from rat brain results in the synthesis of functional receptors *(R)* that, when activated, act via a guinine binding protein *(G_p)* to activate phospholipase *C (PLC)*. Inositol trisphosphate *(IP_3)*, one of the by-products of hydrolysis of phosphatidyl inositol bisphosphate *(PIP_2)* by phospholipase C, stimulates the release of Ca^{2+} from internal stores *(ER)*. The rise in intracellular Ca^{2+} then activates an endogenous Ca^{2+}-dependent Cl^- channel. *DAG*, diacylglycerol; *PKC*, protein kinase C.

3 Extracellular Recording Technique

3.1 Principle of Extracellular Recording Technique

The electrical activity of an excitable cell can be studied by means of an electrode placed onto the cell membrane. The electrode itself can be either a fine wire insulated almost to its tip or a capillary tube filled with salt solution. This technique allows for the distinction between spontaneously firing cells and quiescent cells, and supplies information about electrical activity from sponteanously firing cells (firing rates, interspike intervals, and conduction velocity) in basal condition and after neuromodulator application. This electrophysiological technique may investigate a population of neurons (multiunit activity, MUA), but the signals from a single neuron (unit activity) can also be identified. During the 1950s, several papers were published giving detailed descriptions of electrophysiological responses from populations of motoneurons recorded extracellularly (see for a review, Lloyd 1951). Although extracellular recording techniques also include mass recording methods using D.C. and E.E.G. techniques, only the multiunit and single unit recording methods will be discussed here.

3.2 Multiunit Recordings

The MUA technique can be useful to record from a large number of elements displaying uniform electrical behavior. However, information obtained from MUAwould currently seem to be of little value in the study of neuronal activity from excitable tissue such as hypothalamic neurons, for two essential reasons: The first reason derives from methodological considerations. A multiunit recording cannot be considered as a spatial summation. It collects the action potentials of a great number of neurons and fibers, so that the contribution of any given neuron to the MUA depends on distance and orientation from the recording electrode, and on physiological characteristics which are beyond the control of the experimenter. The second reason is based on morphological considerations. The hetereogeneous structure of the hypothalamus and the proximity within a given area of neurons belonging to different physiological regulatory systems make the data of MUA recordings difficult to interpret. Thus the single cell approach to understanding the function of hypothalamic neurons is preferred.

3.3 Single Unit Recordings

From this technique, electrophysiological characteristics such as action potentials or firing patterns can be studied at the single cell level. Moreover, electrophysiological unit techniques may also be useful to obtain a better understanding of anatomical organization of neuronal circuitry (Dyer 1977). Thus, the stimulation of an axon or a nerve ending generates antidromically an action potential in the cell body. This property has been widely used in neuroendocrinological study in order to identify many neurosecretory neuronal pathways. Kandel used this procedure to identify neurosecretory cells in the preoptic nucleus of the goldfish by stimulating nerve endings in the posterior pituitary gland (Kandel 1964). Following the application of this technique to mammals by Yagi et al. (1966), a direct demonstration of the ability of neurosecretory axons to conduct action potentials was first obtained by antidromic stimulation; when the neurohypophysis is stimulated electrically, action potentials can be triggered in the cell bodies of hypothalamic neurons (Fig. 5). Alternatively, electrical stimulation of the pituitary stalk evokes orthodromically a compound action potential down to the neurohypophysis (Fig. 5), both in vivo (Dyball and Leng 1987) and in vitro (Dreifuss et al. 1971; Nordmann and Stuenkel 1986). This axonal action potential appears to be of the classical sodium-potassium-dependent type since it is abolished by reducing the concentration of sodium or adding tetrodotoxin to the medium (Dreifuss et al. 1971). From magnocellular neurons of the supraoptic and paraventricular nuclei, it has been shown that action potentials display a complex waveform when recorded extracellularly, showing evidence for sodium and calcium spike components at different membrane sites (Mason and Leng 1984); the most likely site for the Ca^{2+} component is the cell dendrite. Thus, the electrical activity of these neurosecretory cells is an essential and rapid means for the cell

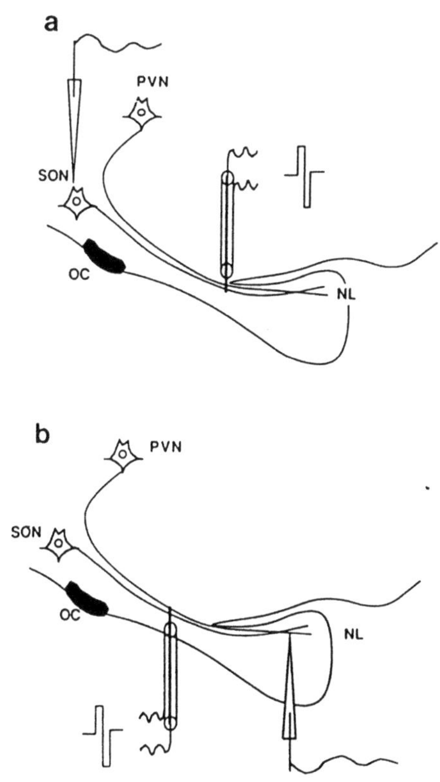

Figure 5a, b. Conduction of action potentials along the hypothalamo-neurohypophyseal tract. **a** Antidromic conduction. Using a dorsal approach to the hypothalamus, the neural stalk was electrically stimulated, and action potentials were recorded extracellularly within the supraoptic nucleus *(SON)* in a urethane-anasthetized rat. **b** Orthodromic conduction. Using a ventral approach to the base of the brain, the neural stalk was electrically stimulated, and action potentials were recorded extracellularly from the neurohypophysis (*NL*, neural lobe) in a urethane anesthetized rat. *PVN*, paraventricular nucleus; *OC*, optic chiasma. (Adapted from Poulain and Theodosis 1988, with permission)

body to control the rate of release of secretory products into the general circulation in respect to afferent solicitations (Poulain and Theodosis 1988).

However, this technique also has its limitations. In single unit studies, recording of samples is often biased in favor of large cells which are more readily accessible, or cells displaying an active firing pattern, thereby giving a distorted view of the neuronal population under investigation. Moreover, stable long-term recordings from a single cell are difficult to accomplish.

To overcome some limitations of single cell analysis, a sensitive technique was developed by Baertschi and Dreifuss (1979) for measuring continously the summated neural activity of the rat hypothalamo-neurohypophyseal tract. Using this technique, compound potentials with a peak amplitude of 4-9 mV were recorded from the fiber tract in response to electrical stimulation of the neurohypophysis.

3.4 Drug Administration During Extracellular Recordings

Microiontophoresis is a technique whereby drugs and other ionized particles in solution can be ejected in very small amounts from glass micropipettes. The ejection is accomplished by application of a voltage causing the electrode to become polarized. Ions and charged molecules will migrate toward or away from the source of the imposed electrical field depending on their net charge (Fig. 6). This pharmacological technique can be used to determine the effects of various substances upon firing parameters of both central and peripheral neurons and muscle (Kelly et al. 1975; Hicks 1984). If the pipette assembly is positioned close to a neuron so that extracellular recordings of its activity can be made through another electrolyte-filled barrel, drugs may be ejected and their pharmalogical effects inferred by resulting changes in the rate and/or pattern of firing (Hicks 1984). Although the amount of drug actually ejected is somewhat less than expected due to complex interactions occurring at the pipette tip, several pharmacological studies have been performed because the delivery remains proportional to the magnitude of current applied (Hall et al. 1979).

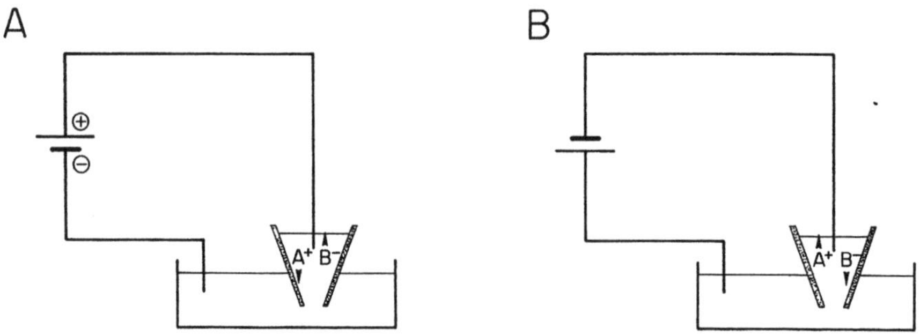

Fig. 6. A, B. Schematic diagram of a microiontophoresis pipette which contains a salt A^+B^-, showing the direction of current flow necessary to eject (*A*) and retain (*B*) the ion A^+

Another method of drug administration is by pressure. Quantitative studies (Dufy et al. 1979) have revealed that administration by pressure (microperfusion) shows certain advantages over the microiontophoresis technique: (1) the accurate determination of concentration within the micropipette, and therefore of the maximum drug concentration administered to the neuron under study; (2) the ejection of relatively large quantities of solution (an advantage that is especially useful in certain intracellular experiments); and (3) the administration of noncharged materials and substances such as the marker enzyme, horseradish peroxidase (HRP). On the other hand, the pressure method has certain detracting features, such as the relative imprecision of the actual amount of substance delivered from the pipette per pulse, the frequent shattering or blocking of micropipette tips, and drug leakage which necessitates the mechanical displacement of the tip away from the site of recording.

4 Intracellular Recording Technique

4.1 History

In 1945, a wire electrode was inserted into the cytoplasm of a squid giant axon to measure its resting and action potentials (Hodgkin and Huxley 1945). In order to extend the possibility of intracellular recordings to smaller cells, Ling and Gérard (1949) performed intracellular recordings using glass microelectrodes of 0.5-μ*M* tip diameter. Since then, most electrophysiological studies performed in invertebrates (e.g. nematodes, annelids, crustaceans, molluscs, etc.) or vertebrates (e.g. fishes, mammals, etc.) have been realized with glass microelectrodes (for reviews see Tauc 1967; Gerschenfeld 1973).

4.2 Principle of the Intracellular Recording Technique

Intracellular recordings are used to obtain information about electrical processes which occur in the cell membrane of excitable cells, and their modulation by external or internal signals. Glass microelectrodes are pulled from fine capillary tubes and have a resistance varying from 30 to 100 MΩ when filled with a salt solution as a conducting fluid (3 *M* potassium chloride or 4 *M* potassium acetate). The tip of the microelectrode (less than 0.2μm in diameter) is inserted into the cell, and drug application can be effectuated by microiontophoresis or pressure ejection (as reported above).

In current-clamp experiments, the microelectrode measures the potential difference between the inside and the outside of the cell and can also be used for passing transmembrane current to bring about charges in the cell membrane potential. Alternatively, when a conductance change occurs in an area of membrane that is small compared with the space constant of a cell, it is possible to control the membrane potential in that area, using either a single microelectrode both to monitor potential and to pass current (single microelectrode recordings) or two microelectrodes (double microelectrode recordings). This technique, called voltage-clamp, was first used sucessfully by Takeuchi and Takeuchi (1959) to record currents generated by acetylcholine at the motor end plate and has been extended to the measurement of membrane ionic currents underlying membrane potential changes in all excitable tissues. From these measurements it is possible to obtain a description of the ion conductances and transient conductance changes that occur in response to electrical and chemical stimuli.

A variety of techniques can be used to study the electrical properties of individual cells, but in terms of sensitivity, time resolution, and the isolation of different conductances, the voltage-clamp method is unsurpassed. In the last few years, extensions of voltage-clamp concept, (e.g., noise measurement) have been used to demonstrate how permeability change is mediated by the conformational transitions of discrete ionic channels, which constitute the fundamental molecular basis for electrophysiological events. An excellent book, entitled *Voltage and Patch Clamping with Microelectrodes*, edited by Smith et al. (1985), covers most aspects of the voltage-clamp technique.

5 Patch-Clamp Recording Technique

5.1 History

Conventional intracellular microelectrode methods for current measurement are associated with a background noise of at least 100 pA (10^{-10} A). The current flowing when a single channel opens is only a small fraction of the background noise (on the order of only 1-5 pA). Neher and Sakmann (1976) solved this problem by the patch-clamp method. Instead of inserting a microelectrode into a cell, they pressed a microelectrode tip (diameter less than 1 μm) onto its surface, effectively isolating a patch of membrane. In 1952, Fatt and Katz had previously showed the possibility of recording from a discrete area of membrane. They demonstrated that the current flowing through the area of membrane at endplates in frog muscle could be discretely recorded by positioning a microelectrode very near the synaptic junction. The currents were measured by the voltage drop they induced across resistance developed by the close apposition of the microelectrode tip with the membrane. A variant of this technique was developed by Strickholm (1961) in which a larger-tipped glass pipette was pressed against the synaptic connection. Using this method it is possible to obtain lower noise and higher resolution recordings of postsynaptic events such as miniature endplate currents.

In the initial experiments of Neher and Sakmann (1976) the seal resistance was less than 100 MΩ (10^8 Ω). One great improvement in the method occurred when it was found that gentle suction of the pipette greatly increased the resistance of the seal between the pipette tip and the cell membrane (Hamill et al. 1981). Because of the extremely high restistance of the seal, in the range of $10^1 -_n 100$GΩ (10^{10}- 10^{11} Ω), the current noise of the system is very low (less than 1 pA). From ionic channel study, ionic fluxes in the channels are detected as electric current flow in the recording circuit, and since the recording sensivity is very high, the opening and closing of one channel appears as clear step changes in the record (for a review see Auerbach and Sachs 1984). For example, acetylcholine-sensitive channels have been shown to elicit elementary current step corresponding to over 10^7 ions flowing per second in the open channel, from muscle membrane (Hamill and Sakman 1981). Although many variants of the patch-clamp technique have now been performed, its full potential is not realized. Our understanding of the physiological and pharmacological properties of ionic channels is being expanded daily by use of this technique.

5.2 Fundamental Concepts

If a glass pipette is sealed tightly against a cell, then any current flowing through the patch of membrane enclosed by the pipette will flow through the electronic circuit connected to the pipette rather than leak out around the area of its attachment to the membrane. In addition, any current applied to the membrane patch will flow through the patch rather than the seal. These observations have had a tremendous impact on electrophysiology for two reasons. First, the seal is electrically tight; resistance of the seal is often in excess of 10 GΩ, therefore random current flow through the seal is sufficiently reduced to permit the current through an indivdual membrane channel to be clearly recorded above the background noise. Secondly, the seal is mechanically tight; the cell membrane will rupture or tear before the seal breaks. This tightness permits the use of a variety of procedures to isolate a particular patch or region of membrane and permits manipulating of experimental conditions far more than was preciously possible (Hamill et al. 1981).

The patch area included in the recording pipette can be estimated from the magnitude of capacity currents arising from external voltage clamps. Assuming that the equivalent curcuit for a sealed patch is a parallel resistance/capacitance (RC) network and that the specific membrane capacity is 1 μF/cm^2, capacitive currents imply that a patch has an area between 1 and 10 μm^2, an area that is equivalent to a hemisphere with a radius of 0.5–1 μm.

The nature of the seal between the membrane and the glass remains obscure. However, we can identify four types of interaction that should participate in the glass-membrane seal. The first is ionic bonds between positive charges on the membrane and negative charges carried by SiO_2 on the glass surface. The second consists of salt bridges formed by divalent cations (Mg^{2+}, Mn^{2+}, Ca^{2+} etc.) between negatively charged groups on the membrane (e.g., phosphate) and glass. A third is hydrogen bonds between nitrogen or oxygen atoms in the phospholipids and oxygen atoms on the glass surface. Finally, it could be speculated that if the glass and the bilayer surfaces are extremely close, Van der Waals forces could also be involved.

5.3 Recording Configurations in Patch-Clamp Experiments

The patch-clamp technique is versatile. There are at least six different configurations which are schematized in Fig. 7 When a microelectrode is applied onto a cell, the seal between the tip and the outer surface of the cell membrane has, under suitable conditions (fire-polished, clean microelectrode tip and clean membrane surface), a high electrical resistance and is mechanically very stable. The seal isolates the patch both electrically and chemically. This first configuration has been called "cell-attached". The electrically isolated membrane patch can be pulled off the cell in such a way that the inside of the plasma membrane faces the bath solution ("inside-out") or alternatively so that the outside faces the bath solution ("outside-out"). By breaking the patch membrane in the cell-attached recording conformation, the solution in the pipette interior gains direct access to the cell interior and cell dialysis is carried out under conditions where the currents across the whole cell membrane can be measured ("whole-cell"). These methods are well decribed in an original paper (Hamill et al. 1981) and a book (Sakmann and Neher 1983). Equilibration between the cell interior and the bath solution can be attained whilst single-channel currents are recorded ("open cell-attached") by making holes in the plasma membrane outside the isolated patch area with the help of detergents like saponin (Petersen and Petersen 1986). On the other hand, equilibration of the cell interior with the pipette solution can be achieved by partially permeabilizing the patch membrane included in the pipette with ATP (Lindau and Fernandez 1986) or nystatin (Horn and Marty 1988) to obtain the "slow whole-cell" recording configuration (Fig. 7).

5.4 Advantages and Disadvantages of Configurations for Patch-Clamping

There are numerous advantages with the use of this method. In many ways patch-clamp experiments are much easier to perform than those using microelectrodes or other types of recording configurations. When an appropriate cell preparation has been found, the seal can be readily made. In addition to the ease of use, the conditions to which the membrane is exposed can be accurately controlled since the ionic gradient across the cell membrane is determined. This permits the physiology and pharmacology of a given channel to be examined at the molecular level. Descriptions above are general considerations of the patch-clamp technique, but each configuration offers advantages and limitations which can be summarized as follows.

5.4.1 Cell-Attached Patch

This is the easiest configuration which allows one to record ionic currents of a small membrane patch without any disturbances of the intracellular medium or cell structure. In several studies, this configuration is chosen in order to preserve cytoplasmic biochemistry and thus to record single-channel currents under relatively normal physiological conditions (see for a review, Farley 1988).

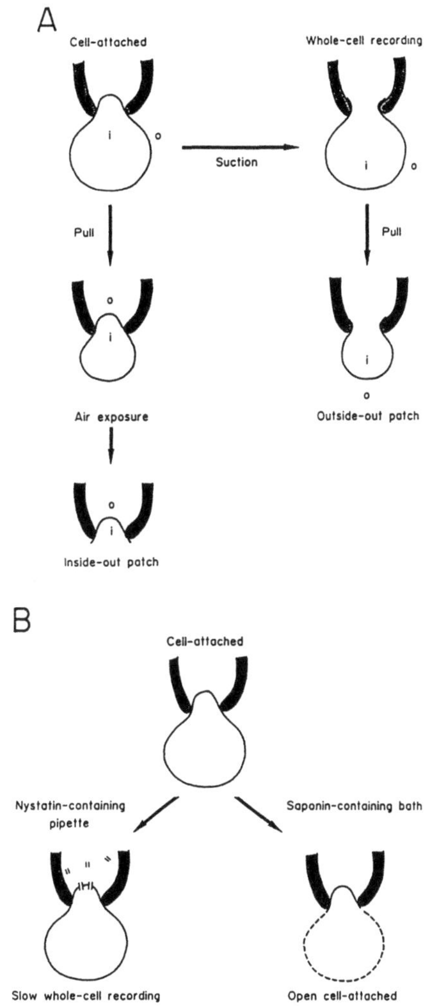

Fig. 7 A, B. Different configurations for patch clamping. All methods start with a clean pipette pressed against an intact cell to form a gigaohm seal between the pipette and the membrane it touches. **A** Single-channel activity can be recorded in this cell-attached configuration. Additional manipulations permit the same pipette to be used to clamp a whole cell or to excise a patch of membrane in inside-out or outside-out configurations. **B** Some variant configurations can be performed using a nystatin-containing pipette (slow whole cell) or saponin-containing bath (open cell-attached)

Moreover, this configuration distinguishes between direct and second messenger-mediated neurotransmitter action on channel activity by studying the effects of the neurotransmitter after bath or pipette administrations (Soejima and Noma 1984). Hence, we have shown in experiments performed in the cell-attached configuration that activation of single K^+ channels of cultured pituitary lactotrophs occurred when dopamine was administered in the bath rather than in the pipette (Fig. 8). However, during cell-attached configuration, the internal medium is undefined and cannot be controlled. The resting intracellular potential is also not precisely known; nor, therefore, is the membrane potential across the patch (unless a separate voltage-measuring electrode is inserted into the cell).

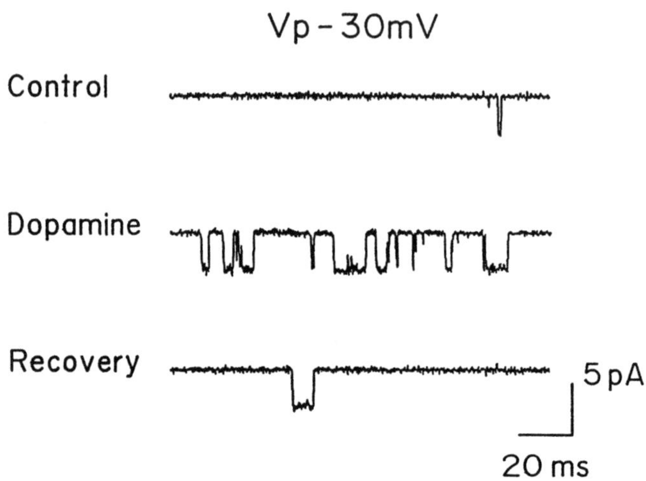

Fig. 8. Cell-attached patch recordings of single-channel activity in cultured lactotrophs before (*Control*), during (*Dopamine*), and after bath application of dopamine (*Recovery*). Note that the inward-directed channel current in the cell-attached patch is reversibly activated by 10-n*M* dopamine. The bathing solution contained 5-m*M* K^+ while the patch electrode contained 140-m*M* K^+. The patch electrode was held at -30 mV (*Vp*). Data were sampled from analogue tape recordings at 16-bit resolution and low-pass filtered at 2.5 kHz

A variant of this technique has been termed the "loose patch-clamp" (Almers et al. 1983). A large-tip pipette (10-20μm) is loosely pressed against the cell membrane. The membrane patch is voltage clamped and the currents flowing through the membrane can be measured (Stühmer et al. 1983). This configuration allows mapping of the density of channels in the cell membrane and is useful for high-resolution focal recording from extreme nerve endings (Forda et al. 1982).

5.4.2 Inside-Out-Patch

In this configuration for single-channel current recordings, the cytoplasmic or inside surface of the patch membrane is facing the bath. The voltage across the membrane is that between the pipette interior and the bath, thus it is strictly under the control of the experimenter and can be changed as desired. Furthermore, the free access to the cytoplasmic surface of the membrane is the most significant feature of this configuration. This latter advantage permits one to control the "cytoplasmic" concentration of permeanions, second messenger (Camardo et al. 1983), enzymes, or other agents that could modify the activity of channels. Unfortunately, the inside-out patch configuration has one serious limitation as the membrane patch is without its normal cytoskeleton, thus many metabolic control mechanisms which could affect channel activity are perturbed.

5.4.3 Outside-Out patch

This configuration allows single-channel current recordings. The most obvious uses of this technique are experiments in which the concentration of ions, drugs, or a neurotransmitter on the outer membrane surface has to be changed. However, it was observed that stability and maintenance of channel activity, when they are recorded in outside-out configuration, are much lower than in the case of the cell-attached configuration (Farley 1988). This problem may arise because the membrane cytoskeleton and metabolic control mechanisms are disrupted.

5.4.4 Whole-Cell Clamp

Although patch pipettes were originally developed for the recording of single channel events, they can be of great advantage for more conventional recordings in the whole-cell mode. For this configuration, the pipette is left attached to the cell after formation of a seal, but the membrane patch under it is ruptured by an additional suction in the pipette. This technique allows measurement of total membrane currents and/or voltage from the whole cell (Hamill et al. 1981; Fenwick et al. 1982). Compared with other cell recording techniques using a dialysis pipette (Kostyuk and Krishtal 1977; Lee et al. 1978), it is characterized by the very high resistance of the pipette-cell seal (>10 GΩ). Therefore, the method was initially called "tight-seal whole-cell recording", but in the following description, we use the shorter term "whole cell recording" (WCR).

One of the major characteristics of WCR is the access resistance. Because pipette resistance is comparatively low, the clamp circuitry can change the membrane potential quite rapidly: settling times of 50-200 μsec are common (Corey 1983). This is faster than the speed encountered using conventional two-microelectrode voltage clamps to change the membrane potential of small cells, and thus permits the study of rapidly activating and deactivating conductances. Another consequence of the low-resistance connection between pipette and cell interior is that the cytoplasm is rapidly exchanged with the pipette solution. This can be an advantage in some conditions: For the study of ionic current, the ability to control ionic composition of solution on both sides of the membrane helps to isolate particular conductances; and for loading cells with different substances, the WCR configuration method offers a more gentle way of modifying the intracellular milieu. It has been shown that in typical mammilian cells (10-15 μm diameter) intracellular K^+ is exchanged with small mobile cations present in the pipette within 5-10 s (Marty and Neher 1983). Likewise, larger molecules diffuse into the cells with a time constant in the range of 10 s–1 min depending on pipette resistance and on the root of molecular weight (Pusch and Neher 1988). This diffusional exchange offers the possibility of loading cells with substances of interest. In this way, the intracellular concentration of free calcium has been controlled by means of Ca/EGTA buffers. Other second messengers such as cAMP and cGMP have also been loaded into cells (for a review, see Neher 1988). After second messengers themselves, many other substances which participate in second messenger generation and function have been incorporated. These include the kinases or their regulatory subunits (Hescheler et al. 1986), non-hydrolyzable analogues of GTP (Breitwieser and Szabo 1985; Lledo et al. 1990a), and GTP-binding proteins (Hescheler et al. 1987). In this respect, we have recently performed intracellular injection of an affinity-purified polyclonal antibody directed against the α subunit of GTP-binding proteins (G proteins) using patch electrodes containing antibody. We have shown, in lactotrophs in primary culture, that G_0 proteins mediate the dopamine-induced decrease in voltage-dependent Ca^{2+} currents (Fig. 9) whereas $G_{i(3)}$ proteins are involved in the dopamine-induced increase in voltage-dependent K^+ currents (Lledo et al. 1992). These latter findings have been confirmed using specific antisense oligodeoxyribonucleotide which were injected into recorded cells using the patch-clamp electrode. An antisense oligonucleotide complementary to the mRNA coding for $G_0\alpha$forms a hybrid with the endogenous sense sequence, thus reducing the efficiency of translation, stability, or transport of the mRNA, and therefore, inhibits synthesis of $G_0\alpha$. Fifty-two hours after injection of the oligonucleotide, dopamine failed to induce a significant inhibition of voltage-dependent Ca^{2+} currents in cells which responded in this way before injections. This suggests that transcription of the G_0 protein subtype was specifically blocked.

Another advantage of this configuration is the study of particular cells that are far too small for conventional two-microelectrode clamp: pituitary cells, erythrocytes, and even synaptosomes have been clamped with patch pipettes. In addition, on pulling away the recording pipette the cell membrane reseals, leaving the cell essentially intact, except that its internal solution has been replaced with the pipette solution. Thus successive WCRs with different internal solutions can be performed on a given cell by changing the recording pipette. Moreover, WCR provides an assay for the secretory process itself through membrane capacitance

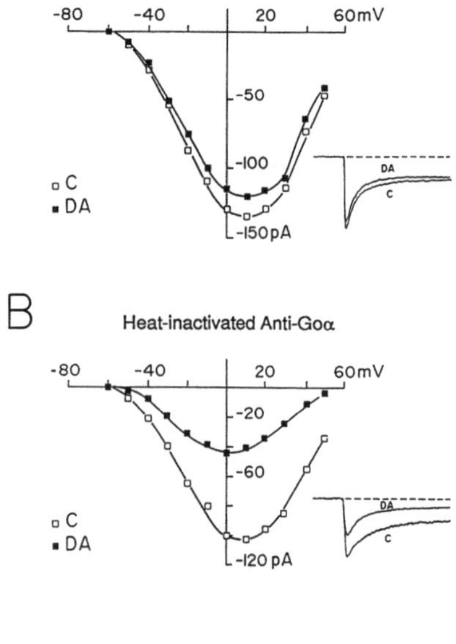

Fig. 9 A, B. Current-voltage relations for the peak inward calcium current recorded from pituitary cells. Calcium currents elicited from a holding potential of -80 mV were recorded in the absence (*C*) and presence of dopamine (*DA*; 10 n*M*). **A** The cell had been loaded with antibody raised against the α subunit of the G_0 protein. **B** As a control experiment, the cell had been injected with heat-inactivated anti-G_0 antiserum (65°C during 30 min). Note that the dopamine-induced response was dramatically attenuated in pituitary cells injected with anti-G_o antiserum

measurements. The resolution of the WCR is so great that fusion events of single vesicles can be captured; the resolution is 0.5 fF (5×10^{-16}F) corresponding to the membrane area of a vesicle 0.13 μm in diameter. Neher and Marty (1982) measured membrane capacitance of bovine chromaffin cells which have been well studied as a model of Ca^{2+}-dependent exocytosis. These experiments constitute the best electrophysiological evidence for vesicle fusion in exocytosis (for a review, see Penner and Neher 1989).

On the other hand, the disadvantage of WCR is that the internal solution is necessarily exchanged; soluble cytoplasmic enzymes, messengers, and cofactors rapidly diffuse out of the cell toward the pipette. This problem of washout limits the use of whole-cell clamp. In order to prevent wash-out, two methods have been elaborated. the first consists in performing WCR with a pipette containing a crude cytosolic extract (Dufy et al. 1986). The second method, in which a cell-attached patch is partially permeabilized (Lindau and Fernandez 1986; Horn and Marty 1988) is slow whole-cell recording. It provides electrical continuity between the pipette and the cytoplasm with minimum alteration of the cytoplasmic environment, which is necessary for the maintenance of some drug responses and/ or channel activity.

6 Conclusions

With so much evidence in its favor, the relationship between secretion and cell membrane electrical activity can no longer be contested. With electrophysiological techniques, we can directly study the effects of different second messengers on various conductances and so understand the mechanisms of action of various stimulatory and inhibitory substances. The actual conductances involved of Na^+, K^+, Ca^{2+}, and Cl^- ions can also be studied in detail in order to understand the nature of the molecules or channels responsible for these conductances and their mechanisms of action. Electrophysiological techniques are now largely associated with molecular biology, which permits the selective expression, or inhibition of expression of given molecules, such as elements of the second messenger systems, receptors, and channels, as well as directed mutations of the operational areas in molecules. The combination of these techniques will lead to detailed understanding of the molecular interactions responsible for the control of liberation of hormones.

Acknowledgements: The authors particularly wish to thank Mrs. D. Verrier and Mr. B. Dupouy for preparing cell cultures and Dr. V. Homburger for providing us with the antibodies directed against G-proteins. This work was supported by a grant from the „Fondation pour la Recherche Médicale“.

References

Adams WB, Benson JA (1985) The generation and modulation of endogenous rhythmicity in the *Aplysia* bursting pacemaker neuron R15. Prog Biophys Mol Biol 46:1–49

Akaike N, Kaneda M, Hori N, Krishtal (1988) Blockade of *N*-methyl-D-aspartate response in enzyme-treated rat hippocampal neurons. Neurosci Lett 87:75–79

Almers W, Stanfield PR, Stuhmer W (1983) Lateral distribution of sodium and potassium channels in frog skeletal muscle: measurements with a patch-clamp technique. J Physiol (Lond) 336:261–284

Amédée T, Ellie E, Dupouy B, Vincent JD (1991) Voltage-dependent calcium and potassium channels in Schwann cells cultured from dorsal root ganglia of the mouse. J Physiol (Lond) 441:35–56

Arnauld E, Vincent JD, Dreifuss JJ (1974) Firing patterns of hypothalamic supraoptic neurons during water deprivation in monkeys. Science 185:535–537

Arnauld E, Dufy B, Vincent JD (1975) Hypothalamic supraoptic neurones: rates and patterns of action potential firing during water deprivation in the unanaesthetized monkey. Brain Res 100:315–325

Auerbach A, Sachs F (1984) Patch clamp studies of single ionic channels. Annu Rev Biophys Bioeng 13:269–302

Baertschi AJ, Dreifuss JJ(1979) The antidromic compound action potential of the hypothalamoneurohypophysial tract, a tool for assessing posterior pituitary activity in vivo. Brain Res 171:437–451

Bancroft FC (1981) GH cells: functional clonal lines of rat pituitary tumor cells. In: Liss AR (ed) Functionally differentiated cell lines. Academic, New York, pp 47–59

Breitwieser GA, Szabo G (1985) Uncoupling of cardiac muscarinic and β-adrenergic receptors from ion channels by a guanine nucleotide analogue. Nature 317:538–540

Camardo JS, Shuster MJ, Siegelbaum SA, Kandel ER (1983) Modulation of a specific potassium channel in sensory neurons of Aplysia by serotonin and cAMP-dependent protein phosphorylation. Cold Spring Harbor Symp Quant Biol 48:213–220

Corey DP (1983) Patch clamp: current excitement in membrane physiology. Neurosci Comment 1:99–110

Denechaud M, Israel JM, Mishal Z, Vincent JD (1987) Influence of cell cycle phases on the electrical activity and hormone release in a transformed line of anterior pituitary cells. Life Sci 40:2377–2384

Denef C, Hautekeete E, de Wolf A, van der Schuren B (1978) Pituitary basophils from immature male and female rats: distribution of gonadotrophs and thyrotrophs as studied by unit gravity sedimentation hormone-releasing hormone. Endocrinology 103:724–735

Dingledine R (ed) (1984) Brain slices. Plenum, New York

Dingledine R, Dodd J, Kelly JS (1980) The in in vitro brain slice as a useful neurophysiological preparation for intracellular recording. J Neurosci Methods 2:323–362

Dreifuss JJ, Kalnins I, Kelly JS, Ruf KB (1971) Action potentials and release of neurohypophyseal hormones in vitro. J Physiol (Lond) 215:805–817

Dufy B, Vincent JD, Fleury H, Pasquier P, Gourdji D, Tixier-Vidal A (1979) Membrane effects of thyrotropin-releasing hormone and estrogen shown by intracellular recording from pituitary cells. Science 204:509–511

Dufy B, Vincent JD, Gourdji D, Tixier-Vidal A (1980) Electrophysiological study of prolactin-secreting pituitary cells in culture. Prog Reprod Biol 6:31–43

Dufy B, Israel JM, Zyzek E, Vincent JD (1982) An electrophysiological study of cultured human pituitary cells. Mol Cell Endocrinol 27:179–190

Dufy B, MacDermott A, Barker JL (1986) Rundown of GH_3 cell K^+ conductance response to TRH following patch recording can be obviated with GH_3 cell extract. Biophys Biochem Res Commun 137:288–396

Dyball REJ, Leng G (1987) Action potential recordings from the rat neural lobe in vivo. J Physiol (Lond) 394:122P

Dyer RG (1977) Localisation of neurosecretory cells by electrophysiological methods. In: Vincent JD, Kordon C (eds) Cell biology of hypothalamic neurosecretion. Centre National de la Recherche Scientifique, Paris, pp 323–335

Edwards FA, Konnerth A, Sakmann B, Takahashi T (1989) A thin slice preparation for patch-clamp recordings from neurons of the mammalian central nervous system. Pflugers Arch 414:600–612
Farley JM (1988) Patch clamp recording methodology. In: Boulton AA, Baker G, B, Walz W (eds) Neuromethods; the neuronal microenvironment. Humana, Clifton pp, 363–419
Fatt P, Katz B (1952) Spontaneous subthreshold activity at motor nerve endings. J Physiol (Lond) 117:109–128
Fedoroff S (1977) Primary cultures, cell lines and cell strains: terminology and characteristics. In: Fedoroff S, Hertz L (eds) Cell, tissue and organ cultures in neurobiology. Academic, New York, pp 265–286
Fenwick EM, Marty A, Neher E (1982) A patch-clamp study of bovine chromaffin cells and their sensitivity to acetylcholine. J Physiol (Lond) 331:577–597
Forda SR, Jessell TM, Kelly JS, Rand RP (1982) Use of the patch electrode for sensitive high resolution extracellular recording. Brain Res 249:371–378
Gerschenfeld HM (1973) Chemical transmission in Invertebrate central nervous systems and neuromuscular junctions. Physiol Rev 53:1–119
Gray R, Johnston D (1985) Rectification of single GABA-gated chloride channels in adult hippocampal neurons. J Neurophysiol 54:134–142
Gray R, Johnston D (1987) Noradrenaline and β-adrenoreceptor agonists increase activity of voltage-dependent calcium in hippocampal neurons. Nature 327:620–622
Gurdon JB, Lane CB, Woodland HR, Marbaix G (1971) Use of frog eggs and oocytes for the study of messenger RNA and its translation in living cells. Nature 233:177–182
Hall JG, Hicks TP, McLennan H, Richardson TL, Wheal HV (1979) The excitation of mammalian central neurons by amino acids. J Physiol (Lond) 286:29–39
Hall M, Howell SL, Schulster D, Wallis MA (1982) Procedure for the purification of somatotrophs isolated from rat anterior pituitary glands using Percoll density gradient. J Endocrinol 94:257–266
Hamill OP, Sakmann B (1981) Multiple conductance states of single acetylcholine receptor channels in embryonic muscle cells. Nature 294:462–464
Hamill OP, Marty A, Neher E, Sakmann B and Sigworth FJ (1981) Improved patch-clamp technique for high resolution current recording from cells and cell-free membrane patches. Pflugers Arch 391:85–100
Harrison RG (1907) Observations of the living developing nerve fiber. Proc Soc Exp Biol Med 4:140–143
Hatfield JM, Hymer WC (1985) Flow cytometric immunofluorescence of rat anterior pituitary cells. Cytometry 6:137–142
Hayward JN, Vincent JD (1970) Osmosensitive single neurones in the hypothalamus of unanaesthesized monkeys. J Physiol (Lond) 210:947–972
Hescheler J, Kameyama M, Trautwein W (1986) On the mechanism of muscarinic inhibition of the cardiac Ca current. Pflugers Arch 407:182–189
Hescheler J, Rosenthal W, Trautwein W, Schultz G (1987) The GTP-binding protein, G_o, regulates neuronal calcium channels. Nature 325.445–447
Hicks TP (1984) The History and development of microiontophoresis in experimental neurobiology. Prog Neurobiol 22:185–240
Hille B (ed) (1984) Ionic channels of excitable membranes. Sinauer, Sunderland, 426pp
Hodgkin AL, Huxley AF (1945) Resting and action potentials in single nerve fibres. J Physiol (Lond) 104:176–195
Horn R, Marty A (1988) Muscarinic activation of ionic currents measured by a new whole-cell recording method. J Gen Physiol 92:145–159
Hymer WC, Hatfield JM (1984) Separation of cells from the rat anterior pituitary gland. In: Pretlow TG, Pretlow TP (eds) Cell separation: methods and selected applications, vol. 3 Academic, New York, pp 163–194
Israel JM, Vincent JD (1990) The pituitary gland as excitable tissue. Front Neuroendocrinol 11:339–362
Israel JM, Denef C, Vincent JD (1983) Electrophysiological properties of normal somatotrophs in culture: an intracellular study. Neuroendocrinology 37:193–199

Israel JM, Jaquet P, Vincent JD (1985) The electrical properties of isolated human prolactin-secreting adenoma cells and their modification by dopamine. Endocrinology 117:1448–1455

Israel JM, Kirk C, Vincent JD (1987) Electrophysiological responses to dopamine of rat hypophysial cells in lactotroph-enriched primary cultures. J Physiol (Lond) 390:1–22

Jahnsen H, Laursen AM (1983) Brain slices. In: Barker JL, MacKelvy JF (eds) Current Methods in cellular neurobiology, vol 3: electrophysiological techniques. Wiley, New York

Jahnsen H, Llinàs R (1984) Electrophysiological properties of guinea-pig thalamic neurones: an in vitro study. J Physiol (Lond) 344:205–226

Kandel ER (1964) Electrical properties of hypothalamic neuroendocrine cells. J Gen Physiol 47:691–717

Kaneko S, Kato K, Yamagishi S, Sugiyama H, Nomura Y (1987) GTP-binding proteins G_i and G_o transplanted onto Xenopus oocytes by rat brain messenger RNA. Mol Brain Res 3:11–17

Kato M, Lledo PM, Vincent JD (1991) Blockade by lithium ions of potassium channels in rat anterior pituitary cells. Am J Physiol 261:C218–C223

Kelly JS, Simmonds MA, Straughan DW (1975) Microelectrode techniques. In: Bradley PB (ed) Methods in brain nervous system. Wiley, London, pp 337–377

Kidokoro Y (1975) Spontaneous calcium action potentails in a clonal pituitary cell line and their relationship to prolactin secretion. Nature 258:741–742

Koizumi K, Yamashita H (1971) Studies of antidromically identified neurosecretory cells of the hypothalamus by intracellular and extracellular recordings. J Physiol (Lond) 221:683–705

Koizumi K, Yamashita H (1972) Studies on neurosecretory cells in the mammalian hypothalamus. In: Kao FF, Koizumi K, Vassalle M (eds) Research in physiology. Aulo Gaggi, Bologna,pp 352– 387

Konnerth A (1990) Patch-clamping in slices of mammalian CNS. Trends Neurosci 13:321–323

Kostyuk PG, Krishtal OA (1977) Separation of sodium and calcium currents in the somatic membrane of mollusc neurons. J Physiol (Lond) 117:500–544

Kubo T, Fukuda K, Mikami A, Maeda A, Takahashi H, Mishina M, Haga T, Haga K, Ichiyama A, Kangawa K, Kojima M, Matsuo H, Hirose T, Numa S (1986) Cloning, sequencing and expression of complementary DNA encoding the muscarinic acetylcholine receptor. Nature 323:411– 416

Kuffler SW, Nicholls JG, Martin AR (eds) (1984) From neuron to brain, 2nd edn. Sinauer, Sunderland 650 pp

Kukstas LA, Verrier D, Zhang J, Chen C, Israel JM, Vincent J-D (1990) Evidence for a relationship between lactotroph heterogeneity and physiological context. Neurosci Lett 120:84–86

Lane CD (1983) The fate of genes, messengers, and proteins introduced into *Xenopus* oocytes. Curr Top Dev Biol 18:89–116

Lee KS, Akaike N, Brown AM (1978) The suction pipette method for internal perfusion and voltage clamp in small excitable cells. J Neurosci Methods 2:58–78

Legendre P, Westbrook G (1990) The inhibition of single *N*-methyl-D-aspartate-activated channels by zinc ions on cultured rat neurons, J Physiol (Lond) 429:429–449

Legendre P, Cooke IM, Vincent JD (1982) Regenerative response of long duration recorded intracellularly from dispersed cell cultures of fetal mouse hypothalamus. J neurophysiol 48:1121–1141

Lendahl U, McKay DG (1990) The use of cell lines in neurobiology. TINS 13:132–137

Lindau M, Fernandez JM (1986) IgE-mediated degranulation of mast cells does not require opening of ion channels. Nature 319:150–153

Ling G, Gérard RW (1949) The normal membrane potential of frog sartorius fibres. J Cell Physiol 34:383–405

Lledo P-M, Israel JM, Vincent J-D (1990a) A guanine nucleotide-binding protein mediates the inhibition of voltage-dependent calcium currents by dopamine in rat lactotrophs. Brain Res 528:143–147

Lledo P-M, Legendre P, Zhang J, Israel JM, Vincent J.D (1990b) Effects of dopamine on voltage-dependent potassium currents in identified rat lactotroph cells. Neurodenocrinology 52:545–555

Lledo P-M, Guerineau N, Mollard P, Vincent J-D, Israel JM (1991a) Physiological characterization of two functional states in subpopulations of prolactin cells from lactating rats. J Physiol (Lond) 437:477–494

Lledo P-M, Israel JM, Vincent J-D (1991b) Chronic stimulation of D_2 dopamine receptors specifically inhibits calcium but not potassium currents in rat lactotrophs. Brain Res 558:143–147
Lledo P-M, Homburger V, Bockaert J, Vincent J-D (1992) Differential G Protein-mediated coupling of D_2 dopamine receptors to K^+ and Ca^{2+} currents in rat anterior pituitary cells. Neuron 8:455–463
Llinàs R, Sugimori M (1980) Electrophysiological properties of in vitro Purkinje cell somata in mammalian cerebellar slices. J Physiol (Lond) 305:171–195
Llinàs R, Yarom (1981) Electrophysiology of mammalian inferior olivary neurones in vitro. Different types of voltage dependent ionic conductances. J Physiol (Lond) 315:549–567
Llinàs R, Yarom Y, Sugimori M (1981) Isolated mammalian brain in vitro: new technique for analysis of electrical activity of neuronal circuit function. Fed Proc 40:2240–2245
Lloyd DPC (1951) Electrical signs of impulse conduction in spinal motoneurones. J Gen Physiol 35:255–288
Luque EH, Munoz de Toro M, Smith PF, Neill JD (1986) Subpopulations of lactotrophes detected with the reverse hemolytic plaque assay show differential responsiveness to dopamine. Endocrinology 118:2120–2124
MacDonald JC, Adelman JP, Douglass J, North RA (1989) Expression of a cloned rat brain potassium channel in *Xenopus* oocytes. Sciences 244:221–224
MacVicar BA (1984) Infrared video microscopy to visualize neurons in the in vitro brain slice preparation. J Neurosci Methods 12:133–139
MacVicar BA, O'Beirne M (1988) Electrophysiological methods for studying ionic currents in brain slices and cell cultures. In: Boulton AA, Baker GB, Walz W (eds) Neuromethods; the neuronal microenvironment. Humana, Clifton, pp 545–587
Marty A, Neher E (1983) Tight-seal whole-cell recording. In: Sakmann B, Neher E (eds) Single channel recording. Plenum, New York, pp 107–121
Mason WT, Leng G (1984) Complex action potential waveform recorded from supraoptic and paraventricular neurons of the rat; evidence for sodium and calcium spike components at different membranes sites. Exp Brain Res 56:135–143
McIlwain H (1961) Techniques in tissue metabolism. Chopping and slicing tissue samples. Biochem J 27:213–218
Mody I, Salter MW, MacDonald JF (1989) Whole-cell voltage-clamp recordings in granule cells acutely dissociated from hippocampal slices of adult or aged rats. Neurosci Lett 96:70–75
Neher E (1988) The use of the patch clamp technique to study second messenger-mediated cellular events. Neuroscience 26:727–734
Neher E, Marty A (1982) Discrete changes of cell membrane capacitance observed under conditions of enhanced secretion in bovine adrenal chromaffin cells. Proc Natl Acad Sci USA 79:6712–6716
Neher E, Sakmann B (1976) Single-channel currents recorded from membrane of denervated frog muscle fibres. Nature 260:799–802
Neill JD, Frawley LS (1983) Detection of hormone release from individual cells in mixed populations using a reverse hemolytic plaque assay. Endocrinology 112:1135–1137
Nelson PG, Lieberman M (eds) (1981) Excitable cells in tissue culture. Plenum, New York
Noda M, Ikeda T, Kayano T, Suzuki H, Takeshima H, Kurasaki M, Takahashi H, Numa S (1986) Existence of distinct sodium channel messenger mRNAs in rat brain. Nature 320:188–192
Nordmann JJ, Stuenkel EL (1986) Electrical properties of axons and neurophysial nerve terminals and their relationship to secretion in the rat. J Physiol (Lond) 380:521–539
Novin D Sundsten JW, Cross BA (1970) Some properties of antidromically activated units in the paraventricular nucleus of the hypothalamus. Exp Neurol 26:330–341
Oron Y, Gillo B, Straub RE, GershengornMC (1987) Mechanism of membrane electrical response to thyrotropin-releasing hormone in *Xenopus* oocytes injected with GH_3 pituitary cell messenger ribonucleic acid. Mol Endocrinology 1:918–925
Ozawa S (1985) TRH-induced membrane hyperpolarization in rat clonal anterior pituitary cells. Am J Physiol 248:E64–E69
Penner R, Neher E (1989) The patch-clamp technique in the study of secretion. TINS 12:159–163
Petersen OH, Petersen CCH (1986) The patch-clamp technique: recording ionic currents through single pores in the cell membrane. News Physiol Sci 1:5–8

Poulain DA, Theodosis DT (1988) Coupling of electrical activity and hormone release in mammalian neurosecretory neurons. In: Ganten D, Pfaff D, Pickering B (eds) Stimulus-secretion coupling in neuroendocrine systems. Springer, Berlin Heidelberg New York, pp 73–100 (current topics in neuroendocrinology, vol 9)

Poulain DA, Vincent JD (1987) Mammalian neurosecretory cells: electrical properties in vivo and in vitro. Curr Top Membr Transp 31:313–331

Pusch M, Neher E (1988) Rates of diffusional exchange between small cells and a measuring patch pipette. Pflugers Arch 411:204–211

Sakmann B, Neher E (eds) (1983) Single channel recording. Plenum, New York

Sakmann B, Edwards FA, Konnerth A, Takahashi T (1989) Patch-clamp techniques used for studying synaptic transmission in slices of mammalian brain. Q J Exp Physiol 74:1107– 1118

Schofield PR, Darlison MG, Fujita N, Burt DR, Stephenson FA, Rodriguez F, Rhee LM, Ramachandran J, Reale V, Glencorse TA, Seeburg PH, Barnard EA (1987) Sequence and functional expression of the $GABA_A$ receptor shows a ligand-gated receptor super-family. Nature 328:221–227

Smith TG, Lecar JH, Redmann SJ, Gage PW (eds) (1985) Voltage and patch clamping with microelectrodes. American Physiological Society, Bethesda

Snutch TP (1988) The use of *Xenopus* oocytes to probe synaptic communication. Trends Neurosci 11:250–256

Soejima M, Noma A (1984) Mode of regulation of the ACh-sensitive K-channel by the muscarinic receptor in rabbit atrial cells. Pflugers Arch 400:424–431

Soreq H (1985) The biosynthesis of biologically active proteins in mRNA-microinjected *Xenopus* oocytes. CRH Crit Rev Biochem 18:199–238

Strickholm A (1961) Impedance of a small electrically isolated area of the muscle cell surface. J Gen Physiol 44:1073–1088

Strumwasser F (1965) The demonstration and manipulation of a circadian rhythm in a single neuron. In: Aschoff J (ed) Circadian clocks. North-Holland, Amsterdam, pp 442–462

Stühmer W, Roberts WM, Almers W (1983) The loose patch clamp. In: Sakmann B, Neher E (eds) Single channel recording. Plenum, New York, pp 123–132

Takeuchi A, Takeuchi M (1959) Active phase of frog's end-plate potential. J Neurophysiol 22:395–411

Tauc L (1967) Transmission in Invertebrate and Vertebrate ganglia. Physiol Rev 47:521–593

Theodosis DT, Legendre P, Vincent JD, Cooke I (1983) Immunocytochemically identified vasopressin neurons in culture show slow, calcium-dependent electrical response. Science 221:1052–1054

Vincent JD (1983) Approches électrophysiologiques des rhythmes hypothalamiques liés à la reproduction. J Physiol (Paris) 44:309–312

Vincent JD, Dufy B (1982) Electrophysiological correlates of secretion in endocrine cells. In: Conn MP (ed) Cellular regulation of secretion and release. Academic, New York, pp 107–145

Vincent JD, Hayward JN (1970) Activity of single cells in the osmoreceptor-supraoptic nuclear complex in the hypothalamus of the waking rhesus monkey. Brain Res 23:105–108

Vincent JD, Arnauld E, Bioulac B (1972) Activity of osmosensitive single cells in the hypothalamus of the behaving monkey during drinking. Brain Res 44: 371–384

Vincent JD, Poulain D, Arnauld E (1980) Electrophysiology of vasopressinergic neurons in the hypothalamus. In: Marsan A, Traczyk WZ (eds) Neuropeptides and neural transmission, Raven, New York, pp 281–291

Walton K, Llinàs R (1982) Electrophysiological activity recorded from the cerebral cortex of an isolated mammalian brain in vitro. Soc Neurosci Abstr 8:935

Yagi K, Azuma T, Matsuda K (1966) Neurosecretory cell: capable of conducting impulse in rats. Science 154:778–779

Yamashita H, Koizuma K, Brooks CM (1970) Electrophysiological studies of neurosecretory cells in the cat hypothalamus. Brain Res 20:462–466

Yamamoto C, Chujo T (1978) Visualization of central neurons and recording of action potentials. Exp Brain Res 31:299–301

Yamamoto C and MacIlwain H (1966) Electrical activities in thin sections from the mammalian brain maintained in chemically defined media in vitro. J Neurochem 13:1333–1343

MIX
Papier aus verantwortungsvollen Quellen
Paper from responsible sources
FSC® C105338

If you have any concerns about our products,
you can contact us on
ProductSafety@springernature.com

In case Publisher is established outside the EU,
the EU authorized representative is:
Springer Nature Customer Service Center GmbH
Europaplatz 3, 69115 Heidelberg, Germany

Printed by Libri Plureos GmbH
in Hamburg, Germany